Emerging Applications and Implementations of Metal-Organic Frameworks

Shimaa Mohamed Elsaeed
Egyptian Petroleum Research Institute, Egypt

Elsayed Zaki
Egyptian Petroleum Research Institute, Egypt

Abdel-Azim Abdel-Azim
Egyptian Petroleum Research Institute, Egypt

IGI Global
PUBLISHER of TIMELY KNOWLEDGE

A volume in the Advances in Chemical and Materials Engineering (ACME) Book Series

Published in the United States of America by
IGI Global
Engineering Science Reference (an imprint of IGI Global)
701 E. Chocolate Avenue
Hershey PA, USA 17033
Tel: 717-533-8845
Fax: 717-533-8661
E-mail: cust@igi-global.com
Web site: http://www.igi-global.com

Library of Congress Cataloging-in-Publication Data

Names: Elsaeed, Shimaa Mohamed, 1977- editor. | Zaki, Elsayed, 1984- editor. | Abdel-Azim, Abdel-Azim, 1947- editor.
Title: Emerging applications and implementations of metal-organic frameworks / Shimaa Mohamed Elsaeed, Elsayed Zaki, and Abdel-Azim Abdel-Azim, editors.
Description: Hershey, PA : Engineering Science Reference, an imprint of IGI Global, [2021] | Includes bibliographical references and index. | Summary: "This book contains research on emerging applications and implementations of metal-organic frameworks"-- Provided by publisher.
Identifiers: LCCN 2020014001 (print) | LCCN 2020014002 (ebook) | ISBN 9781799847601 (hardcover) | ISBN 9781799857242 (paperback) | ISBN 9781799847618 (ebook)
Subjects: LCSH: Metallic composites--Industrial applications. | Polymeric composites--Industrial applications. | Organometallic polymers--Industrial applications.
Classification: LCC TA418.9.C6 E427 2021 (print) | LCC TA418.9.C6 (ebook) | DDC 624.1/7730284--dc23
LC record available at https://lccn.loc.gov/2020014001
LC ebook record available at https://lccn.loc.gov/2020014002

This book is published in the IGI Global book series Advances in Chemical and Materials Engineering (ACME) (ISSN: 2327-5448; eISSN: 2327-5456)

British Cataloguing in Publication Data
A Cataloguing in Publication record for this book is available from the British Library.

For electronic access to this publication, please contact: eresources@igi-global.com.

Advances in Chemical and Materials Engineering (ACME) Book Series

J. Paulo Davim
University of Aveiro, Portugal

ISSN:2327-5448
EISSN:2327-5456

Mission

The cross disciplinary approach of chemical and materials engineering is rapidly growing as it applies to the study of educational, scientific and industrial research activities by solving complex chemical problems using computational techniques and statistical methods.

The **Advances in Chemical and Materials Engineering (ACME) Book Series** provides research on the recent advances throughout computational and statistical methods of analysis and modeling. This series brings together collaboration between chemists, engineers, statisticians, and computer scientists and offers a wealth of knowledge and useful tools to academics, practitioners, and professionals through high quality publications.

Coverage

- Composites
- Biomaterials
- Multifuncional and Smart Materials
- Computational methods
- Toxicity of Materials
- Metallic Alloys
- Fracture Mechanics
- Statistical methods
- Artificial Intelligence Methods
- Numerical Techniques

The Advances in Chemical and Materials Engineering (ACME) Book Series (ISSN 2327-5448) is published by IGI Global, 701 E. Chocolate Avenue, Hershey, PA 17033-1240, USA, www.igi-global.com. This series is composed of titles available for purchase individually; each title is edited to be contextually exclusive from any other title within the series. For pricing and ordering information please visit http://www.igi-global.com/book-series/advances-chemical-materials-engineering/73687. Postmaster: Send all address changes to above address.

Titles in this Series

For a list of additional titles in this series, please visit: http://www.igi-global.com/book-series/advances-chemical-materials-engineering/73687

Normal Partitions and Hierarchical Fillings of N-Dimensional Spaces
Gennadiy Vladimirovich Zhizhin (Russian Academy of Natural Sciences, Russia)
Engineering Science Reference • © 2021 • 280pp • H/C (ISBN: 9781799867685) • US $195.00

Applications of Nanomaterials in Agriculture, Food Science, and Medicine
Mohd Amin Bhat (Sathyabama University, India) Irshad Ahmad Wani (Government Degree College, Anantnang, India) and Shah Ashraf (National Institute of Technology, Srinagar, India)
Engineering Science Reference • © 2021 • 442pp • H/C (ISBN: 9781799855637) • US $215.00

Technology Development for Adsorptive Heat Energy Converters Emerging Research and Opportunities
Kostyantyn M. Sukhyy (Ukrainian State University of Chemical Engineering, Ukraine) Elena A. Belyanovskaya (Ukrainian State University of Chemical Engineering, Ukraine) and Mikhailo P. Sukhyy (Ukrainian State University of Chemical Engineering, Ukraine)
Engineering Science Reference • © 2021 • 305pp • H/C (ISBN: 9781799844327) • US $195.00

Nanotechnologies and Clusters in the Spaces of Higher Dimension Emerging Research and Opportunities
Gennadiy Vladimirovich Zhizhin (Independent Researcher, Russia)
Engineering Science Reference • © 2021 • 286pp • H/C (ISBN: 9781799837848) • US $195.00

Advanced Surface Coating Techniques for Modern Industrial Applications
Supriyo Roy (Haldia Institute of Technology, India) and Goutam Kumar Bose (Haldia Institute of Technology, India)
Engineering Science Reference • © 2021 • 342pp • H/C (ISBN: 9781799848707) • US $215.00

Handbook of Research on Advancements in Supercritical Fluids Applications for Sustainable Energy Systems
Lin Chen (Institute of Engineering Thermophysics, Chinese Academy of Sciences, China & University of Chinese Academy of Sciences, China)
Engineering Science Reference • © 2021 • 821pp • H/C (ISBN: 9781799857969) • US $415.00

Inspiration and Design for Bio-Inspired Surfaces in Tribology Emerging Research and Opportunities
Hisham Abdel-Aal (Drexel University, USA)
Engineering Science Reference • © 2021 • 367pp • H/C (ISBN: 9781799816478) • US $195.00

IGI Global
PUBLISHER of TIMELY KNOWLEDGE

701 East Chocolate Avenue, Hershey, PA 17033, USA
Tel: 717-533-8845 x100 • Fax: 717-533-8661
E-Mail: cust@igi-global.com • www.igi-global.com

Table of Contents

Detailed Table of Contents

In recent years, metal-organic frameworks (MOFs)-derived composites have emanated as a significant class of substantial materials with specific functional characteristics such as flexibility, high porosity, and diverse composition. Specific chemical modifications can also endow MOFs with specific functionality by offering the possibility of manufacturing all new generation of sensing devices. As on comparison with pure MOFs, the mix up of MOFs with matrix materials(e.g., metal nanoparticles, quantum dots, molecular species, enzymes, silica, and polymers) or functional species not only exhibits enhanced properties, but also widens its applications to modern field of heterogeneous catalysis, gas separation, potential hydrogen storage material and many others due to its high adsorption nature and excellent reversibility kinetics as in result reveals its various undefined attributes, such as reproducible syntheses, amenability to scale-up and chemical modification due to interactions of the functional matrix or species with the MOFs structures.

Different metal ions and anions are crucial for the life of organisms, but their excessive intake or deficiency cause various physiological abnormalities or diseases. Toxic metal ions such as Hg2+, Cr6+, Cd2+, explosive nitroaromatic compounds, and volatile organic compounds are widely used in agriculture, rocket fuels, and industry, but they cause severe effects on human health and environmental change. Hence, sensing and identification of these analytes are important for life, health, and environment protection. Recently, metal-organic frameworks (MOFs) have been extensively explored as luminescent sensors due to their unique crystallinity, structural diversity, tunable porosity, and functional groups. This chapter reviewed the origin of luminescent of MOFs and its applications as luminescent sensors for detecting different environmentally toxic analytes such as metal ions, anions, nitroexplosives, and volatile organic compounds. In addition, the detection mechanism of MOFs and its structural dependent properties are also illustrated.

An analysis of the contribution that metal-organic frameworks (MOFs) have made to the development of energy storage devices over the past two decades such as rechargeable batteries and supercapacitors is presented here. This chapter reviews the different versions of electrode manufacturing based on metal-organic frameworks to be used in the design and manufacture of rechargeable batteries and supercapacitors. The MOFs examined in this chapter include those based on MOF-derived materials, MOF-based composites, and conductive MOFs. Despite the significant progress that has been achieved so far, many tasks must be made to reach total security so that performance parameters required for optimal performance of rechargeable batteries and supercapacitors in commercial, industrial, and military applications. Therefore, innovative conceptions of the actions that must be performed are explored in this chapter.

Metal-organic frameworks (MOFs) are versatile materials that are of interest due to their application and properties. MOFs are highly crystalline and porous materials; they are composed of organic bridging ligands, acting as linkers, and a three-dimensional (3D) network of metal ions that are secondary building units. Since the MOFs have a high surface area, high porosity, tunable topography, and their structures are quite diverse, these materials are used in process of separation/purification, gas/energy storage, drug delivery, catalysis, and chemical sensors. Since the MOFs can be modified to selectively adsorb chemical species, they can be used as sensitive layer for modification of sensors. This process allows the sensor to detect the target analyte. Quartz crystal microbalances (QCMs) are highly sensitive mass sensors. In this chapter, the authors review the literature related to QCMs modified with MOFs. In particular, the relationship between target analyte, class of MOF, and instrument used for measurement of frequency variations.

Defluoridation of drinking water is a subject of great concern due to the side effects of exposure to fluorine such as fluorosis, damage to vital organs such as the brain, liver, and kidney, among others. Fluorine finds itself in water resources both by geological and anthropogenic processes. Adsorption still remains the popular method of defluoridation and water treatment in general due to its inherent advantages such as simplicity and low-cost. This chapter discusses at length the up-to-date data on the preparation and adsorption capacities of metal-organic framework (MOF) materials as emerging adsorbents for sequestration of fluoride from water. Their synthesis, efficiency, and suggested adsorption mechanisms are examined. It is demonstrated that MOFs are low-cost, efficient, and environmentally benign alternative adsorbents for the removal of fluoride from aqueous solution.

Clean and safe drinking water is one of the most important and essential components on earth for human survival. However, due to the various anthropogenic activities, the surface and groundwater have been contaminated with heavy metals and radionuclides, and it has become a serious problem of concern globally. These non-biodegradable heavy metal ions from water tends to accumulate in the soil and living organisms and causes serious health hazards in humans. Thus, the finding of recyclable, water-stable MOFs, and multi-functional MOFs (hybrid MOFs) with superior adsorption capability have emerged as a novel material for the removal of heavy metal species (such as Cr(III)/Cr(VI), Cu(II), Hg(II), Cd(II), As(III)/As(V), Pb(II), radionuclides U(VI), Se(IV)/(VI), Cs(I) from aqueous solution. This chapter deals with the recent development of MOFs and their application in heavy metal removal from water.

Over the past few years, metal-organic frameworks have gained considerable attention. Metal-organic frameworks are widely regarded as promising materials for separation, catalysis, molecular recognition, and storage applications. The applications and potentials of engineered forms of metal-organic frameworks such as thin films and membranes supported on various substrates in the area of water purification are discussed in this chapter. Metal-organic frameworks applications in separation and purification techniques with metal-organic frameworks-based membrane matrix is underlined at their promising levels. Metal-organic frameworks polymer composite membranes, including polymer-supported metal-organic frameworks membranes, are identified and the recent development are discussed. This chapter will provide a comprehensive assessment of recent studies on the removal of various contaminants from water using metal-organic frameworks-based membranes.

Metal-organic frameworks (MOFs) are based on metals and organic linkers; they possess large surface areas, suitable pore size and shape, wide range of chemical composition, and functionalized pore surface,

Preface

Metal-organic frameworks or MOFs can be defining as a kind of materials constituted by metal ions and coordination complexes with organic bonds forming networks with one-, two-, or three-dimensional structures containing potential voids that can be chemically exploited.

Metal-organic frameworks can be used in many topics related to our applications in world today. applied in rechargeable batteries and supercapacitors, structure and luminescent properties of the metal-organic frameworks (MOFs).

Some of the recent findings on the development of MOF-based drug delivery systems, focusing on non-steroidal anti-inflammatory drugs, anticancer drugs, and antimicrobials. MOFs used in drug controlled release and also, illustrates the perspectives, insights, and critical reflections on the controlled release of drugs from MOFs, and the biological evaluations of MOFs being used in CR of drugs.

Adsorption technique has achieved a great interest in removal of heavy metals due to its effectiveness, economic nature, simple operating conditions, and wide pH range.

The synthesis methods for preparation of MOFs and their application as adsorbents for the removal of fluorides from aqueous solutions.

the MOFs porosity is better than displayed by zeolites, since those are constructed from tethaedral units which can give rise to a finite number of Secondary building units (SBU), while MOFs can form SBUs from metal atom or cluster and one-, two- or three-dimensional extended inorganic substructures.

This book is meant for students studying MOFs and those working on adsorbents and other allied units industries but without a formal educational background in metal-organic frameworks (MOFs).

This book consists of nine chapters.

Conclusion of this topic in book is application of MOFs in recent industries as energy storage and overcome water scarcity.

Shimaa Elsaeed
Egyptian Petroleum Research Institute, Egypt

Elsayed Zaki
Egyptian Petroleum Research Institute, Egypt

Abdel-Azim Abdel-Azim
Egyptian Petroleum Research Institute, Egypt

Acknowledgment

This work has been done at the Egyptian Petroleum Research Institute (EPRI).

Chapter 1
Preparation of Composite Metal-Organic Frameworks

Abhinay Thakur
Lovely Professional University, India

Shveta Sharma
Lovely Professional University, India

Ashish Kumar
Lovely Professional University, India

ABSTRACT

In recent years, metal-organic frameworks (MOFs)-derived composites have emanated as a significant class of substantial materials with specific functional characteristics such as flexibility, high porosity, and diverse composition. Specific chemical modifications can also endow MOFs with specific functionality by offering the possibility of manufacturing all new generation of sensing devices. As on comparison with pure MOFs, the mix up of MOFs with matrix materials(e.g., metal nanoparticles, quantum dots, molecular species, enzymes, silica, and polymers) or functional species not only exhibits enhanced properties, but also widens its applications to modern field of heterogeneous catalysis, gas separation, potential hydrogen storage material and many others due to its high adsorption nature and excellent reversibility kinetics as in result reveals its various undefined attributes, such as reproducible syntheses, amenability to scale-up and chemical modification due to interactions of the functional matrix or species with the MOFs structures.

1. INTRODUCTION

In recent decades, the synthesis of metal-organic frameworks (MOFs) has gained tremendous attention due to the potential of achieving a wide range of esthetically appealing structures that exhibits considerable interest to applications in a range of porous materials fields. Based on pore size, shapes these MOFs have gained a promising traditional application in the field of storage, catalysis and separation

DOI: 10.4018/978-1-7998-4760-1.ch001

as of the host-guest interactions(Stock & Biswas, 2012). A metal-organic framework material (MOF) consisting of two main components which are a metal ion (or metal ion cluster) and an organic molecule (or bridging ligands).

Metal ions + Organic units (bridging ligands) → MOF materials

Organic ligands may be defined as species which donate to metal ions multiple lone pairs of e^-, while metal ions having vacant orbital shells accept these lone pairs of electrons results in formation of non-organic frame material(X. Liu, Sun, Hu, & Wang, 2016).

With such functionality, MOFs are used as integrated with the structural, electrical, magnetic and catalytic properties developed by selecting suitable metal ions and organic ligands (Adhikari & Lin, 2016). With researcher's efforts, various synthetic techniques have been developed to form the MOFs having various crystal structures, surface chemistries and pore size/shapes. On comparison with other porous materials such as mesoporous silica and zeolites, it deliver better chemical tailoring capabilities such as separation, molecular storage and catalysis attributed to their existence of various functional groups inside their frameworks. For applications like catalysis, MOFs are necessary which are generated by synthetic room temperature techniques, because MOFs synthesized at low temperatures terms out to be extremely thermal stable. Initially the researchers from the coordination and solid-state / zeolite chemistry fields carried out the research of MOFs (Lin, Adhikari, Lin, & Tu, 2016). Now days, MOFs are also attracting attentions of material scientists. Several scientists have introduced a post-synthetic modification of MOFs via a sequence of chemical reactions to incorporate the functionalities and creation of composite materials that appear to impart features in MOFs via the incorporation of functional species like organic colours, minor bio-molecules, NWs, NPs, polymers and nanofibers through the composite formation(Yang, 2005). MOF composites can be formed by two means: a) by deposition of MOFs to create the thin films on a two-dimensional planar substratum; b) by combining MOFs with the other components to create mixed structures.

The preparations and applications of MOF films occurred on various substrates have been applied by many chemists in various fields. This chapter will therefore concentrate on a concise overview of MOF composite materials preparation (Ben, Lu, Pei, Xu, & Qiu, 2012). Techniques utilized for MOF composites synthesis according to their formulation phase have been discussed throughout the chapter. Figure 1 showing various applications of MOFs composites.

Figure 1. Applications of MOFs composites in various fields

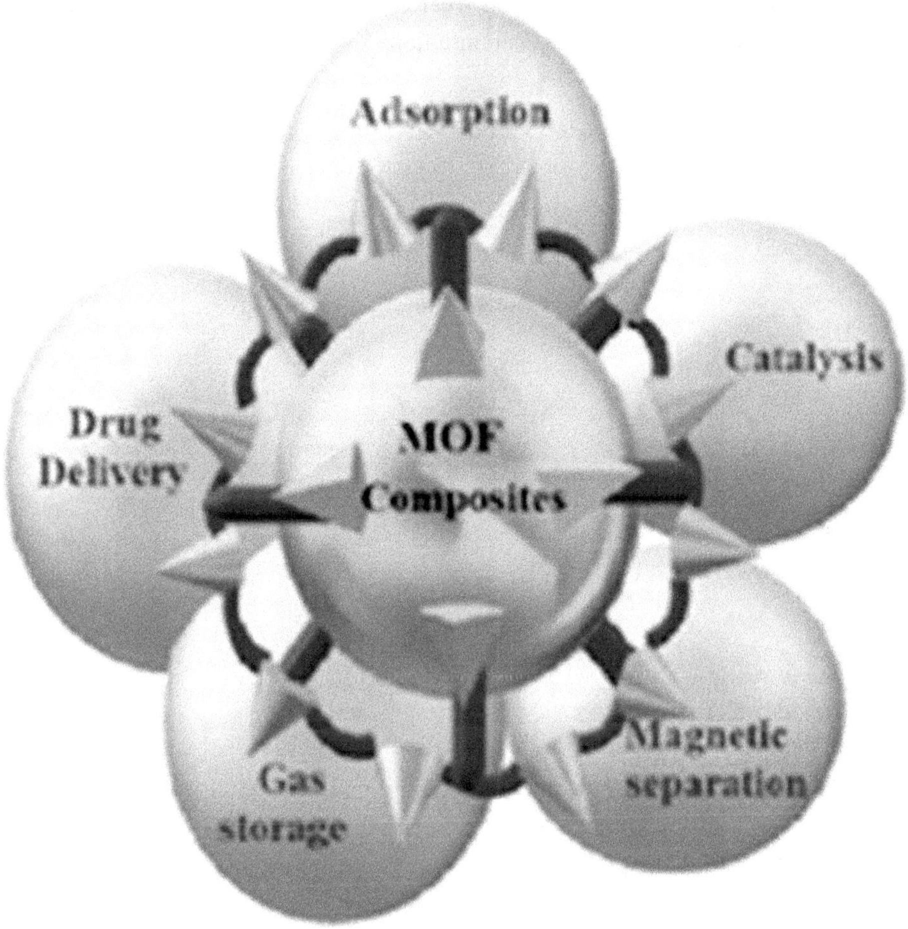

2. MOF COMPOSITES

The word "Composite" can be described as a solid material consisting of two or even more substances that contributes to attain desirable characteristics to the structure as a whole without losing their own identity. In fact, there have been always one substance that dominates the material as the matrix. MOF composites usually developed where the functional species and the matrix materials combine together. The structural fragility of MOFs has contributed their processability in manufacture (Perez, Balkus, Ferraris, & Musselman, 2009). MOF composites were therefore developed for applications like purification, small molecule separation and catalysis where MOF clusters were incorporated in one continuous, flexible matrix (Garrido, Mar, & Castillo, 2010). The matrices under such structures also often allow the MOFs to be reused and recovered, but they also improve the formability of MOF substances. In addition to functional organisms, MOFs itself can also function as matrices for applications like heterogeneous catalysis and drug delivery that correspond with functionalities conveyed both to MOFs and to the active species. MOFs are regarded as stabilizing matrix materials in these applications that provide enclosed regions for all functional species. The specific structures of MOFs attributes to the potential diffusion and accumulation of equipped species in the MOFs and also alter the reaction mechanisms, leading to new catalytic properties(Ostermann, Cravillon, Weidmann, & Smarsly, 2011).Figure 2 shows preparation of MOFs composites by indulging cyclopentadiene in MOF structure.

Figure 2. Formation pf MOFs composites (MOF + Cyclopentadiene)

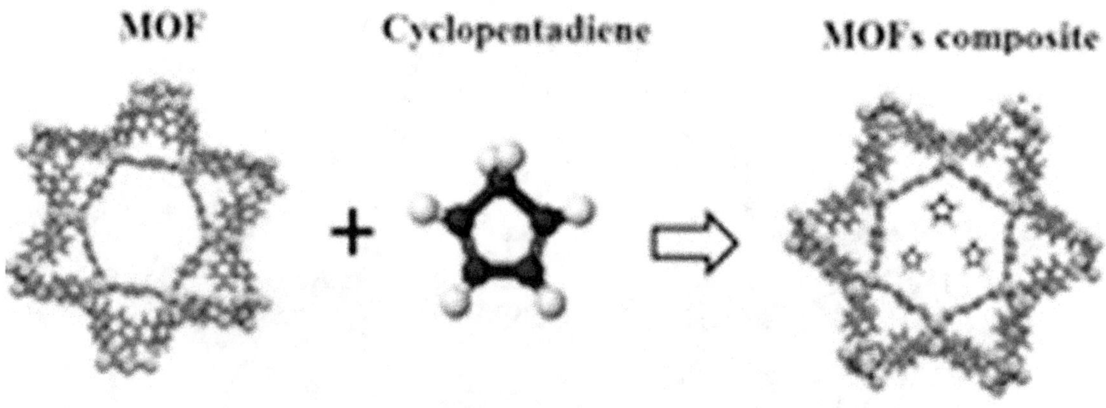

2.1. MOF–Carbon Composites

Several researches focused on the development and application of MOF composites along with carbon-based materials. MOF- Carbon composites have been identified by the different structures of carbon-based materials like those of activated carbons (ACs), carbon nanotubes (CNTs) and carbon fibres such as vast chemical stability, grafting preparation and large surface area. A study of MOF – carbon composites has recently been published by many researchers to outline synthesis protocols, features and future use in numerous applications (Chem, 2012). Where they discussed about the various process to make MOF-Carbon composites such as one pot synthesis, ex situ synthesis, stepwise approach synthesis etc.(X. Liu et al., 2016).

2.2. MOF–CNT Composites

Carbon nanotubes are deliberated as tubular structures comprises of carbon atoms structured in a cylindrical shape with sufficient high aspect ratios. Such structures provide CNT with extraordinary physical and mechanical strength, along with excellent electrical and thermal properties which are further useful in various industrial applications. As alignment of carbon molecules, they could form rather single- wall CNTs or multiple-wall CNTs (MWCNTs)(Nath, Mohamedali, Henni, & Ibrahim, 2018). Due to these characteristics, CNTs are found as one of the best materials for improving the properties of MOFs by manufacturing MOF – CNT composites (Ishida, Nagaoka, Akita, & Haruta, 2008).

2.3. MOF–AC Composite

ACs are regarded as extremely porous carbon-based material with immense surface area, making it best candidate for separation applications because of its high chemical, low cost and thermal stability(Aijaz & Xu, 2014). Some researchers have imbedded ACs into MOF-199 framework by utilizing in situ hydrothermal synthesis for H_2S adsorption applications. ACs has inflated the composite surface area and formed other robust structures with improved adsorptive capabilities. The synthesis was performed following a similar protocol as for MOF–CNT composites in which ACs are combined with MOF precursor solution under sonication to assist the dispersion of ACs in the solution by hydrothermal synthesis like in a case where Pd was permeated into AC and later on doped into MOF-74 by performing a complete carbonization in the presence of an intermediate sucrose to attain Pd-MOF-74 composites (Adhikari & Lin, 2016). By implementing same procedure Pd-MIL-53(Cr) through Pd/AC intermediate have been prepared for hydrogen storage applications (Lin et al., 2016).

3. SYNTHESIS OF MOFS COMPOSITES

3.1 MOF Composites Formation by Solution Blending Approach

Mechanical stirring was applied to the mixture of polymers to emphasize the solution blending method and then functional species to increase homogeneity of incorporated materials in solution beyond solvent vaporization in result leaving a polymer-based composite as residue[14–16]. As a lot of polymers could really easily dissolve in liquid solvents making it appealing in synthesis of polymer-based composites (Yang, 2005). Many MOF/polymer membranes like HKUST-1/PMMA(Ben et al., 2012), MOF- 5/ matrimid(Perez et al., 2009) and ZIF-8/PSU36(Garrido et al., 2010) have been produced by this method. The ease in the formation of MOF/polymer in low cost makes it insistence high for numerous industrial applications. The MOF lattice variation resulted from host-guest interaction of MOF structures during the removal of solvents results in a weak adhesion among polymers and MOF particles (Manuscript, n.d.). However, the host-guest interactions of MOF structures results in the variation of MOF lattice when solvents removal results in a weak adhesion between polymers and MOF particles. Moreover, MOF/ polymer membranes formed are mostly defective with uncontrollable voids. These issues are undesirable to the membranes used for gas separation and liquid filtration. Further, undesirable issues regarding the use of MOF/polymer composite membranes for liquid filtration and gas separation have been studied by the researchers worldwide. Figure 3 deliberating formation of MOF composites by solution blending procedure; blending of polymeric solution with CNT dispersion together.

Figure 3. Solution blending of polymer and CNT dispersion solution to form MOF composite

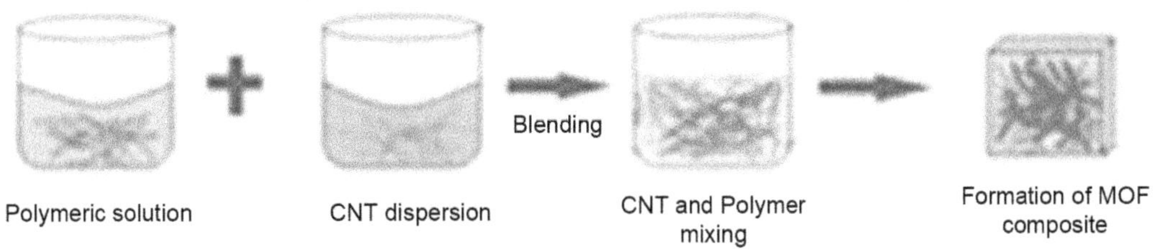

3.2. Formation of MOF Composites by Electrospinning

Electrospinning has been employed as one of the best process to attained fiber structures in polymer engineering for MOF/polymer composites synthesis. This method can be represented as the high voltage indulgent applied to a liquid droplet to generate a charge on it, therefore the high molecule-to-liquid cohesion leads to the configuration of a highly charged liquid jet and the rapid vaporization of the solvent throughout the liquid jet flight leads to the configuration of fiber composites with such a broad range of MOF / polymer combinations(Qin et al., 2015). In addition, the diameters of fibers and the MOF loadings can be adjusted by differing the polymer to MOF ratio and polymer concentration in spun mixtures. Many such MOF composites such as ZIF-8/PVP (Ostermann et al., 2011), MIL-101(Fe)/PS and ZIF-8/PS(Chem, 2012) fibrous composites have been obtained by this technique. Such composites, combined with the simple electrospinning method with the diverse functions of MOFs, show different applications in gas separation field and can also be used to extract toxic gasses present in atmosphere (Petit & Bandosz, 2009). Figure 4 showing organic fibers getting metals ions on their flip surfaces by the ease of electrospinning method.

Figure 4. MOFs organic fibers attained by electrospinning technique

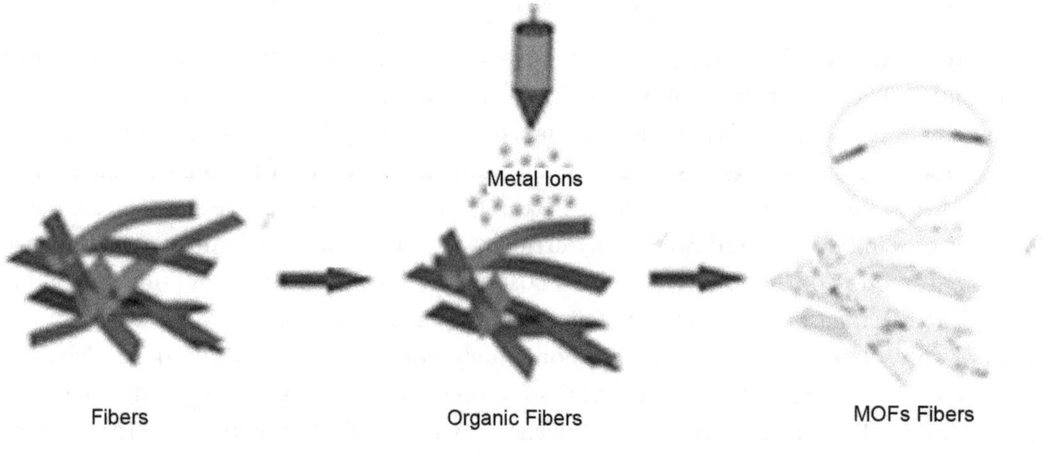

3.3. Solid Grinding Method

This approach allows the integration of MNPs within MOF matrix (Ishida et al., 2008). For this process metal precursor like $(Me)_2Au(acac)$ with MOF powder is obtained by a pestle and mortar. The derived compound later on reduced in presence of hydrogen gas at atmospheric level at inflated temperatures in results attaining desired MOF–MNP composite(Gascon & Kapteijn, 2010). Figure 5 demonstrating the occurrence of Au clusters on PCP by solid grinding process under reduction in hydrogen gas at 120^0 C.

Figure 5. Au cluster formation on PCP by solid grinding process

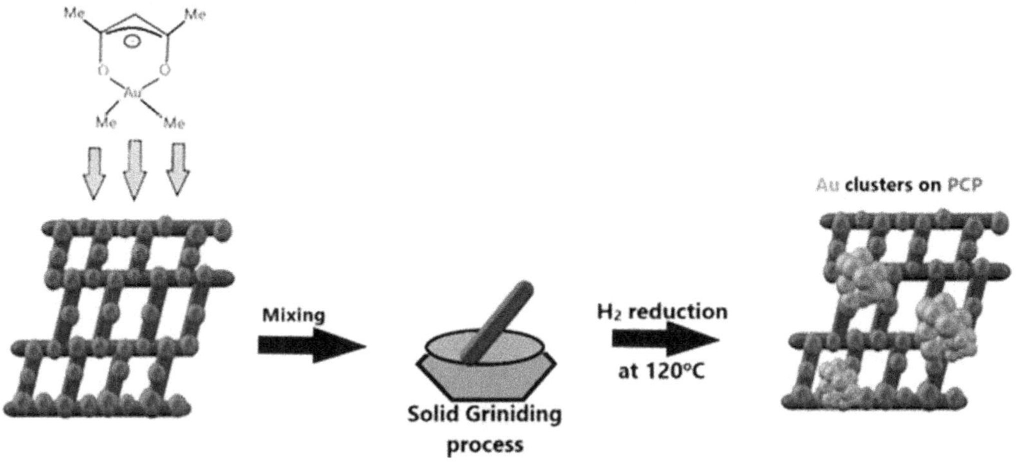

3.4. Gas Infiltration Method

This process is also classified as chemical vapor deposition which declared as a probable technique for indulging metal precursor within the MOFs by Fischer and his colleagues (X. W. Liu, Sun, Hu, & Wang, 2016). In this procedure, solid MOFs are arranged under vacuum conditions inside a sealed Schlenk tube and later on exposed in presence of volatile metal precursor vapors like Au(CO)Cl for Au at sufficient metal precursor temperature and vapor pressure in static strain to integrate metal of preference into MOF pores(Górka, Fulvio, Pikus, & Jaroniec, 2010). Targeted MOF – nanoparticle composites are then created by reducing the metal loaded solid MOF with hydrogen gas or simple thermal decomposition.

3.5. Encapsulation of the Pre-Synthesized Structures into MOFs

Various nano-objects feature such as their shape, size and composition. The impregnation approaches lacks of tenability on these discussed parameters for the incorporated structures, therefore widens the applications of MOF composites(Aijaz & Xu, 2014). New approaches have been developed to encapsulate the nano-objects pre-synthesized along the MOF shells post-formation on their surfaces (W. Liu et al., 2017). The MOF shells can also be developed through MOF solution post-growth or chemical modification of pre-accumulated metal compound structures into MOFs (Ahmed & Jhung, 2014).

3.6. One-Pot Synthesis

One-pot synthesis is an excellent technique that saved resources and time by confining reactants to successive chemical reactions in a single reactor and minimizing a long process of separation and purifying of intermediate chemicals(Tsuruoka, Kawasaki, Nawafune, & Akamatsu, 2011). This synthesis can be used to synthesize all forms of MOF–carbon composites. Typically, carbon-based materials are introduced to the solution comprising well-dissolved MOF precursors and so the hydrothermal reactions are conducted with the same constraints as the pure MOF synthesis. The synthesis of ZIF-8/CNTs [26] and

NH2–MIL 101(Al)/CNTs[27] composites was recorded by various researchers using this process, where CNTs were solubilized with hydrogen peroxide to improve dispersion in polar solvents and functional groups on anchoring metal nodes interfaces. Figure 6 illustrates the MOF composites formation when MOF precursor transformed into the MCM -41 under microwave irradiation and results in composite formation. This diagram illustrates the one pot synthesis.

Figure 6. One pot synthesis to attain MOF composites through MOF precursor

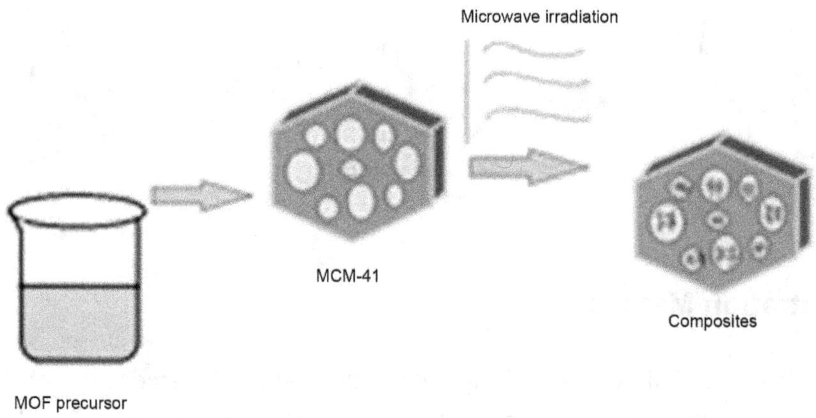

3.7. Stepwise Synthesis Approach

For instance, sometimes it is difficult to synthesize MOF composite and preparing a thick and deformity-free membrane by the one pot synthesis. However, difficulties in forming any modifications to carbonaceous substrates without functional groups on interfaces, the heterogeneous crystallization of MOFs on carbonaceous substrates surface via one-pot synthesis becomes an enormous problem(Petit & Bandosz, 2011). So, step-by-step synthesis methods like seeded growth technique also considered as a preferred solution.

3.7.1. Seeded Growth Method

This is further known as secondary development in which is commonly used for MOF–carbon membranes synthesis as it fabricate membranes. This method allows in attaining for a defect-free MOF membrane by more effective control of orientation [29–32]. During the procedure, MOF crystals are first synthesized and deposited on the supports as seeds, which will greatly promote the MOF crystallization. Several techniques such as rubbing, wiping, spin coating and heating are of vital importance and could be use in such method.

3.7.2. Layer by Layer Assembly

This method, a superior approach to the conventional solvo thermal growth in maintaining the MOFs coating thickness and structure penetration via circulating growth cycles which endorsed by many researchers to combine MOFs with carbon-based materials (Petit, Burress, & Bandosz, 2011). Some researchers constructed an Al_2O_3/HKUST-1 fiber core-shell structure having smooth surface and high coverage property at low-temperature in which firstly introduced ALD to facilitate MOF nucleation. Later on, using the layer by layer assembly method, MOF films finally fabricated on the cotton fiber mats and PET(M. Bechelany,* M. Drobek, C. Vallicari, A. Abou Chaaya, 2015).

4. MOF COMPOSITES APPLICATIONS

4.1. Gas Adsorption and Storage

Aroused by the rapidly rising rate of atmospheric CO_2, one of modern major environmental issues, the adsorption / capture of CO_2 becomes a significant application of MOF composites. MOF composites have been well researched because of the synergistic impact of reactive CO_2 adsorption on the active sites developed at the interface between matrix and MOFs. A further important use of MOF composites is storage of hydrogen. Low- or no-carbon energy sources are anticipated to decrease CO_2 levels in the atmosphere, and hydrogen is among the most worthy candidates. MOFs result in higher absorption of hydrogen at low temperatures, they have little benefit in storing hydrogen at room temperature relative to other porous materials like those of active carbons and zeolites. The spillover effect entails the dissociative chemisorption of hydrogen to atomic hydrogen on the layer of metal NPs and its corresponding migration through diffusion to adjacent substances. The metal NP / MOF composites combine this property with the high surface-volume ratio of MOFs and are interesting materials for enhancing hydrogen efficiency.

4.2. Drug Delivery

MOFs has become extensively documented as drug carriers whose key concerns are the delivery mechanism of medicines through endogenous MOFs and the toxicity of MOFs. The use of MOFs and magnetic NPs becomes an alternative option to this concept. The magnetic NPs can move in human bodies through blood or lymphatic vessels when combined with target-specific biomolecules and identify preferred biological goals. By then, magnetic NPs can also steer therapeutic agents to target by fixating the area on external magnetic field.

4.3. Catalysis

MOFs and related composites appeared with potential in catalysis as the innovative materials. The use of MOF in the field of catalysis have been done by the use of species like POMs (usually anions) that consisting of transition metal oxyanions which are linked together by shared oxygen atoms to make huge, closed 3D frameworks. The low specific surface area, poor catalytic stability and high solubility in aqueous solutions limit their catalytic formulations. In some cases semiconductor/MOF composites have also been used in improving catalytic properties.

4.4. Electrical Applications

A convergence of increased MOF porosity with electronic or protonic conductivity is scheduled to create a variety of new technologies like fuel cells and electrical sensors. MOFs are generally poor conductors due to their non-redox-active nature of metal nodes and perhaps almost organic ligands being used in MOF synthesis does not allow the electron transmission. MOFs with enhanced permeability synthesis strategies depend over the use of redox-active metal centers, electron donor which limitates due to redox changes and nature to enable their conductivity. To improve MOFs electrical conductivity, several substances such as GO (use in reduced rGO) form and imidazole or histamine has been used for the MOF composites synthesis.

5. CONCLUSION

With so many decades of advancement and improvement, studies of MOFs have been shifted from synthesis and design of a functional crystal structures to the MOFs application and commercialization among with their derivatives. Composite design is one of the best and alternative ways to broaden the application scopes of MOFs from gas absorption, separation and catalysis to new fields such as vapor sensing, electrical applications, drug delivery and even ahead of its applications. As during the formation of composites, the MOF structures may be devastated and action of MOFs under admirable processing conditions. The MOFs should be maintained with ordered framework structure, high surface area and fully accessibility of their pores to the substrates before and after the formation of MOF composites, the materials should be characterized correctly to confirm their stability. Although much effort has been dedicated to the study of composite formation strategies, some drawbacks of these approaches restrict their potential utilizations. However, the MOF structures after the compounding process may degrade due to the oxidation of the functional groups or metal nodes as both of which can convince the collapse of the micro pores. Therefore, expanding the metal loading and avoiding their aggregation in MOFs in concert become one of the main challenges on the synthesis of metal MOF composites.

REFERENCES

Adhikari, A. K., & Lin, K. S. (2016). Improving CO2 adsorption capacities and CO2/N2 separation efficiencies of MOF-74(Ni, Co) by doping palladium-containing activated carbon. *Chemical Engineering Journal, 284*(January), 1348–1360. doi:10.1016/j.cej.2015.09.086

Ahmed, I., & Jhung, S. H. (2014). Composites of metal-organic frameworks: Preparation and application in adsorption. *Materials Today, 17*(3), 136–146. doi:10.1016/j.mattod.2014.03.002

Aijaz, A., & Xu, Q. (2014). *Catalysis with Metal Nanoparticles Immobilized within the Pores of Metal – Organic Frameworks*. Academic Press.

Bechelany, M., Drobek, M., Vallicari, C., Abou Chaaya, A., Julbe, A., & Miele, P. (2015). Highly crystalline MOF-based materials grown on electrospun nanofibers†. *Nanoscale, 7*(13), 5794–5802. Advance online publication. doi:10.1039/C4NR06640E PMID:25759092

Ben, T., Lu, C., Pei, C., Xu, S., & Qiu, S. (2012). Polymer-Supported and Free-Standing Metal-Organic Framework Membrane. *Polymer-Supported and Free-Standing Metal–Organic Framework Membrane, 18*(3), 1–5. doi:10.1002/chem.201201574 PMID:22807003

Chem, J. M. (2012). *Electrospinning technology applied in zeolitic imidazolate framework membrane synthesis*. doi:10.1039/c2jm35401b

Chen, L., & Xu, Q. (2019). Metal-Organic Framework Composites for Catalysis. *Matter, 1*(1), 57–89. doi:10.1016/j.matt.2019.05.018

Doherty, C. M., Buso, D., Hill, A. J., Furukawa, S., & Kitagawa, S. (2014). Using Functional Nano- and Microparticles for the Preparation of Metal À Organic Framework. *Accounts of Chemical Research, 47*(2). Advance online publication. doi:10.1021/ar400130a PMID:24205847

Garrido, L., Mar, L., & Castillo, L. F. (2010). *CO 2 Transport in Polysulfone Membranes Containing Zeolitic Imidazolate Frameworks As Determined by Permeation and PFG NMR Techniques*. doi:10.1021/ma902303e

Gascon, J., & Kapteijn, F. (2010). Metal-organic framework membranes-high potential, bright future? *Angewandte Chemie International Edition, 49*(9), 1530–1532. doi:10.1002/anie.200906491 PMID:20127924

Górka, J., Fulvio, P. F., Pikus, S., & Jaroniec, M. (2010). Mesoporous metal organic framework-boehmite and silica composites. *Chemical Communications, 46*(36), 6798–6800. doi:10.1039/c0cc01578d PMID:20717603

Hindocha, S. (2017). The chemistry of metal-organic frameworks: Synthesis, characterization, and applications. *Johnson Matthey Technology Review, 61*(2), 138–141. doi:10.1595/205651317X695091

Ishida, T., Nagaoka, M., Akita, T., & Haruta, M. (2008). Deposition of gold clusters on porous coordination polymers by solid grinding and their catalytic activity in aerobic oxidation of alcohols. *Chemistry (Weinheim an der Bergstrasse, Germany), 14*(28), 8456–8460. doi:10.1002/chem.200800980 PMID:18752232

Lin, K., Adhikari, A. K., Lin, K., & Tu, M. (2016). Hydrogen storage capacity enhancement of MIL-53 (Cr) by Pd loaded activated carbon doping. *Journal of the Taiwan Institute of Chemical Engineers*. doi:10.1016/j.jtice.2016.02.033

Liu, W., Shao, Q., Ji, G., Liang, X., Cheng, Y., Quan, B., & Du, Y. (2017). Metal–organic-frameworks derived porous carbon-wrapped Ni composites with optimized impedance matching as excellent lightweight electromagnetic wave absorber. *Chemical Engineering Journal, 313*, 734–744. doi:10.1016/j.cej.2016.12.117

Liu, X., Sun, T., Hu, J., & Wang, S. (2016). Composites of metal–organic frameworks and carbon-based materials: preparations, functionalities and applications. *Journal of Materials Chemistry A: Materials for Energy and Sustainability, 4*(October), 3584–3616. doi:10.1039/C5TA09924B

Liu, X. W., Sun, T. J., Hu, J. L., & Wang, S. D. (2016). Composites of metal-organic frameworks and carbon-based materials: Preparations, functionalities and applications. *Journal of Materials Chemistry. A, Materials for Energy and Sustainability, 4*(10), 3584–3616. doi:10.1039/C5TA09924B

Melek, Y., & Aytun, O. U. (2007, August). An energy benchmarking model based on artificial neural network method utilizing US Commercial Buildings Energy Consumption Survey (CBECS) database. *International Journal of Energy Research*, *31*, 135–147. doi:10.1002/er

Nath, D., Mohamedali, M., Henni, A., & Ibrahim, H. (2018). Composites of Metal-Organic Frameworks (MOFs): Synthesis and Applications in Separation and Catalysis. *Metal-Organic Frameworks*, (iii), 251–295. doi:10.1002/9783527809097.ch9

Ostermann, R., Cravillon, J., Weidmann, C., & Smarsly, B. M. (2011). *Metal – organic framework nanofibers via electrospinning w z*. doi:10.1039/C0CC02271C

Perez, E. V., Balkus, K. J. Jr, Ferraris, J. P., & Musselman, I. H. (2009). Mixed-matrix membranes containing MOF-5 for gas separations. *Journal of Membrane Science*, *328*(1–2), 165–173. doi:10.1016/j.memsci.2008.12.006

Petit, C., & Bandosz, T. J. (2009). MOF-graphite oxide composites: Combining the uniqueness of graphene layers and metal-organic frameworks. *Advanced Materials*, *21*(46), 4753–4757. doi:10.1002/adma.200901581

Petit, C., & Bandosz, T. J. (2011). Synthesis, characterization, and ammonia adsorption properties of mesoporous metal-organic framework (MIL(Fe))-graphite oxide composites: Exploring the limits of materials fabrication. *Advanced Functional Materials*, *21*(11), 2108–2117. doi:10.1002/adfm.201002517

Petit, C., Burress, J., & Bandosz, T. J. (2011). The synthesis and characterization of copper-based metal-organic framework/graphite oxide composites. *Carbon*, *49*(2), 563–572. doi:10.1016/j.carbon.2010.09.059

Qin, J. S., Du, D. Y., Guan, W., Bo, X. J., Li, Y. F., Guo, L. P., Su, Z.-M., Wang, Y.-Y., Lan, Y.-Q., & Zhou, H. C. (2015). Ultrastable Polymolybdate-Based Metal-Organic Frameworks as Highly Active Electrocatalysts for Hydrogen Generation from Water. *Journal of the American Chemical Society*, *137*(22), 7169–7177. doi:10.1021/jacs.5b02688 PMID:25933041

Shekhah, O., Wang, H., Kowarik, S., Schreiber, F., Paulus, M., Tolan, M., Sternemann, C., Evers, F., Zacher, D., Fischer, R. A., & Wöll, C. (2007). Step-by-step route for the synthesis of metal-organic frameworks. *Journal of the American Chemical Society*, *129*(49), 15118–15119. doi:10.1021/ja076210u PMID:18020338

Stock, N., & Biswas, S. (2012). Synthesis of metal-organic frameworks (MOFs): Routes to various MOF topologies, morphologies, and composites. *Chemical Reviews*, *112*(2), 933–969. doi:10.1021/cr200304e PMID:22098087

Tella, A. C., & Aaron, I. Y. (2013). ChemInform Abstract: Syntheses and Applications of Metal-Organic Frameworks Materials. *ChemInform*, *44*(11). doi:10.1002/chin.201311277

Yang, Y. (2005). *High-efficiency solution processable self-organization of polymer blends*. doi:10.1038/nmat1500

Zhao, M., Lu, Q., Ma, Q., & Zhang, H. (2017). Two-Dimensional Metal–Organic Framework Nanosheets. *Small Methods*, *1*(1–2), 1600030. Advance online publication. doi:10.1002mtd.201600030

Chapter 2
Luminescent Metal-Organic Frameworks-Based Sensors for Environmentally Toxic Analytes

Khemnath Patir
Behali Degree College, India

ABSTRACT

Different metal ions and anions are crucial for the life of organisms, but their excessive intake or deficiency cause various physiological abnormalities or diseases. Toxic metal ions such as Hg2+, Cr6+, Cd2+, explosive nitroaromatic compounds, and volatile organic compounds are widely used in agriculture, rocket fuels, and industry, but they cause severe effects on human health and environmental change. Hence, sensing and identification of these analytes are important for life, health, and environment protection. Recently, metal-organic frameworks (MOFs) have been extensively explored as luminescent sensors due to their unique crystallinity, structural diversity, tunable porosity, and functional groups. This chapter reviewed the origin of luminescent of MOFs and its applications as luminescent sensors for detecting different environmentally toxic analytes such as metal ions, anions, nitroexplosives, and volatile organic compounds. In addition, the detection mechanism of MOFs and its structural dependent properties are also illustrated.

INTRODUCTION

Metal-Organic frameworks (MOFs) are class of highly porous crystalline solids constructed from metal ions (primary building unit or PBU) or metal clusters (secondary building unit or SBU) and organic ligands (organic linkers) via coordination bonds, Figure 1a (Keeffe & Yaghi, 2012). Depending on the binding sites of the organic ligands, the MOFs may be arranged themself to one, two or three dimensional networks (Tranchemontagn et al., 2009). Owing to its inherent tunable pores and high porosity, the MOFs have been successfully applied in gas storage and separation, heterogeneous catalysis, drug delivery, sensing etc (Collin & Zhou, 20017; Li et al., 2011; Lee et al., 2009; Hu et al., 2014; Horcajada et al., 2010).

DOI: 10.4018/978-1-7998-4760-1.ch002

Figure 1. Schematic illustration of formation of metal-organic frameworks, assembled from organic linkers and metal ions

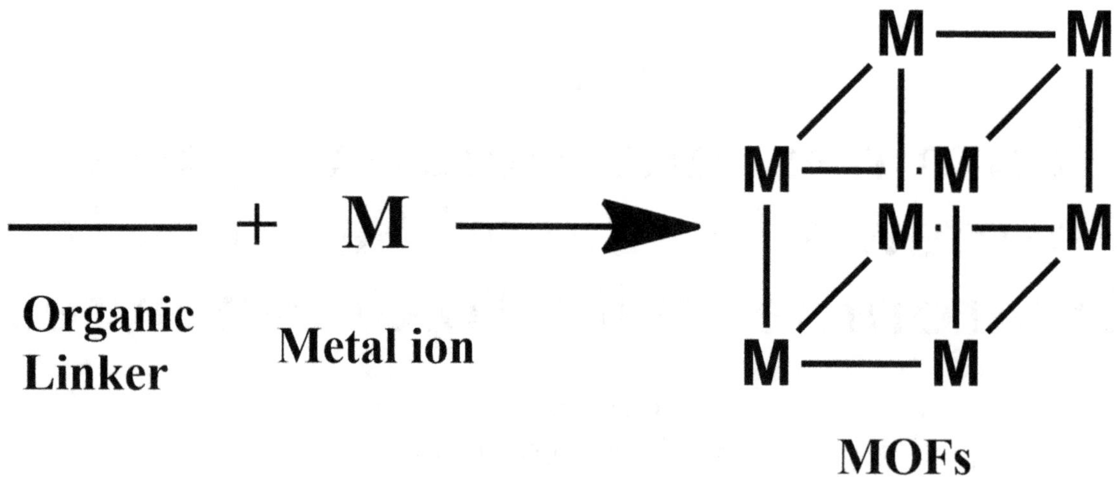

With the increasing population and industry development, environmental pollution has become a serious problem for the ecosystem and public health. Many types of pollutants, e.g., water and air pollution of metal ions, anions, nitroexplosives and organic compounds are associated with health risks (Lustig et al., 2017). Hence, sensing and identifying these analytes are important for life, health, environment protection and industry. Chromatography and its coupled techniques are widely used methods for determining environmentally toxic analytes (Petrovic et al., 2010; Nuijs et al., 2011). However, chromatography-based techniques require costly equipment, complex samples preparation and time consuming. Thus, new sensing techniques, which possess the advantages of high sensitivity, rapid detection and suitability for onsite, real-time, and continuous monitoring of environmental toxic analytes, are necessary. Luminescent materials possess all these requirements. Hence, luminescent materials can be used for detecting environmentally toxic analytes.

Luminescence is defined as the emission of light after absorption of energy from suitable source such as X-rays, electron beam, ultraviolet light (UV). There are mainly two types of luminescence: (1) Fluorescence is spin allowed radiative transition from lowest singlet excited state S_1 to the singlet ground state S_0, (2) Phosphorescence is the spin forbidden radiative transition from the triplet state T_1 to the ground state S_0 (Lakowicz, 2006). Recently, luminescent metal-organic frameworks (LMOFs) are widely explored porous materials for sensing applications. In comparison to other reported luminescent materials, LMOFs are more attractive candidate as analytes sensors due to their inherent merits such as structural and functional components, effective interaction between the analytes and MOFs. The MOFs have inherent crystallinity which is beneficial in understanding the LMOFs-analyte sensing interactions at molecular level in the detection process. In addition, suitable selection of the metals and organic ligands offers different compositions and porosities which in turn facilate LMOFs-analytes interactions to promote the recognition capability of specific targets (He et al., 2019). For example, the incorporation of fluorescent ligands to MOFs can significantly vary the luminescent properties of LMOFs and therefore is promising candidate for sensing applications. Furthermore, the porosity of LMOFs also offers a suitable host to analytes and enhances sensing performance. In addition, the tunable porosities of the

LMOFs also make them suitable host for adsorbing and releasing guest molecules, thereby resulting in specific target sensing performance (Allendorf et al., 2009). The pore structures of the MOFs can be tune by functionalization with different functional groups such as $-HSO_3$, $-NH_2$, $-CONH_2$, pyridyl etc which improve the sensing performance of the analytes (Evans et al., 2014). Hence, pore features (size, shape and environment) of LMOFs can be tune which is beneficial for sensing performance. Numerous LMOFs with wide-ranging structures are reported and applied for sensing of various analytes due to its tunability and functionalization. The origin of emission from LMOFs is assigned by two processes: ligand to metal charge transfer (LMCT) and metal to ligand charge transfer (MLCT). Both MLCT and LMCT types are observed in MOFs which consists of transition metal ions in d^{10} configuration such as Zn (II), Cd (II), Cu (I), Ag (I) etc (Tang et al., 2013; Chen et al., 2003; Tsuge et al., 2016; Jin et al., 2013). The detection processes of the analytes with LMOFs are either fluorescence "turn off" or "turn on" process via energy transfer (electron transfer). Hence, exploring the structure-properties relationship of LMOFs is of great significance for the rational designed and modification of LMOFs.

In this chapter, structure and luminescent properties of the metal-organic frameworks (MOFs) are discussed. In addition, recent applications of luminescent metal organic frameworks (LMOFs) for sensing various analytes such as metal ions, anions, nitroexplosives and volatile organic compounds are also discussed. The different types of mechanism of LMOFs involved in sensing of analytes are also illustrated. In addition, dependent of structure tuning and functionalization of LMOFs on the sensing performance of analytes are also discussed with examples. This book chapter would be useful as reference for the synthesis of new MOFs and its sensing applications.

BACKGROUND

LMOFs have been applied for detection of analytes due to their unique shape selective character and large surface area which enable the reversible adsorption and release of target molecules (Tranchemontagn et al., 2009). Moreover, metals-organic ligands components of the MOFs can be varied which is helpful for tunable fluorescence performance. Detection of different analytes with LMOFS is discussed below with examples.

Metal Ions

Iron plays an important role in biological systems. However, excessive intake or deficiency of iron causes physiological disorder (Carter et al., 2014). Hence, LMOFs have been applied for detection of iron. Bu and coworkers reported the detection of Fe^{3+} with a chiral lanthanide-based MOF, {[Tb (BPTC-2)]-(Me_2NH_2). $(Solvents)_x$}, (H_4BPTC-2 = biphenyl-3, 3′, 5, 5′-tetra carboxylic acid) (Zhao et al., 2016). It exhibits high selectivity for detection of Fe^{3+} by fluorescence "turn off" response and mechanism is assigned as competitive absorption process between ligands and Fe^{3+}. Qian and coworkers constructed a 2D MOF nanosheets, [Ti_2 $(HDOBDC)_2$], (H_2DOBDC = 2, 5-dihydroxyterephthalic acid) and applied as a Fe^{3+} sensor with high sensitivity and rapid response (Zu et al., 2016). The sensing efficiency of MOFs nanosheets is significantly enhanced due to increase in surface area, making active sites more accessible in comparison to the bulk MOFs. Sun and coworkers developed three types of Cd-based MOFs, {[Cd_2Na (L3)$(BDC)_{2.5}$].9H_2O}, (BDC = terephthalic acid),{[Cd_2(L3)(2,6-NDC$)_2$].DMF.5H_2O}, (2,6 NDC = 2, 6-naphthalenedicarboxylic acid, DMF = N, N′-dimethylformamide) and {[Cd_2(L3)$(BPDC)_2$].DMF.

$9H_2O$}, (BPDC = 4, 4′-diphenyldicarboxylic acid), where L3 = N′-(4-(1H-1, 2, 4-triazole-1-yl) benzyl) -N′-2-amino-ethyl) ethane-1, 2-diamine. These MOFs act as an effective sensor for Fe^{3+} by fluorescence quenching process (Zhao et al., 2017). The mechanism of sensitivity of these MOF towards Fe^{3+} is due to the presence of open channels and chelating NH_2 sites in the MOFs. Yan groups reported the synthesis of fluorescent metal-organic framework MIL-53(Al) and applied for selective detection of Fe^{3+} (Yang et al., 2013). Here, the cation Fe^{3+} is exchange with Al^{3+} from the metal organic framework, which results in the transformation of strong fluorescent MIL-53(Al) to weak fluorescent MIL-53(Fe). The limit of detection of Fe^{3+} by this system was determined to be 0.9 µM.

Cu^{2+} cation is a significant environmental pollutant and an essential trace element in biological systems (Kozlowski et al., 2009; Que et al., 2008). However, lower or higher concentration of Cu^{2+} can cause adverse effects on biological systems such as inhibit cell metabolism or gastrointestinal disturbance and damage to liver or kidney. Thus, the efficient, straightforward and real time detection of trace amounts of Cu^{2+} in biological systems is helpful to diagnose and treat the diseases induce by copper metabolism disorders. Cheng groups reported the construction of fluorescent BPIEI-CQDs/ZIF-8, MOFs based composites and it is found to be effective for selective detection of Cu^{2+} due to the adsorption property of the MOFs (Lin et al., 2014). The detection limit was found to be 80 pM. Shao groups constructed a 3D MOF, $[Cd_2(DTP)_2(bibp)_{1.5}]_n$, ($H_2$DPT = 4′-(3, 5-dicarboxylphenoxy) phenyl)-4, 2′:6′, 4′,-terpyridine), (bibp = 1, 3-di(1H-imidazol-1-yl)propane) and demonstrated as effective and selective Cu^{2+} sensor in DMF (N, N-dimethylformamide) (Yang et al., 2017). The mechanism of sensing is ascribed as electron transfer from MOFs to the Cu^{2+} ions. Yan and coworkers detected trace amounts of Cu^{2+} in aqueous medium using MIL-53-L5 [(L5 = 2-(pyrenes-1-imine) terephthalic acid] nanocrystals via fluorescence turned off process (Liu et al., 2016). The luminescent components-pyrene group of the MOFs is effectively quenched by interactions with the Cu^{2+} ions.

Aluminium (Al) metal and its compounds exist in environment and human daily life such as aluminum alloy has been used as cookware, the compounds of aluminum also serve as clinical drugs. However, the excessive intake of Al is harmful to liver, kidney and bones of human body specially can induce Alzheimers disease and osteoporosis. Hence, detection of Al^{3+} is necessary. Bu and co-workers designed a new 3D MOF, $[Zn_2$ (O-BTC) (4, 4′-BPY)$_{0.5}$ $(H_2O)_{1.5}$· $(DMA)_{0.5}$]n and was demonstrated as effective Al^{3+} sensor by fluorescence turn-on process (Wang et al., 2016). It was based on 2-hydroxybenzene-1, 3, 5-tricarboxylic acid (HO-H_3BTC) and 4, 4′-bipyridine (4, 4′-BPY) ligands in which the pyridine rings from 4, 4′-BPY are parallelly arranged in the pores of the framework. It show weak fluorescence in ethanol, but with addition of the Al^{3+}, the emission intensity increases significantly. 4, 4′-BPY acts as guest molecules which facilitate the electron transfer process. The pore structure of the MOFs also plays important role in the sensing performance. Sun and co-workers reported a 3D porous MOF decorated with exposed pyrimidyl Lewis base sites, $[Co_2$ (dmimpym) $(nda)_2]_n$ (dmimpym = 4, 6-di(2-methylimidazol-1-yl)-pyrimidine, H_2nda = 1, 4-naphthalenedicarboxylic acid) and applied for detecting Al^{3+} (Chen et al., 2017). Here, pyrimidyl N atom in the pyridine ligands of the MOF and Al^{3+} formed bond which facilate the electron transfer process resulting fluorescence intensity enhancement. Zhai and coworkers designed an excited state intramolecular proton transfer (ESIPT)-based MOF, Mg-TPP-DHBDC, (TPP = 6′-(pyridin-4-yl)-4, 2′:4′, 4″-terpyridine, DHBDC = 2, 5-dihydroxybenzene-1, 4-dicarboxylic acid) for detection of Al^{3+} (Li et al., 2019). The intramolecular hydrogen bonds between -OH and -COOH of DHBDC ligand activate the ESIPT process. On increasing concentration of Al^{3+} to the MOF, it first turn-off and then turn-on fluorescence was observed. The mechanism of fluorescence response is due to the effective coordination bonds of the -OH of DHBDC ligand and Al^{3+}. Hence, the sensor displays high sensitivity and selectivity towards Al^{3+} in the range of 0-15 μm. Fang groups reported a new Ln-MOF, $[Tb_4\text{-}(\mu_6\text{-}L4)_2\ (\mu\text{-}HCOO)\ (\mu_3\text{-}OH)_3\ (\mu_3\text{-}O)(DMF)_2(H_2O)_4]_n$ composed of Tb^{3+} secondary building units (SBUs) and [2-(5-carboxypyridin-3-yl)terephthalic acid] (H_3L4) ligands which was found to exhibits a high sensitivity and low detection limit of Ce^{2+} (Zhang et al., 2018). The weak interactions of pyridine nitrogen atoms of the ligand within Ln-MOF and the Ce^{2+} enhanced the fluorescent quenching and hence, improved the sensitivity. Liu and coworkers synthesized MOF composite, [TbK (BPDSDC) (DMF) $(H_2O)_2$], (H_4BPDSDC, biphenyl-3, 3′-disulfonyl-4, 4′-dicarboxylic acid) and is applied for sensing Cr^{3+} (H. Wang et al., 2016). The mechanism of sensing by this system is that the sulfonate groups within the pores recognized sizes for Cr^{3+}.

Mercury ion (Hg^{2+}) is highly toxic substances and post risks to human health. Hence, many researchers have investigated the utility of MOFs for detecting Hg^{2+}. For tuning the host-guest interactions between MOFs and analytes, most of the MOFs containing nitrogen centers, amine groups and alkyne group have been constructed and investigated for detecting Hg^{2+} (Samanta et al., 2018; Pankajakshan et al., 2019). Ghosh and co-workers reported a butyne-functionalized UiO-66 fluorescent MOF and applied for detecting Hg^{2+} in aqueous medium via fluorescence turn-off process (Samanta et al., 2018). The MOF displays rapid and high selectivity for Hg^{2+}, Figure 2. The limit of detection was found to be 10.9 nmol L^{-1}. Mandal and coworkers prepared a MOF, {[Zn (4, 4′-AP) (5-AIA)]-$(DMF)_{0.5}$}n with nitrogen-containing ligands, [4, 4′-AP = 4, 4-azopyridine, 5-AIA = 5-aminoisophthalic acid], which is able to detect Hg^{2+} (Pankajakshan et al., 2019). In this system, Hg (II) interacts with –N=N– group of 4, 4′-AP of the framework which is responsible for the fluorescence quenching of the MOF. Chen groups designed a turn-on fluorescent Hg^{2+} sensor based on Eu-MOF nanoparticles (Feng et al., 2013). The mechanism of sensing of Hg^{2+} with MOF is through an inner filter effect. Gu and coworkers prepared PCN-224 (PCN, porous coordination network) from zirconium clusters and porphyrinic linker TCPP (TCPP = Tetra (4-carboxyphenyl) porphyrin) which was found to effective for Hg^{2+} detection in HEPES (HEPES = 4-(2-hydroxyethyl)-1-piperazineethanesulfonic acid) buffer solution (Yang et al., 2016). The Hg^{2+} ions bind at the porphyrin center, resulting a fluorescent quenching effect in the visible region. In addition, this MOF-based sensor was regenerated by adding a potassium iodide solution to the system.

Huang and coworkers synthesized fluorene-based MOFs and demonstrated as Pb^{2+} sensor by fluorescence enhancement (Li et al., 2016). Xing and coworkers prepared an indium-based MOF, $[In_2(OH)_2(H_2TTHA)(H_2O)_2]$, ($H_6$TTHA = 1,3,5-triazine-2,4,6-triamine hexaacetic acid), which can detect UO_2^{2+} and Ru^{3+} cations by fluorescence turn off (Du et al., 2016). The mechanism of sensitivity is dependent on the pyrophosphate and phosphate groups of the ligands. Shi groups reported the synthesis of lanthanide and manganese based $\{[Eu(PDA)_3\ Mn_{1.5}(H_2O)_3]_3.5H_2O\}$ and $\{[Tb(PDA)_3\ Mn_{1.5}(H_2O)_3]_3.5H_2O\}$, (PDA = pyridine- 2,6-dicarboxylic acid) and was found to effective for detection of Zn^{2+} via fluorescence enhancement (X. Zhao et al., 2009). The fluorescence intensity of both this MOF increased with addition of Zn^{2+}. However, other metal ions do not change the fluorescence intensity of the MOF. The mechanism of sensitivity for Zn^{2+} is assigned to be effective intramolecular energy transfer from the ligand to the Eu^{3+}/Tb^{3+}. In addition, Zhao groups also reported $[Eu\ (PDA)_3\ Fe_{1.5}(H_2O)_3].1.5H_2O$ for the detection of Mg^{2+} ions via electron transfer between the host-guest interaction (B. Zhao et al., 2009).

Figure 2. The extent of the fluorescence response of UiO-66-butyne toward various metal ions, (F0 = fluorescence intensity of UiO-66-butyne in absent of analytes, F = fluorescence intensity of UiO-66-butyne in present of analytes, 1= Hg2+, 2 = Ba2+, 3 = Cd2+, 4 = Co2+, 5 = Cr3+, 6 = Cu2+, 7 = Fe3+, 8 = Mg2+, 9 = Mn2+, 10 = Ni2+, 11 = Sr2+, 12 = Zn2+)

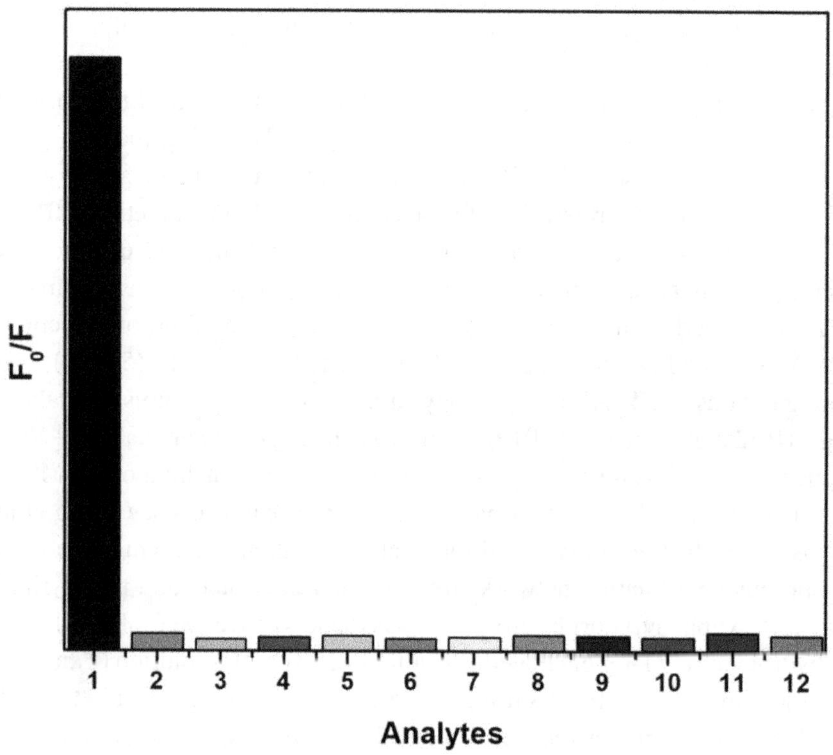

Cadmium ions (Cd^{2+}) are toxic ions found in air, soil and water. It accumulates in the lungs, livers, kidneys and bones of the human body, poisons the cardiovascular and immune systems, causing certain cancers (Marzo et al., 2013). Therefore, detection of Cd^{2+} is important to prevent diseases. Yan et al reported that Eu (III) functionalized Zr-based MOF act as a superior fluorescent probe for detection of Cd^{2+} in an aqueous medium (Xu & Yan, 2016). The mechanism of sensing is due to effective enhancement of energy transfer between the organic ligand and the central Eu^{3+} system by the Cd^{2+} ions.

Anions

Recently, luminescent MOFs have been employed for detection of different types of anions. Phosphate ions (PO_4^{3-}), one type of anions when present in water reduces or eliminate dissolved oxygen, resulting in a harmful effect on the water ecosystem (Kolliopoulos et al., 2015). Therefore, determination of PO_4^{3-} is of great importance. Qian and coworkers constructed Tb (III) based MOF, TbNTA (NTA

= nitrilotriacetate) and applied as highly selective detection of PO_4^{3-} via fluorescence turn-off process (Xu et al., 2011). However, all the other ions F^-, Cl^-, Br^-, I^-, NO^{3-}, NO^{2-}, HCO^{3-}, CO_3^{2-} and SO_4^{2-} have no effect on the fluorescence intensity of TbNTA. The mechanism of selectivity is due to the prevention of energy transfer from ligand to the Tb^{3+} by phosphate ions, resulting in fluorescence quenching. Lu groups reported the preparation of MOF/ZnO composite which is non-fluorescent (Zhao et al., 2014). However, after addition of PO_4^{3-} into the MOF/ZnO system, the fluorescence is turn on. The fluorescence intensity is found to be directly dependent on the concentration of the phosphate ions and is not affected by other ions. The limit of detection of PO_4^{3-} is determined to be 53 nM using this MOF/ZnO system. Xu groups synthesized 3D MOF, Eu-H_3BTB, (H_3BTB = 1, 3, 5-benzenetribenzoate), which is exploited as a reusable PO_4^{3-} sensor (Zu et al., 2015). The fluorescence intensity of the MOF remained same upto five testing, indicating that interaction between the Eu-H_3BTB and PO_4^{3-} is weak and fluorescence regeneration may be due to the removal of the PO_4^{3-}. Liu and coworkers reported Tb^{3+} doped dual emitting fluorescence Tb-Zn-MOF by the encapsulation of the Tb^{3+} into the Zn-MOF, {[Zn_4-($L7^{3-}$)$_2$ (O^{2-}) (H_2O)$_2$]. 4EtOH}$_n$,{(L7 = 4, 4′, 4″-{[(1, 3, 5-triazine-2, 4, 6-triyl) tris (sulfanediyl))] tribenzoic acid)}(Ji et al., 2018). It shows fluorescence quenching behaviour for PO_4^{3-} detection with rapid response and a low limit of detection of 0.1 ppm, attributed to the weak interaction between PO_4^{3-} and Tb^{3+}. The relative ratio of the fluorescence intensity (F_L/F_{Tb}) is linearly dependent on the concentration of PO_4^{3-}.

Chlorine is powerful oxidant used in many production processes such as water disinfection, deodorization and bleaching. However, it is harmful to human health in excess amounts. Hence, detection of free chlorine in drinking water is of intense research. Lv and coworkers synthesized NH_2-MIL-53(Al) MOF for detection of chloride ions (Lu et al., 2016). The mechanism of detection involved the effective fluorescence quenching of MOF by the energy transfer of the NH----OCl via hydrogen bonding. The limit of detection was found to be 0.04 μM. The recovery test performed in real tap water and swimming pool water samples showed that the recoveries were 97 to 101%. Cyanide ion (CN^-) is water contaminant and toxic pollutants occur in nature. Ghosh and coworkers reported a MOF based materials (ZIF-90) for selective and sensitive fluorescent sensor for detection of CN^- in aqueous medium with detection limit of 2 μM lower than 2 mM in drinking water permitted by WHO (Karmakar et al., 2016). The mechanism of sensing is assigned as the nucleophilic addition reaction of the cyanide groups with the free aldehydes of the MOFs converting to dicyanovinyl group, which results in significant change in fluorescence intensity.

Hexavalent chromium (Cr^{6+}) is one of the most toxic heavy metal ions (Mon et al., 2018). Recently, many MOFs have been reported for detection of chromium containing ions. Bu groups synthesized Zn based MOF featuring triangular channels, {[Zn_3(bpanth)-(oba)$_3$]. 2DMF}$_n$ (bpanth = 9, 10 bis(4-pyridyl) anthracene, H_2oba = 4, 4′-oxybis(benzoic acid) (Yao et al., 2018). This MOF system is applied as rapid and selective fluorescent sensor for $Cr_2O_7^{2-}$/CrO_4^{2-} with detection limit of 0.7 ppm and 0.3 ppm respectively for $Cr_2O_7^{2-}$ and CrO_4^{2-}. The mechanism of sensing is due to the competitive resonance energy transfer between the organic ligands and Cr^{6+}. The strong interactions between metal clusters and analytes determined the high sensitivity of MOFs. Li and coworkers reported a Zr (IV)-MOF by using H_3BTBA ligand, [$Zr_6O_4(OH)_7(H_2O)_3$(BTBA)$_3$], (BTBA = 4, 4′, 4″-(1H-benzo [d]imidazole-2, 4, 7-triyl) tribenzoic acid) (He et al., 2018). The fluorescence of the Zr (IV) MOF is selectively quenched in the presence of Cr^{6+} due to the competitive absorption and the host-guest interactions derived from the coordination between Zr_6 node of MOF and $Cr_2O_7^{2-}$ anions, Figure 3. The limit of detection of $Cr_2O_7^{2-}$ was determined to be 1.5 μM. Zhao and coworkers developed efficient $Cr_2O_7^{2-}$ fluorescent sensor based on 2D MOF, [Cd (TPTZ) (H_2O)$_2$ (HCOOH) (IPA)$_2$], (TPTZ = 4-[4-(1H-1, 2, 4-triazol-1-yl) phenyl] phenyl)-1H-1, 2, 4-triazole, IPA = 1, 3-benzene dicarboxylic acid) (Wang et al., 2015). The toxic anions are capture in the MOF framework through ion exchange and it caused larger fluorescent quenching for $Cr_2O_7^{2-}$ in comparison to the other anions. In another work, Zheng groups synthesized Cd (II)-based MOF, {[Cd (L3) (BPDC)]. 2 H_2O}, (L3 = 4, 4′-(2, 5-bis-methylthio)-1, 4-phenylene) bipyridine, H_2BPDC = 4, 4′-biphenyl dicarboxylic acid) for $Cr_2O_7^{2-}$ sensing (S. Chen et al., 2017). The $Cr_2O_7^{2-}$ anions harvest more excitation energy than MOFs due to wider absorption bands, resulting in hindering the energy absorption of the framework. Yan and coworkers doped Eu^{3+} cations into the channels of the MIL-121, {[Al (OH) (H_2BTEC)]. (H_2O)}, (H_2BTEC = benzene 1, 2, 4, 5-tetracarboxylic acid) to differentiate F^- and $Cr_2O_7^{2-}$ (Hao & Yan, 2016). In the presence of F^- and $Cr_2O_7^{2-}$ anions, the fluorescence color of the MIL-121- Eu^{3+} changed from red to blue or black. The fluorescence quenching of the MIL-121-Eu^{3+} in present of F^- is due to the metal based system while quenching for $Cr_2O_7^{2-}$ is attributed to ligand based system.

Permanganate ion (MnO_4^-) an oxidizing agent is used in analytical chemistry, agriculture, municipal water treatment, human health and safety, medicine, ground water remediation etc. However, high concentration of MnO_4^- causes side effects and produced risk to aquatic ecosystem. Yan groups synthesized Eu (III) functionalized In-MOF and demonstrated as fluorescent sensor for detection of MnO_4^- (Wu et al., 2017). The mechanism of sensing of MOFs towards MnO_4^- is due to the decreased in energy transfer from ligands to the lanthanide ions. In another work, Yan and coworkers reported heterometallic MOF based on alkaline earth and lanthanide {[Ba_3 $La_{0.5}$ (μ_3-L4)$_{2.5}$ (H_2O)$_3$ (DMF)].(DMF)$_3$}, (L4 = p-terphenyl-3, 4″, 5-tricarboxylic acid, DMF = N, N′-dimethylformamide) for detecting MnO_4^- (Ding et al., 2016). This MOF system display high selectivity and sensitivity towards MnO_4^-. The MnO_4^- anions easily diffused into the pores of the MOF, causing the fluorescence of the MOF to quench.

Excess accumulation of F^- ions in human body may results in X-shaped or O-shaped legs, hunchbacks or arms that can only be bent but not stretched. Li and coworkers synthesized [Eu (O-cpia) (phen) MOF], (O-cpia = 5-(2-carboxy-phenoxy)-1, 3-benzenedicarboxylic acid) and demonstrated as high selectivity ratiometric detection of PO_4^{3-} and F^- in aqueous media (Wang et al., 2018). The detection mechanism involves the interaction between the analyte and MOF, causing the absorption peaks to weaken, which results in lower energy transfer from ligand to the metal. Hence, the fluorescence intensity of the MOF is significantly reduced. Wang groups designed a Eu-based MOF complex, [Eu (III) tris (2-thenoyl-trifluoroacetonate], (Eu-(TTA)$_3$-1), which is able to detect different anions including F^-, acetate and dehydrogen phosphate (Zheng et al., 2012). Different anions were detected based on the different in their fluorescence emission peaks. F^-, acetate and dehydrogen phosphate show emission peaks at purple, green and blue region respectively. Yin and coworkers synthesized Eu-MOF from 5-boronisophthalic acid (5-bop) and Eu^{3+}, which is found to be suitable for sensing F^- ions (Yang et al., 2017). This MOF system is found to be effective for selective detection of F^- ions. The introduction of boric acid was able to tune the energy level of the ligand, producing two emission centers at single excitation wavelength of 275 nm. This originated from the combination of T_1 energy level of boric acid functionalized ligands and the energy gap between the T_1 and Eu^{3+}. As the concentration of the F^- increased, the fluorescence intensity of 5-bop center increased while that of Eu^{3+} decreased along with color changed from red to blue. The dual fluorescence emission of Eu-MOF was used to detect F^- ions with a low limit of detection of 2 μ mol L^{-1}. The interactions between F^- and the boron ligand contribute to the change of the fluorescence properties.

Figure 3. Fluorescence quenching percentage of Zr (VI)-MOF in aqueous medium with different anions, (1 = Cr2O72-, 2 = NO2-, 3 = PO43-, 4 = MnO42-, 5 = I-, 6 = CO32-, 7 = Br-, 8 = NO3-, 9 = Cl-, 10 = SO42-, 11 = ClO42-, 12 = F-, 13 = IO3-, 14 = AcO-, 15 = WO42-)

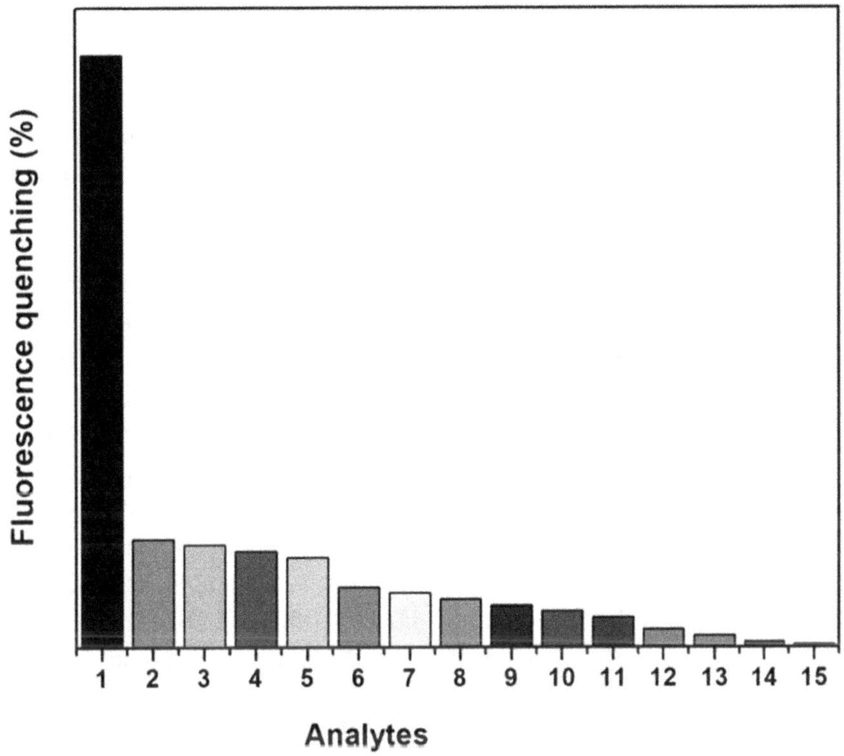

The fluorescence turn-on detection of F^- was achieved by Lobkovsky groups using a lanthanide-based MOF, {[Tb (BTC)]. (CH_3OH)}, (H_3BTC = 1, 3, 5-benzenetricarboxylic acid) (Chen et al., 2008). The addition of F^- enhanced the emission intensity of the MOF much more significantly than other anions such as Br^-, Cl^-, SO_4^{2-} and CO_3^{2-} due to the formation of strong hydrogen bond between F^- and the O-H group of terminal methanol molecules. Another turn-on mode MOF based fluoride probe was constructed by Bein and coworkers (Hinterholzinger et al., 2013). The MOF composite, NH_2-MIL-101 (Al), (MIL = Materials of Institute Lavoisier), first binds with fluorescein dyes in the pores to form a non-emissive host-guest system. Then the F^- ions competitively coordinate to the aluminum clusters, resulting in the decomposition of the MOF framework and ultimately the release of the luminescent dyes.

Iodide ions (I^-) from industry and nuclear waste are toxic and can induce serious health problems. Hence, few groups have reported the detection of I^- using MOFs. Zhang and coworkers constructed two cationic heterometallic-organic frameworks with formula of {[Ln_2Zn-$(L6)_3(H_2O)_4$] $(NO_3)_2$.$12H_2O$}, (Ln = Eu, Tb, L6 = 4, 4′-dicarboxylate-2, 2′ bipyridine anion) (Shi et al., 2015). These heterometallic-organic frameworks are found to be sensitive and efficient as fluorescent probes for I^- in aqueous medium with fast response and low limit of detection of 0.001 ppm. It was demonstrated that I^- confined in the channels can be oxidized to I_3^- with the help of cationic framework, resulting in the fluorescence quenching. By this mechanism the MOF display high selectivity towards I^- than the other anions including Cl^-, F^-, $H_2CO_2^-$, NO_3^-, $H_2PO_4^-$, HSO_3^-, HCO_3^-, CO_3^{2-}, SO_4^{2-}, SO_3^{2-} and PO_4^{3-} via fluorescence turn-off process. The sensing experiments can be recycled for upto ten times by immersing the MOFs in a saturated KNO_3 solution.

Organic Compounds

Nitro-explosives are environment pollutants which are related to modern bombs in criminal intent or terrorist attacks. It causes strong eye or skin irritation and damage the respiratory system. Therefore, methods for detection and determination of nitro-explosives have attracted intense research. Many luminescent MOFs have reported for sensing nitro-explosives. The mechanism of fluorescence sensing of MOFs are assigned as π-π interactions, binding reaction, photoinduced electron transfer between the MOF and the analytes. The host structure and pore sizes also contribute the sensing properties of MOFs. In 2009, Li and coworkers for the first time reported the sensing of nitro-explosives with MOFs (Lan et al., 2009). A microporous luminescent MOF, $[Zn_2(bpdc)_2(bpee)]$, (bpdc = 4, 4′-biphenyldicarboxylate, bpee = 1, 2-bipyridylethene) was prepared and is applied for detection of 2, 4 dintitrotoulene (DNT) and 2, 3-dimethyl-2, 3-dinitrobutene (DMNB) with fast response. In this system, the microporous structure of the MOF favour the π-π interactions between the aromatic ligands and analytes, leading to fluorescence quenching by electron transfer. The tuning of pore sizes and electronic structure of the MOFs can target host-guest interactions which is useful for explosive sensing applications. Bu and coworkers synthesized 3B MOF, $[NH_2(CH_3)_2]_2[Cd_{17}(L8)_{12}(\mu_3\text{-}H_2O)_4(DMF)_2(H_2O)_2]$, ($H_3L8$ = 2, 4, 6-tris(1-(3-carboxyl-phenoxy)yl-methyl] mesitylene) and demonstrated as fluorescent probes for electron deficient nitrocompouds in solution or vapour phase including nitrobenzene (NB), 1, 4-dinitrobenzene (2, 4-NB), 1, 3-dinitrobenzene (1, 3-NB) and 2, 4-dinitrotoulene (2, 4-NT) (Tian et al., 2014). The limit of detection of nitrobenzene was found to be 0.135 ppm. The mechanism of sensing is due to the effective electron transfer from electron rich ligand to the electron withdrawing nitro-aromatic explosives. Zhao and coworkers constructed a tetraphenylethylene derivative based new MOF, $[Zn_2(H_2L9)_2(4, 4'\text{-Bpy})_2(H_2O)_3(H_2O)]$, (L9 = 4, 4′, 4″, 4‴-(1, 4-phenylene bis-(2-phenylethene-2, 1,1-triyl) tetrabenzoic acid), which displays high selectivity and sensitivity for 2, 4, 6-trinitrophenol (TNP) by fluorescence turn off response (Deng et al., 2017). The limit of detection of TNP was determined to be 110 ppb. The authors assigned the mechanism of sensing to fluorescence resonance energy transfer (FRET), since there is as significant spectral overlap between the absorption spectrum of TNP and emission spectrum of the MOF. In another work, Du and coworkers developed dual emissive Rhodamine 6G (Rh6G) based MOF, $\{(NH_2Me_2)[Zn_3(\mu_3\text{-}OH)(tpt)(TZB)_3](DMF)_{12}\}_n$, (tpt = 2, 4, 6-tri(4-pyridyl)-1,3,5-triazine, H_2TZB = 4-(1H-tetrazol-5-yl) benzoic acid) which was found to be effective for selective detection of TNP in the presence of other nitro-aromatic molecules, Figure 4 (Chen et al., 2017a).[68] In this system, the electrostatic interactions between the TNP and Lewis base of the ligands induce the change of the fluorescence properties. Liu and coworkers explored two MOFs, TMU-31 and TMU-32 for sensing of nitroaromatic compounds (Tehrani et al., 2017). The fluorescence quenching efficiency of the MOFs follows the order of 1, 3-dinitrobenzene (m-DNB) > 2, 4-dinitrobenzene (o-DNB), > nitrobenzene (NB), > nitromethane (NM), 2, 4, 6-trinitrotoulene (TNT). The study reveals that urea groups inside the pore cavity of the MOFs act as binding sites for nitroanalytes through N-H---O hydrogen bonding and π-π interactions.

The fluorescence enhancement (turn on) MOF based sensor for the detection of explosives are also reported. Retailleau groups reported three MOFs, TMU-40 (Zn), TMU-40 (Cd) and TMU-40 (Co) using 5, 6-di(pyridin-4-yl)-1, 2, 3, 4-tetrahydropyrazine ligand with aggregation induced emission (AIE) properties (Rouhani et al., 2018). These MOFs have the capabilities of detecting energetic phenol derivative by fluorescence turn on process. The π-π interaction between the phenol and electron poor ring of the ligands leads to rigidity of the MOF framework to increase the fluorescence intensity.

Figure 4. Relative fluorescence intensity of dye to ligand of Rh6G-MOF after addition of 200 ppm of various analytes, (1 = Rh6G, 2 = Benzene, 3 = Toluene, 4 = Bromobenzene, 5 = Chlorobenzene, 6 = p-Xylene, 7 = m-Xylene, 8 = o-Xylene, 9 = Pyridine, 10 = 2, 6-DNB, 11 = m-DNB, 12 = NB, 13 = PNT, 14 = TNP)

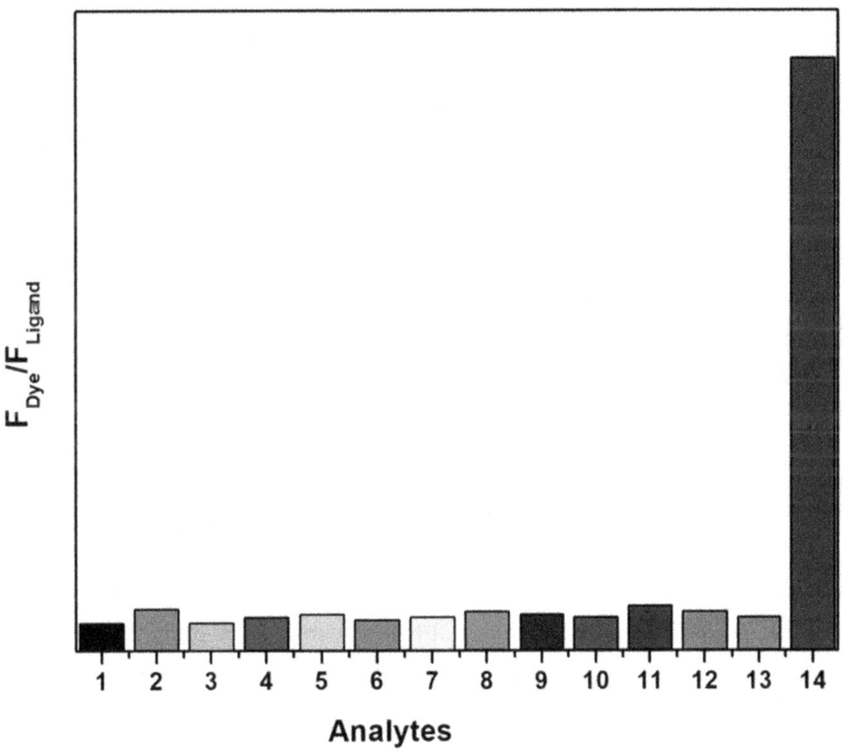

Wang and coworkers synthesized three MOFs namely TABD-MOF (Mg), TABD-MOF (Ni) and TABD-MOF (Co), [(TABD-COOH = 4, 4′-(Z Z)-1, 4-diphenylbuta-1, 3-diene-1, 4 diyl) dibenzoic acid] for detection of heterocyclic compounds (Guo et al., 2014). The study reveals that all these MOFs efficiently probe the 5-membered ring heterocyclic compounds in a fluorescence turn on mode. Since the C=N and N=N bonds in the heterocyclic ring competitively coordinated to the metal clusters and leads to the destruction of the MOF structure and subsequently AIE linkers are released which form emissive aggregates in solution. These MOFs can detect 5-nitro-2, 4-dihydro-3H-1, 2, 4-triazole-3-one (NTO) in both solution and solid phase. The test show color change as well as turn on emission.

Volatile organic compounds (VOCs) such as benzene and its derivatives are toxic pollutants and cause environmental problems and climate change. Recently, Yan groups incorporated Eu^{3+} ions into nanocrystalline MOF, $[Zr_6 (\mu_3\text{-}O) (OH)_4 (bpy)_{12}]$, leveled as bpy-UiO, (bpy = 2, 2-bipyridine-5, 5-dicarboxylic acid) which could detect aromatic VOCs with high performance (Zhou & Yan, 2016). In this system, the MOF and VOCs binding event could be recognized with 2D readouts that combined the intensity ratio of the ligand-based emission to Eu^{3+} emission (I_L/I_{Eu}). Since the fluorescence characteristics of the MOFs changes with VOCs properties, the identities of the VOCs can be recognized. Li and coworkers reported a Zn-MOF, $[Zn_2(bpdc)_2(bpee)]$, (bpdc = 4, 4′-biphenyldicarboxylate, bpee = 1, 2-bipyridylethene) as a sensor for benzoquinone (BQ) via a fluorescence turn off process (Lan et al., 2009). When the MOF is exposed to BQ, 94% of its fluorescent was quenched due to the electron transfer. Some MOF-based sensors with turn on fluorescence response have also been developed for sensing VOCs. Zhao and coworkers constructed a layered MOF, named NUS-1 (NUS = National University of Singapore), from Zn_4O-like SBUs and aggregation induced emissive (AIE) ligands, H_2DPEB (H_2DPEB = 4, 4′-(2, 2-diphenylethene-1, 1-diyl) dibenzoic acid) for sensing VOCs (Zhang et al., 2014). The motion of the dangling phenyl rings is restricted by C-H---π interactions in the DPEB crystals, while it is free in a MOF matrix. The free rotation of the dangling phenyl rings quenches the florescence of the as synthesized NUS-1, reducing the quantum yield to 17% (versus 79% in crystalline H_2DPEB). In order to test NUS-1 for VOCs sensing, they first evacuated the MOF to remove free solvent molecules, producing the activated species NUS-1a. After soaking in various VOCs, the crystalline NUS-1a displays obvious turn-on fluorescence under UV radiation in the evidence of noticeable intensity changes and peak shifts in the emission spectrum. NUS-1a shows an emission maximum bathochromically shifted 18 nm from that of pristine NUS-1a in benzene solution, and a hypochromic shift of 28 nm in mesitylene. Such large wavelength shifts can be ascribed to the different conformations of the dangling phenyl rings caused by the sensor-analyte interactions. When sensing molecules, such as benzene or cyclohexane, which tend to lock phenyl rings in a coplanar conformation, the degree of π electron conjugation in the MOF increases, leading to a red shift. However, the phenyl rings are locked into a perpendicular conformation in solvents like tetrahydrofuran (THF) or toluene, which leads to a decrease of conjugation and a correspondingly blue shifts were observed. Du and coworkers constructed a MOF by assembly of Tb^{3+}, H_2oba (4, 4′-oxybis (benzoic acid) and Hatz (3-amino-1, 2, 4-triazole) to realize the sensing of p-xylene (Chen et al., 2017b). This MOF display a significant fluorescent enhancement due to the photoinduced electron transfer from p-xylene to H_2oba ligand. Duan groups designed a white light emitting MOF composite by encapsulating green emitting $[Ir (CF_3\text{-}ppyF_2) (bpy)]^+$ and red emitting $[Ru\text{-}(bpy)_3]^{2+}$, (CF_3-$ppyF_2$ = 5-trifluomethyl) into blue emitting Zn-MOF, $\{(Me_2NH_2)[Zn_2(L10)(H_2O)].4DMF\}$, ($H_5L10$ = 2, 5-(6-(3-carboxyphenylamino)-(1, 3, 5-triazine-2, 4-diyldiimin-O)-diterephthalic acid (Zhao et al., 2018). This MOF composite was applied as a multi-ratiometric probe to detect fluorobenzene, where color change from white to orange can be observed by naked eyes. Meng and coworkers reported a luminescent 2D Cu-MOF, $[Cu_6 (L11)_6(3H_2O)\text{-}DMSO]$, (L11 = 5, 6 diphenyl-1, 2, 4-triazine-3-thiol, DMSO = dimethylsulphoxide), which displays strong fluorescence quenching response in presence of benzene and toluene via energy transfer process (Bai et al., 2006).

Jena groups presented a MOF, $\{[Cu(L13)(I)]_{2n}\cdot 2nDMF\cdot nMeCN\}$, (L13 = 4′-(4-methoxyphenyl)-4,2′:6′,4″-terpyridine) to realize the probing solvents, toxic halobenzenes, N-heterocycles, amines and nitro-aromatic explosive vapors with a rapid fluorescence response (Khatua et al., 2015). In this process, single-crystal-to-single-crystal (SC-SC) phase transition processes were realized and the framework integrity remains similar. The color change could be clearly observed by naked eyes. Formaldehyde is a widely used in industry and organic transformation. However, long-term exposure of formaldehyde causes different diseases such as asthma, cancer. Huang and coworkers synthesized highly porous dual emissive Eu/Zr-MOF composite which is highly sensitive towards formaldehyde (Li et al., 2017). On excitation at 365 nm, the ratio of intensity of the two emissions centers of Eu/Zr-MOF composite were found to be increases almost linearly with the amount of formaldehyde. This sensing mechanism was attributed to the binding interaction between the free amino group in Eu/Zr-MOF and formaldehyde. The electron transfer from the amino group increased with the positively charged formaldehyde containing the lone pair electrons, leading to a sharp enhancement in luminescence at 465 nm, while the characteristic emission of Eu^{3+} at 615 nm increased slightly.

FUTURE RESEARCH DIRECTIONS

Although MOF-based materials have shown promising results, future works is still required to enhance the sensor performance in terms of sensitivity, selectivity, stability and reusability. First, the function of the MOFs in sensing platform should be better understood which will be helpful in designing the sensor and optimized its performance. Second, MOFs pore structure is crucial for many sensing applications and hence new synthesis or surface functionalization techniques are necessary to tune the MOF structure and enhance its sensing activities and stability. Third, the detection function of the LMOFs as analytes sensors arises not only from the diverse porous structure but also dependent on the interactions among different subsets including inorganic/organic component features, guest molecules etc. As a result depth studies of the structure luminescence relationships as well as host-guest interactions should be explored by both experimental investigation and theoretical calculations. Fourth, most of the established LMOFs sensors are derived from single emission center such as fluorescent ligands or lanthanide ions. Therefore, ratiometric LMOFs sensors that utilize the ratio between the intensity of two or multiple emission centers with the same MOF should be developed. The rational design and construction of the ratiometric sensors based on LMOFs should be developed. Moreover, the stability of MOF-based sensing materials should to be improved, especially for sensors that work under an acid or basic condition. Finally, most of the LMOFs sensors demonstrations were carried out using laboratory conditions. Therefore, extensive research and developments are still needed in order to find its reliability as sensor performance in complex environmental conditions, such as real water samples. In addition, the sensor selectivity and reliability should be improved as they are essential requirements for the practical use of the sensors.

Apart from sensing applications, the MOFs have probabilities in other areas too. Nanoscale luminescent MOFs have potential applications in various fields such as sensing in living cells, drug delivery, medical diagnosis and treatment. Research in these areas can be explored more. In addition, the ideas of incorporating MOFs to optical or electronic devices, such as thin films or membrane can also be explored. Bright future of MOFs based materials can be expected with collaborative works of scientists from different subjects.

Table 1. List of selected LMOFs for sensing applications

Sensing application MOF Analyte Ref.
Cations [Tb (BPTC-2)]-(Me_2NH_2). $(Solvents)_x$ Fe^{3+} 20
[Ti_2 $(HDOBDC)_2$ (H_2DOBDC]) Fe^{3+} 21
[Cd_2Na (L3) $(BDC)_{2.5}$]. $9H_2O$ Fe^{3+} 22
MIL-53(Al) Fe^{3+} 23
BPIEI-CQDs/ZIF-8 Cu^{2+} 26
[Cd_2 $(DTP)_2$ $(bibp)_{1.5}]_n$ Cu^{2+} 27
MIL-53-L5 Cu^{2+} 28
[Zn_2 (O-BTC)(4, 4′-BPY)$_{0.5}$$(H_2O)_{1.5}$·$(DMA)_{0.5}$]n Al^{3+} 29
[Co_2 (dmimpym) $(nda)_2]_n$ Al^{3+} 30
Mg-TPP-DHBDC Al^{3+} 31
[Tb4-(μ_6-L4)$_2$ (μ-HCOO)(μ_3-OH)$_3$(μ_3-O)$(DMF)_2$$(H_2O)_4]_n$ Ce^{2+} 32
[TbK (BPDSDC) (DMF) $(H_2O)_2$] Cr^{3+} 33
UiO-66-butyne Hg^{2+} 34
{[Zn (4, 4′-AP) (5-AIA)]-$(DMF)_{0.5}$}n Hg^{2+} 35
Eu-MOP Hg^{2+} 36
PCN-224 Hg^{2+} 37
Fluorene-MOFs Pb^{2+} 38
[$In_2(OH)_2$(H_2TTHA)$(H_2O)_2$] UO_2^{2+} & Ru^{3+} 39
[$Eu(PDA)_3$ $Mn_{1.5}(H_2O)_3]_3$.$5H_2O$ Zn^{2+} 40
[$Eu(PDA)_3$ $Fe_{1.5}(H_2O)_3$].$1.5H_2O$ Mg^{2+} 41
Eu (III) Zr-MOF Cd^{2+} 43
Anions Tb(III)-MOF PO_4^{3-} 45
MOF/ZnO PO_4^{3-} 46
Eu-H_3BTB PO_4^{3-} 47
{[Zn_4-$(L7^{3-})_2$ (O^{2-}) $(H_2O)_2$]. 4EtOH$\}_n$ PO_4^{3-} 48
NH_2-MIL-53(Al) ClO^- 49
ZIF-90 CN^- 50
{[Zn_3 (bpanth)-$(oba)_3$]. 2DMF$\}_n$ $Cr_2O_7^{2-}$/CrO_4^{2-} 52
[Zr_6O_4 $(OH)_7$ $(H_2O)_3$ $(BTBA)_3$] $Cr_2O_7^{2-}$ 53
[Cd (TPTZ) $(H_2O)_2$ (HCOOH) $(IPA)_2$] $Cr_2O_7^{2-}$ 54
{[Cd (L3) (BPDC)]. 2 H_2O} $Cr_2O_7^{2-}$ 55
MIL-121 $Cr_2O_7^{2-}$ 56
ln MOF MnO_4^- 57
{[Ba_3 $La_{0.5}$ (μ_3-L4)$_{2.5}$ $(H_2O)_3$ (DMF)].$(DMF)_3$} MnO_4^- 58
[Eu(O-cpia) (phen) MOF] F^- 59
(Eu-$(TTA)_3$-1) F^- 60
Eu-MOF-5-bop F^- 61
{[Tb(BTC)]. (CH_3OH)} F^- 62
NH_2-MIL-101 (Al) F^- 63
[Ln_2 Zn-$(L6)_3$ $(H_2O)_4$] $(NO_3)_2$.12 H_2O] I^- 64
Nitro-compounds [Zn_2 $(bpdc)_2$ (bpee)] DNT & DMNB 65
[NH_2 $(CH_3)_2]_2$ [Cd_{17} $(L8)_{12}$ (μ_3-$H_2O)_4$ $(DMF)_2$ $(H_2O)_2$] NB 66
[Zn_2 (H_2L9)$_2$ (4, 4′-Bpy)$_2$ $(H_2O)_3$ (H_2O)] TNP 67
(Rh6G)+{(NH_2 Me_2) [Zn_3 (μ_3-OH) (tpt) $(TZB)_3$] $(DMF)_{12}\}_n$ TNP 68
TMU-31& TMU-32 m-DNB 69
TMU-40 (Zn), TMU-40 (Cd) & TMU-40 (Co) Phenol derivative 70
TABD-MOF (Mg), TABD-MOF (Ni) TABD-MOF (Co) NTO 71
Volatile organic compounds [Zr_6 (μ_3-O) $(OH)_4$ $(bpy)_{12}$] VOCs 72
[Zn_2 $(bpdc)_2$ (bpee)] BQ 65
NUS-1 VOCs 73
Tb^{3+}-MOF (H_2oba + Hatz) p-xylene 74
{(Me_2NH_2)[Zn_2(L10)(H_2O)].4DMF} Fluorobenzene 75
[$Cu_6$$(L11)_6$($3H_2O$)-DMSO] Benzene & Toluene 76
Eu/Zr-MOF Formaldehyde 78

CONCLUSION

Table 1 is the summary of the recent MOFs based sensors applied for detecting different analytes. This chapter discusses structure and luminescent properties of metal-organic frameworks (MOFs). These luminescent MOFs have been demonstrated as sensing materials for different analytes such as cations, anions, aromatic nitroexplosives and volatile organic compounds. It has been demonstrated that the coordination environments of the metal ions, framework structure, size and geometry of the pore and interaction with the analytes induced different luminescent properties of MOFs, which can be monitored by fluorescence changes in terms of intensity/emission frequency shifts. The detection processes of the analytes with LMOFs are either fluorescence "turn off" or "turn on" process. In addition, the sensing performances of LMOFs were found to be further improved on doping with other functional materials. For example, lanthanide-based MOFs and fluorophore modified MOFs are sensing materials in which luminescent sensors act as the guest molecules incorporated in MOFs can emit or induced the luminescence. The mechanism of sensing of MOFs is assigned to be effective host-guest interaction via energy transfer. Since, these analytes are hazardous to environment, detecting them will be helpful for monitoring environmental pollution, which in turn will prevent causing of various diseases.

REFERENCES

Allendorf, M. D., Bauer, C. A., Bhakta, R. K., & Houk, R. J. T. (2009). Luminescent metal-organic frameworks. *Chemical Society Reviews, 38*(5), 1330–1352. doi:10.1039/b802352m PMID:19384441

Bai, Y., He, G., Zhao, Y., Duan, C., Dang, D., & Meng, Q. (2006). Porous material for absorption and luminescent detection of aromatic molecules in water. *Chemical Communications (Cambridge), 43*(14), 1530–1532. doi:10.1039/b517002h PMID:16575450

Carter, K. P., Young, A. M., & Palmer, A. E. (2014). Fluorescent sensors for measuring metal ions in living systems. *Chemical Reviews, 114*(8), 4564–4601. doi:10.1021/cr400546e PMID:24588137

Chen, B., Wang, L., Zapata, F., Qian, G., & Lobkovsky, E. B. (2008). A luminescent microporous metal-organic framework for the recognition and sensing of anions. *Journal of the American Chemical Society, 130*(21), 6718–6719. doi:10.1021/ja802035e PMID:18452294

Chen, D. M., Zhang, N. N., Liu, C. S., & Du, M. (2017). Dual-emitting dye-MOF composite as a self-calibrating sensor for 2, 4, 6-trinitrophenol. *ACS Applied Materials & Interfaces, 9*(29), 24671–24677. doi:10.1021/acsami.7b07901 PMID:28682050

Chen, D. M., Zhang, N. N., Liu, C. S., & Du, M. (2017). Template-directed synthesis of a luminescent Tb-MOF material for highly selective Fe^{3+} and Al^{3+} ion detection and VOC vapor sensing. *Journal of Materials Chemistry. C, Materials for Optical and Electronic Devices, 5*(9), 2311–2317. doi:10.1039/C6TC05349A

Chen, S. G., Shi, Z. Z., Qin, L., Jia, H. L., & Zheng, H. G. (2017). Two New Luminescent Cd (II)-Metal-Organic Frameworks as Bifunctional Chemosensors for Detection of Cations Fe^{3+}, Anions CrO_4^{2-} and $Cr_2O_7^{2-}$ in Aqueous Solution. *Crystal Growth & Design, 17*(1), 67–72. doi:10.1021/acs.cgd.6b01197

Chen, W., Wang, J. Y., Chen, C., Yue, Q., Yuan, H. M., Chen, J. S., & Wang, S. N. (2003). Photoluminescent metal-organic polymer constructed from trimetallic clusters and mixed carboxylates. *Inorganic Chemistry, 42*(4), 944–946. doi:10.1021/ic025871j PMID:12588123

Chen, W. M., Meng, X. L., Zhuang, G. L., Wang, Z., Kurmoo, M., Zhao, Q. Q., Wang, X. P., Shan, B., Tung, C. H., & Sun, D. (2017). A superior fluorescent sensor for Al^{3+} and UO_2^{2+} based on a Co (II) metal-organic framework with exposed pyrimidyl Lewis base sites. *Journal of Materials Chemistry. A, Materials for Energy and Sustainability, 5*(25), 13079–13085. doi:10.1039/C7TA01546A

Collins, D. J., & Zhou, H. C. (2007). Hydrogen storage in metal-organic frameworks. *Journal of Materials Chemistry, 17*(30), 3154–3160. doi:10.1039/b702858j

Deng, Y., Chen, N., Li, Q., Wu, X., Huang, X., Lin, Z., & Zhao, Y. (2017). Highly fluorescent metal-organic frameworks based on a benzene-cored tetraphenylethene derivative with the ability to detect 2, 4, 6-trinitrophenol in water. *Crystal Growth & Design, 17*(6), 3170–3177. doi:10.1021/acs.cgd.7b00131

Ding, B., Liu, S. X., Cheng, Y., Guo, C., Wu, X. X., Guo, J. H., Liu, Y. Y., & Li, Y. (2016). Heterometallic Alkaline Earth-Lanthanide Ba^{II}-La^{II} Microporous Metal-Organic Framework as Bifunctional Luminescent probes of Al^{3+} and MnO_4^-. *Inorganic Chemistry, 55*(9), 4391–4402. doi:10.1021/acs.inorgchem.6b00111 PMID:27088966

Du, N., Song, J., Li, S., Chi, Y. X., Bai, F. Y., & Xing, Y. H. (2016). A Highly Stable 3D Luminescent Indium-Polycarboxylic Framework of UO_2^{2+}, Ru^{3+} and Biomolecules Thiamines. *ACS Applied Materials & Interfaces, 8*(42), 28718–28726. doi:10.1021/acsami.6b09456 PMID:27748584

Evans, J. D., Sumby, C. J., & Doonan, C. J. (2014). Post-synthetic metalation of metal-organic frameworks. *Chemical Society Reviews, 43*(16), 5933–5951. doi:10.1039/C4CS00076E PMID:24736674

Feng, D., Chung, W. C., Wei, Z., Gu, Z. Y., Jiang, H. L., Chen, Y. P., Darensbourg, D. J., & Zhou, H. C. (2013). Construction of ultrastable porphyrin Zr metal-organic frameworks through linker elimination. *Journal of the American Chemical Society, 135*(45), 17105–17110. doi:10.1021/ja408084j PMID:24125517

Guo, Y., Feng, X., Han, T., Wang, S., Lin, Z., Dong, Y., & Wang, B. (2014). Tuning the luminescence of metal-organic frameworks for detection of energetic heterocyclic compounds. *Journal of the American Chemical Society, 136*(44), 15485–15488. doi:10.1021/ja508962m PMID:25325884

Hao, J. N., & Yan, B. (2016). Ln^{3+} post-functionalized metal-organic frameworks for color tunable emission and highly sensitive sensing of toxic anions and small molecules. *New Journal of Chemistry, 40*(5), 4654–4661. doi:10.1039/C5NJ03419A

He, J., Xu, J., Yin, J., Li, N., & Bu, X. H. (2019). Recent advances in luminescent metal-organic frameworks for chemical sensors. *Science China Materials, 62*(11), 1655–1678. doi:10.100740843-019-1169-9

He, T., Zhang, Y. Z., Kong, X. J., Yu, J., Lv, X. L., Wu, Y., Guo, Z. J., & Li, J. R. (2018). Zr (IV)-based metal-organic framework with T-shaped ligand: Unique structure, high stability, selective detection, and rapid adsorption of $Cr_2O_7^{2-}$ in water. *ACS Applied Materials & Interfaces, 10*(19), 16650–16659. doi:10.1021/acsami.8b03987 PMID:29733570

Hinterholzinger, F. M., Rühle, B., Wuttke, S., Karaghiosoff, K., & Bein, T. (2013). Highly sensitive and selective fluoride detection in water through fluorophore release from a metal-organic framework. *Scientific Reports*, *3*(1), 1–7. doi:10.1038rep02562 PMID:24008779

Horcajada, P., Chalati, T., Serre, C., Gillet, B., Sebrie, C., Baati, T., Eubank, J. F., Heurtaux, D., Clayette, P., Kreuz, C., Chang, J.-S., Hwang, Y. K., Marsaud, V., Bories, P.-N., Cynober, L., Gil, S., Férey, G., Couvreur, P., & Gref, R. (2010). Porous metal-organic framework nanoscale carriers as a potential platform for drug delivery and imaging. *Nature Materials*, *9*(2), 172–178. doi:10.1038/nmat2608 PMID:20010827

Hu, Z., Deibert, B. J., & Li, J. (2014). Luminescent metal-organic frameworks for chemical sensing and explosive detection. *Chemical Society Reviews*, *43*(16), 5815–5840. doi:10.1039/C4CS00010B PMID:24577142

Ji, G., Gao, X., Zheng, T., Guan, W., Liu, H., & Liu, Z. (2018). Post synthetic metalation metal-organic framework as a fluorescent probe for the ultrasensitive and reversible detection of PO_4^{3-} ions. *Inorganic Chemistry*, *57*(17), 10525–10532. doi:10.1021/acs.inorgchem.8b00313 PMID:30095895

Jin, F., Zhang, Y., Wang, H. Z., Zhu, H. Z., Yan, Y., Zhang, J., Wu, J. Y., Tian, Y.-P., & Zhou, H.-P. (2013). Diverse structural Ag (I) supramolecular complexes constructed from multidentate dicyanoisophorone-based ligands: Structure and enhanced luminescence. *Crystal Growth & Design*, *13*(5), 1978–1987. doi:10.1021/cg400025j

Karmakar, A., Kumar, N., Samanta, P., Desai, A. V., & Ghosh, S. K. (2016). A post-synthetically modified MOF for selective and sensitive aqueous-phase detection of highly toxic cyanide ions. *Chemistry (Weinheim an der Bergstrasse, Germany)*, *22*(3), 864–868. doi:10.1002/chem.201503323 PMID:26584835

Keeffe, M., & Yaghi, O. M. (2012). Deconstructing the Crystal Structures of Metal Organic Frameworks and Related Materials into Their Underlying Nets. *Chemical Reviews*, *112*(2), 675–702. doi:10.1021/cr200205j PMID:21916513

Khatua, S., Goswami, S., Biswas, S., Tomar, K., Jena, S. H., & Konar, S. (2015). Stable multiresponsive luminescent MOF for colorimetric detection of small molecules in selective and reversible manner. *Chemistry of Materials*, *27*(15), 5349–5360. doi:10.1021/acs.chemmater.5b01773

Kolliopoulos, A. V., Kampouris, D. K., & Banks, C. E. (2015). Rapid and portable electrochemical quantification of phosphorus. *Analytical Chemistry*, *87*(8), 4269–4274. doi:10.1021/ac504602a PMID:25856498

Kozlowski, H., Janicka-Klos, A., Brasun, J., Gaggelli, E., Valensin, D., & Valensin, G. (2009). Copper, iron, and zinc ions homeostasis and their role in neurodegenerative disorders (metal uptake, transport, distribution and regulation). *Coordination Chemistry Reviews*, *253*(21), 2665–2685. doi:10.1016/j.ccr.2009.05.011

Lakowicz, J. R. (2006). *Principles of Fluorescence Spectroscopy*. Springer. doi:10.1007/978-0-387-46312-4

Lan, A., Li, K., Wu, H., Olson, D. H., Emge, T. J., Ki, W., Hong, M., & Li, J. (2009). A luminescent microporous metalorganic framework for the fast and reversible detection of high explosives. *Angewandte Chemie International Edition*, *48*(13), 2334–2338. doi:10.1002/anie.200804853 PMID:19180622

Lee, J., Farha, O. K., Roberts, J., Scheidt, K. A., Nguyen, S. T., & Hupp, J. T. (2009). Metal-organic framework materials as catalysts. *Chemical Society Reviews*, *38*(5), 1450–1459. doi:10.1039/b807080f PMID:19384447

Li, C. M., Huang, J. P., Zhu, H., Liu, L., Feng, Y., Hu, G., & Yu, X. (2017). Dual-emitting fluorescence of Eu/Zr-MOF for ratiometric sensing formaldehyde. *Sensors and Actuators. B, Chemical*, *253*, 275–282. doi:10.1016/j.snb.2017.06.064

Li, J. R., Ma, Y., McCarthy, M. C., Sculley, J., Yu, J., Jeong, H. K., Balbuena, P. B., & Zhou, H. C. (2011). Carbon dioxide capture-related gas adsorption and separation in metal-organic frameworks. *Coordination Chemistry Reviews*, *255*(15), 1791–1823. doi:10.1016/j.ccr.2011.02.012

Li, L., Chen, Q., Niu, Z. G., Zhou, X. H., Yang, T., & Huang, W. (2016). Lanthanide metal-organic frameworks assembled from a fluorene-based ligand: Selective sensing of Pb^{2+} and Fe^{3+} ions. *Journal of Materials Chemistry. C, Materials for Optical and Electronic Devices*, *4*(9), 1900–1905. doi:10.1039/C5TC04320D

Li, Y. P., Zhu, X. H., Li, S. N., Jiang, Y. C., Hu, M. C., & Zhai, Q. G. (2019). Highly selective and sensitive turn off-on fluorescent probes for sensing Al^{3+} ions designed by regulating the excited-state intramolecular proton transfer process in metal-organic frameworks. *ACS Applied Materials & Interfaces*, *11*(12), 11338–11348. doi:10.1021/acsami.8b20410 PMID:30834744

Lin, X. M., Gao, G. M., Zheng, L. Y., Chi, Y. W., & Chen, G. N. (2014). Encapsulation of strongly fluorescent carbon quantum dots in metal-organic frameworks for enhancing chemical sensing. *Analytical Chemistry*, *86*(2), 1223–1228. doi:10.1021/ac403536a PMID:24328057

Liu, C., & Yan, B. (2016). A novel photofunctional hybrid materials of pyrene functionalized metal-organic framework with conformation change for fluorescence sensing of Cu^{2+}. *Sensors and Actuators. B, Chemical*, *235*, 541–546. doi:10.1016/j.snb.2016.05.127

Lu, T., Zhang, L. C., Sun, M., Deng, D., Su, Y., & Lv, Y. (2016). Amino-functionalized metal-organic frameworks nanoplates-based energy transfer probe for highly selective fluorescence detection of free chlorine. *Analytical Chemistry*, *88*(6), 3413–3420. doi:10.1021/acs.analchem.6b00253 PMID:26905374

Lustig, W. P., Mukherjee, S., Rudd, N. D., Desai, A. V., Li, J., & Ghosh, S. K. (2017). Metal-organic frameworks: Functional luminescent and photonic materials for sensing applications. *Chemical Society Reviews*, *46*(11), 3242–3285. doi:10.1039/C6CS00930A PMID:28462954

Marzo, A. M. L., Pons, J., & Merkoci, A. (2013). All-integrated and highly sensitive paper based device with sample treatment platform for Cd^{2+} immunodetection in drinking/tap waters. *Analytical Chemistry*, *85*(7), 3532–3538. doi:10.1021/ac3034536 PMID:23445438

Mon, M., Bruno, R., Ferrando-Soria, J., Armentano, D., & Pardo, E. (2018). Metal-organic framework technologies for water remediation: Towards a sustainable ecosystem. *Journal of Materials Chemistry. A, Materials for Energy and Sustainability*, *6*(12), 4912–4947. doi:10.1039/C8TA00264A

Nuijs, A. L. N. V., Tarcomnicu, I., & Covaci, A. (2011). Application of hydrophilic interaction chromatography for the analysis of polar contaminants in food and environmental samples. *Journal of Chromatography. A*, *1218*(35), 5964–5974. doi:10.1016/j.chroma.2011.01.075 PMID:21316059

Pankajakshan, A., Kuznetsov, D., & Mandal, S. (2019). Ultrasensitive detection of Hg (II) ions in aqueous medium using zinc-based metal-organic framework. *Inorganic Chemistry*, *58*(2), 1377–1381. doi:10.1021/acs.inorgchem.8b02898 PMID:30628772

Petrovic, M., Farre, M., Alda, M. L. D., Perez, S., Postigo, C., Kock, M., Radjenovic, J., Gros, M., & Barcelo, D. (2010). Recent trends in the liquid chromatography-mass spectrometry analysis of organic contaminants in environmental samples. *Journal of Chromatography. A*, *1217*(25), 4004–4017. doi:10.1016/j.chroma.2010.02.059 PMID:20334864

Que, E. L., Domaille, D. W., & Chang, C. J. (2008). Metals in neurobiology: Probing their chemistry and biology with molecular imaging. *Chemical Reviews*, *108*(5), 1517–1549. doi:10.1021/cr078203u PMID:18426241

Rouhani, F., Morsali, A., & Retailleau, P. (2018). Simple one-pot preparation of a rapid response AIE fluorescent metal-organic framework. *ACS Applied Materials & Interfaces*, *10*(42), 36259–36266. doi:10.1021/acsami.8b12404 PMID:30259725

Samanta, P., Desai, A. V., Sharma, S., Chandra, P., & Ghosh, S. K. (2018). Selective recognition of Hg^{2+} ion in water by a functionalized metal-organic framework (MOF) based chemodosimeter. *Inorganic Chemistry*, *57*(5), 2360–2364. doi:10.1021/acs.inorgchem.7b02426 PMID:29461048

Shi, P. F., Hu, H. C., Zhang, Z. Y., Xiong, G., & Zhao, B. (2015). Heterometal-organic frameworks as highly sensitive and highly selective luminescent probes to detect I^- ions in aqueous solutions. *Chemical Communications (Cambridge)*, *51*(19), 3985–3988. doi:10.1039/C4CC09081K PMID:25620743

Tang, Y. Y., Ding, C. X., Ng, W. S., & Xie, Y. S. (2013). Synthesis, structures and photoluminescence of Zn (II), Ag (I), Cu (I) and Co (II) coordination polymers of a tetrapyridyl ligand. *RSC Advances*, *3*(39), 18134–18141. doi:10.1039/c3ra43405b

Tehrani, A. A., Esrafili, L., Abedi, S., Morsali, A., Carlucci, L., Proserpio, D. M., Wang, J., Junk, P. C., & Liu, T. F. (2017). Urea metal-organic frameworks for nitro-substituted compounds sensing. *Inorganic Chemistry*, *56*(3), 1446–1454. doi:10.1021/acs.inorgchem.6b02518 PMID:28085264

Tian, D., Li, Y., Chen, R. Y., Chang, Z., Wang, G. Y., & Bu, X. H. (2014). A luminescent metal-organic framework demonstrating ideal detection ability for nitroaromatic explosives. *Journal of Materials Chemistry. A, Materials for Energy and Sustainability*, *2*(5), 1465–1470. doi:10.1039/C3TA13983B

Tranchemontagn, D. J., Mendoza-Corte, J. L., Keeffe, M., & Yaghi, O. M. (2009). Secondary building units, nets and bonding in the chemistry of metal-organic frameworks. *Chemical Society Reviews*, *38*(5), 1257–1283. doi:10.1039/b817735j PMID:19384437

Tsuge, K., Chishina, Y., Hashiguchi, H., Sasaki, Y., Kato, M., Ishizaka, S., & Kitamura, N. (2016). Luminescent copper (I) complexes with halogenidobridged dimeric core. *Coordination Chemistry Reviews*, *306*, 636–651. doi:10.1016/j.ccr.2015.03.022

Wang, H. H., Zhou, L. J., Wang, Y. L., & Liu, Q. Y. (2016). Terbium-biphenyl-3, 3′-disulfonyl-4, 4′ carboxylate framework with sulfonate sites for luminescent sensing of Cr^{3+}. *Inorganic Chemistry Communications*, *73*, 94–97. doi:10.1016/j.inoche.2016.10.006

Wang, Q. S., Li, J. J., & Li, X. (2018). A luminescent Eu (III)-based metal-organic framework as a highly effective sensor for cation and anion detections. *Sensors and Actuators. B, Chemical*, *258*, 358–364. doi:10.1016/j.snb.2017.11.075

Wang, R., Liu, X., Huang, A., Wang, W., Xiao, Z., Zhang, L., Dai, F., & Sun, D. (2016). Unprecedented solvent-dependent sensitivities in highly efficient detection of metal ions and nitroaromatic compounds by a fluorescent barium metal-organic framework. *Inorganic Chemistry*, *55*(4), 1782–1787. doi:10.1021/acs.inorgchem.5b02693 PMID:26845287

Wang, Y., Cheng, L., Liu, Z. Y., Wang, X. G., Ding, B., Yin, L., Zhou, B. B., Li, M. S., Wang, J. X., & Zhao, X. J. (2015). An Ideal Detector Composed of Two-Dimensional Cd (II)-Triazole Frameworks for Nitro-Compound Explosives and Potassium Dichromate. *Chemistry (Weinheim an der Bergstrasse, Germany)*, *21*(40), 14171–14178. doi:10.1002/chem.201502167 PMID:26291651

Wu, J. X., & Yan, B. (2017). Eu (III)-functionalized In-MOF (In (OH) bpydc) as fluorescent probe for highly selectively sensing organic small molecules and anions especially for $CHCl_3$ and MnO_4^-. *Journal of Colloid and Interface Science*, *504*, 197–205. doi:10.1016/j.jcis.2017.05.054 PMID:28551513

Xu, H., Cao, C. S., & Zhao, B. (2015). A water-stable lanthanide-organic framework as a recyclable luminescent probe for detecting pollutant phosphorus anions. *Chemical Communications (Cambridge)*, *51*(51), 10280–10283. doi:10.1039/C5CC02596F PMID:25940110

Xu, H., Gao, J. K., Qian, X. F., Wang, J. P., He, H. J., Cui, Y. J., Yang, Y., Wang, Z. Y., & Qian, G. D. J. (2016). Metal-organic framework nanosheets for fast-response and highly sensitive luminescent sensing of Fe^{3+}. *Journal of Materials Chemistry. A, Materials for Energy and Sustainability*, *4*(28), 10900–10905. doi:10.1039/C6TA03065C

Xu, H., Xiao, Y. Q., Rao, X. T., Dou, Z. S., Li, W. F., Cui, Y. J., Wang, Z. Y., & Qian, G. D. (2011). A metal-organic framework for selectively sensing of PO_4^{3-} anion in aqueous solution. *Journal of Alloys and Compounds*, *509*(5), 2552–2554. doi:10.1016/j.jallcom.2010.11.087

Xu, X. Y., & Yan, B. (2016). Eu (III) functionalized Zr-based metalorganic framework as excellent fluorescent probe for Cd^{2+} detection in aqueous environment. *Sensors and Actuators. B, Chemical*, *222*, 347–353. doi:10.1016/j.snb.2015.08.082

Yang, C. X., Ren, H. B., & Yan, X. P. (2013). Fluorescent metal-organic framework MIL-53(Al) for highly selective and sensitive detection of Fe^{3+} in aqueous solution. *Analytical Chemistry*, *85*(15), 7441–7446. doi:10.1021/ac401387z PMID:23826852

Yang, J., Wang, Z., Li, Y. S., Zhuang, Q. X., Zhao, W. R., & Gu, J. (2016). Porphyrin MOFs for reversible fluorescent and colorimetric sensing of mercury (II) ions in aqueous phase. *RSC Advances*, *6*(74), 69807–69814. doi:10.1039/C6RA13766K

Yang, L., Lian, C., Li, X., Han, Y., Yang, L., Cai, T., & Shao, C. (2017). Highly selective bifunctional luminescent sensor toward nitrobenzene and Cu^{2+} ion based on microporous metal-organic frameworks: Synthesis, structures, and properties. *ACS Applied Materials & Interfaces*, *9*(20), 17208–17217. doi:10.1021/acsami.7b01758 PMID:28467044

Yang, Z. R., Wang, M. M., & Yin, X. B. (2017). Boric-acid functional lanthanide metal-organic frameworks for selective ratiometric fluorescence detection of fluoride ions. *Analytical Chemistry*, *89*(3), 1930–1936. doi:10.1021/acs.analchem.6b04421 PMID:28208304

Yao, Z. Q., Li, G. Y., Xu, J., Hu, L., & Bu, X. H. (2018). A water-stable luminescent Zn (II) metalorganic framework as chemosensor for high-efficiency detection of Cr^{VI-} anions ($Cr_2O_7^{2-}$ and CrO_4^{2-}) in aqueous solution. *Chemistry (Weinheim an der Bergstrasse, Germany)*, *24*(13), 3192–3198. doi:10.1002/chem.201705328 PMID:29210125

Zhang, M., Feng, G., Song, Z., Zhou, Y. P., Chao, H. Y., Yuan, D., Tan, T. T., Guo, Z., Hu, Z., Tang, B. Z., Liu, B., & Zhao, D. (2014). Two-dimensional metal-organic framework with wide channels and responsive turn-on fluorescence for the chemical sensing of volatile organic compounds. *Journal of the American Chemical Society*, *136*(20), 7241–7244. doi:10.1021/ja502643p PMID:24824627

Zhang, Q., Wang, J., Kirillov, A. M., Dou, W., Xu, C., Yang, L., Fang, R., & Liu, W. (2018). Multifunctional Ln-MOF luminescent probe for efficient sensing of Fe^{3+}, Ce^{3+} and Acetone. *ACS Applied Materials & Interfaces*, *10*(28), 23976–23986. doi:10.1021/acsami.8b06103 PMID:29920195

Zhao, B., Chen, X. Y., Chen, Z., Shi, W., Cheng, P., Yan, S. P., & Liao, D. Z. (2009). A porous 3D heterometal-organic framework containing both lanthanide and high-spin Fe (II) ions. *Chemical Communications*, *21*(21), 3113–3115. doi:10.1039/b821496d PMID:19462104

Zhao, D., Liu, X. H., Zhao, Y., Wang, P., Liu, Y., Azam, M., Ai-Resayes, S. I., Lu, Y., & Sun, W. Y. (2017). Luminescent Cd (II)-organic frameworks with chelating NH_2 sites for selective detection of Fe (III) and antibiotics. *Journal of Materials Chemistry. A, Materials for Energy and Sustainability*, *5*(30), 15797–15807. doi:10.1039/C7TA03849F

Zhao, D., Wan, X. Y., Song, H. J., Hao, L. Y., Su, Y. Y., & Lv, Y. (2014). Metal-organic frameworks (MOFs) combined with ZnO quantum dots as a fluorescent sensing platform for phosphate. *Sensors and Actuators. B, Chemical*, *197*, 50–57. doi:10.1016/j.snb.2014.02.070

Zhao, H., Ni, J., Zhang, J. J., Liu, S. Q., Sun, Y. J., Zhou, H., Li, Y. Q., & Duan, C. Y. (2018). A trichromatic MOF composite for multidimensional ratiometric luminescent sensing. *Chemical Science (Cambridge)*, *9*(11), 2918–2926. doi:10.1039/C8SC00021B PMID:29732075

Zhao, X. L., Tian, D., Gao, Q., Sun, H. W., Xu, J., & Bu, X. H. (2016). A chiral lanthanide metal-organic framework for selective sensing of Fe (II) ions. *Dalton Transactions (Cambridge, England)*, *45*(3), 1040–1046. doi:10.1039/C5DT03283K PMID:26649625

Zhao, X. Q., Zhao, B., Shi, B. W., & Cheng, P. (2009). Structures and luminescent properties of a series of Ln-Ag heterometallic coordination polymers. *CrystEngComm*, *11*(7), 1261–1269. doi:10.1039/b900430k

Zheng, Y. H., Tan, C. L., & Wang, Q. M. (2012). A luminescent lanthanide complex-based anion sensor with electron-donating methoxy groups for monitoring multiple anions in environmental and biological processes. *Spectrochimica Acta. Part A: Molecular and Biomolecular Spectroscopy*, *96*, 387–394. doi:10.1016/j.saa.2012.05.064 PMID:22722072

Zhou, Y., & Yan, B. (2016). A responsive MOF nanocomposite for decoding volatile organic compounds. *Chemical Communications (Cambridge)*, *52*(11), 2265–2268. doi:10.1039/C5CC09029F PMID:26727287

ADDITIONAL READING

Elsaidi, K. S., Mohamed, M. H., Banerjee, D., & Tallapally, P. K. (2018). Flexibility in Metal-organic frameworks: A fundamental understanding. *Coordination Chemistry Reviews*, *358*, 125–152. doi:10.1016/j.ccr.2017.11.022

Hu, Z., Deibert, B. J., & Li, J. (2014). Luminescent metal-organic frameworks for chemical sensing and explosive detection. *Chemical Society Reviews*, *43*(16), 5815–5840. doi:10.1039/C4CS00010B PMID:24577142

Keeffe, M., & Yaghi, O. M. (2012). Deconstructing the Crystal Structures of Metal Organic Frameworks and Related Materials into Their Underlying Nets. *Chemical Reviews*, *112*(2), 675–702. doi:10.1021/cr200205j PMID:21916513

Lei, J., Qian, R., Ling, P., Cui, L., & Ju, H. (2014). Design and sensing applications of metal-organic frameworks. *Trends in Analytical Chemistry*, *58*, 71–78. doi:10.1016/j.trac.2014.02.012

Lustig, W. P., Mukherjee, S., Rudd, N. D., Desai, A. V., Li, J., & Ghosh, S. K. (2017). Metal-organic frameworks: Functional luminescent and photonic materials for sensing applications. *Chemical Society Reviews*, *46*(11), 3242–3285. doi:10.1039/C6CS00930A PMID:28462954

Tranchemontagn, D. J., Mendoza-Corte, J. L., Keeffe, M., & Yaghi, O. M. (2009). Secondary building units, nets and bonding in the chemistry of metal-organic frameworks. *Chemical Society Reviews*, *38*(5), 1257–1283. doi:10.1039/b817735j PMID:19384437

Yang, L., Song, Y., & Wang, L. (2020). Multi-emission metal-organic framework composites for multicomponent ratiometric fluorescence sensing: Recent developments and future challenges. *Journal of Materials Chemistry. B, Materials for Biology and Medicine*, *8*(16), 3292–3315. doi:10.1039/C9TB01931F PMID:31829391

Zhang, Y., Yuan, S., Day, G., Wang, X., Yang, X., & Zhou, H. C. (2018). Luminescent sensors based on metal-organic frameworks. *Coordination Chemistry Reviews*, *354*, 28–48. doi:10.1016/j.ccr.2017.06.007 PMID:30090910

KEY TERMS AND DEFINITIONS

Adsorption: It is processes of removal of toxic analytes from solution with metal organic frameworks by attaching to their pore or surface via host-guest interaction.

Analytes: Cations, anions, nitro, and organic compounds that is of interest in an analytical procedure due to their toxicity.

Environmental Pollution: It is defined as contamination of various components of earth which possesses a serious problem for the ecosystem and public health.

Fluorescence Quenching and Enhancement: Emissive properties of the metal organic frameworks decreases or increases in the presence of analytes such as metal ions, anions, nitro explosives, and organic compounds, etc.

Fluorescence Resonance Energy Transfer: It is an effective energy transfer process between chemical species and sensor due to their significant overlap of absorption and fluorescence spectra which assist in detecting analytes.

Luminescence: It is the emission of light after absorption of energy from suitable source such as x-rays, electron beam, ultraviolet light (UV). It has two types of emission, fluorescence and phosphorescence.

Metal-Organic Frameworks: It is a class of highly porous materials composed of metal ions/metal clusters and organic linkers/organic ligands that may be one, two- or three-dimensional networks.

Photoinduced Electron Transfer: It is a process that involves the transfer of electron between metal organic frameworks and analytes on excitation with suitable light source, changing their optical properties.

Chapter 3
Advances in Rechargeable Batteries and Supercapacitors Based on Metal-Organic Frameworks

Rafael Vargas-Bernal
https://orcid.org/0000-0003-4865-4575
Instituto Tecnológico Superior de Irapuato, Mexico

ABSTRACT

An analysis of the contribution that metal-organic frameworks (MOFs) have made to the development of energy storage devices over the past two decades such as rechargeable batteries and supercapacitors is presented here. This chapter reviews the different versions of electrode manufacturing based on metal-organic frameworks to be used in the design and manufacture of rechargeable batteries and supercapacitors. The MOFs examined in this chapter include those based on MOF-derived materials, MOF-based composites, and conductive MOFs. Despite the significant progress that has been achieved so far, many tasks must be made to reach total security so that performance parameters required for optimal performance of rechargeable batteries and supercapacitors in commercial, industrial, and military applications. Therefore, innovative conceptions of the actions that must be performed are explored in this chapter.

INTRODUCTION

Metal-organic frameworks or MOFs can be defined as a kind of materials constituted by metal ions and coordination complexes with organic bonds forming networks with one-, two-, or three-dimensional structures containing potential voids that can be chemically exploited (Rosseinsky, 2014). The coordination network can be extending unidimensional through cross-links between coordination entities based on different types of links and moieties (Zhang, 2016). The potential advantage of these frameworks is that their pores can be stabilizing during the removal of the invited molecules, and these can be filling

DOI: 10.4018/978-1-7998-4760-1.ch003

with other chemical moieties, which can be exploited to produce conducting solids in electrochemical storage applications such as rechargeable batteries and supercapacitors. For these applications, mixed electrical conduction and redox activity of the frames can lead to electrochemically active materials. The choice of metal and linker establishes the properties and structure of the framework such as size and shape of the pores, which establish how many bonds can be established with the metal and in which orientation. For metal-ion batteries, the material used, as an electrode must be subjecting to an increasing number of adsorption and desorption cycles during its charge and discharge cycles (Tang, 2016). Three charge storage mechanisms are implicit in a rechargeable battery based on MOFs: 1) modification of metal-containing electrodes by means of oxidation states, 2) use of linkers to develop redox processes, and 3) formation of completely reduced metal and alloy complex through the metal ion by means of conversion reactions. Although the electrical conductivity of the metal-organic frameworks is low, its porosity increases the ionic diffusion ratio that is the main characteristic to be used in battery electrode materials. In addition, the frameworks have a high crystallinity, which gives them a mechanical stability, and their tunable structures increase their physicochemical functionalities. The most desired main performance parameters in a rechargeable battery are a maximum charge capacity, long cycling life, fast charge-discharge rate, and a high current density. Frameworks also have a high surface area and higher thermal stability than pure organic materials, and thus increase the efficiency of the use of metal ions (Xu, 2018). Based on these crystalline materials it is possible to develop positive materials (cathodes), negative (anodes) and electrolytes for rechargeable batteries and supercapacitors. In the case of supercapacitors, these are dividing into two types: 1) electrochemical double-layer capacitor and 2) pseudocapacitors. The use of frames in supercapacitors is basing on high electrical conductivity, higher power densities and energy storage capacity, and large surface area accessible to the electrolyte (Wang, 2018). Frameworks in the supercapacitors can be applying for three purposes:

1. Pristine frameworks to store electrical charges by means of physical adsorption on their inner surfaces of the electrolyte ions to exploit the reversible redox reactions of the metal centers;
2. Frameworks to synthesize metal oxides and preserve electrons by means of the transfer of charge between the electrode and electrolyte, and
3. Frameworks to obtain porous carbon-based materials for increasing the capacitance via enrichment of the electrical conductivity.

Different pathways to design electrode materials for supercapacitors and rechargeable batteries based on conductive MOFs include conductive MOFs (Xu, 2018), MOF-based composites (Chen, 2014; Yang, 2018), and MOF-derived materials (Zhang, 2017; Yang, 2018; Kung, 2019; Zhong, 2019). Metal ion batteries based on metal-organic frameworks make use of monovalent cations such as lithium (Li^+), potassium (K^+) and sodium (Na^+), on divalent cations such as zinc (Zn^{2+}), calcium (Ca^{2+}) and magnesium (Mg^{2+}) or on trivalent cations such as aluminum (Al^{3+}). Different metal-organic frames have been proposing as cathodes or anodes for rechargeable batteries and supercapacitors, some examples are:

- A hybrid material constituted of graphene-based composite wrapped in sulfur and microporous carbon obtained from a MOF for lithium-sulfur batteries (Chen, 2014),
- TiO_2 in the form of porous material obtained by annealing Ti-based MOF used in the anode of sodium batteries (Zhang, 2017),

- Cobalt-based MOFs for rechargeable battery anodes based on lithium and sodium ions, as well as for supercapacitors (Hua, 2019), and
- Porous nanocomposite derived from a bi-metallic organic metal framework based on Zn and Ni as the lithium-ion battery anode material (Yang, 2018).

In this chapter, a brief history that describes the evolution that metal-organic frameworks applied in rechargeable batteries and supercapacitors is presenting. Current directions in frameworks research and recent global progress are also analyzing. This study will provide reference to develop new metal-organic frameworks used in electrochemical charge storage systems. Finally, the potential challenges and opportunities are briefly highlighting in order to design metal-organic frameworks in the coming decades.

Background

A rechargeable battery can be defined as a type of electric battery that can be charging, discharging to a load, and recharging a finite number of times. It contains one or more electrochemical cells. These devices are commercially available in many different shapes and sizes, ranging from button cells to connected complex systems to make stable an electrical distribution grid. There are distinct options of electrode and electrolyte materials, which give rise to different costs, environmental impacts, and capacities. They are using in cars, light commercial vehicles (for example electric bicycles, motorized wheelchairs, or golf carts), portable consumer devices, tools, uninterruptible power supplies (UPSs), and battery storage power stations. The relevance of its application means that billions of dollars in research are investing around the world to improve them.

In the charging process, the positive active material (cathode) is oxidizing, producing electrons, and the negative material (anode) is reducing, consuming electrons, as depicted in Figure 1. The electrons establish the flow of current in the external circuit. The electrolyte serves as a simple buffer for internal ion flow between the electrodes or can serve as an active participant in the electrochemical reaction.

A supercapacitor is an energy storage device in the form of electrical capacitance, which can store more energy per unit mass or volume in quantities of ten to a hundred times with respect to electrolytic capacitors. These devices are choosing over batteries due to they are simpler and faster charging, as well as their faster-charging delivery. It is also calling the ultracapacitor or double-layer electrolytic capacitor, as depicted in Figure 2(a). If a supercapacitor is comparing with a capacitor, it has a larger plate area and a smaller plate distance. Its plates are designing with metals and immersed in electrolytes, and these are separating by a thinner insulator. When the plates are charged, an electrical double-layer is producing because opposite electrical charges are created on both sides of an electrical separator. This allows for higher energy density and greater capacitance. In addition, unlike a battery, the supercapacitor has a little wear and tear on long-term use and an unlimited life cycle. In this way, the supercapacitor can be charging and/or discharging an unlimited number of times during its useful life. Therefore, it bridges its properties between rechargeable batteries and electrolytic capacitors. Among the multiple advantages of the supercapacitor are its high-power delivery, high load current, low resistance, simple and fast charging mechanism that is not subject to overload. If its performance is comparing to that of a battery, it has excellent charge and discharge performance at high and low temperatures is highly reliable and has a low impedance. In the group of the disadvantages of a supercapacitor are its high cost, lower voltage limit, and high self-discharge, since it presents lower specific energy and its use in the entire energy spectrum are restricting due to its voltage under conditions of linear discharge. Due to their outstanding electrical properties, supercapacitors can be used in applications where it is required to deliver power and bridge power spaces, and to power battery-free devices. This energy storage device is using in applications in cars such as limited energy storage, regenerative braking, burst-mode power delivery, buses, elevators, cranes, and trains.

Figure 1. Charging process of a rechargeable battery

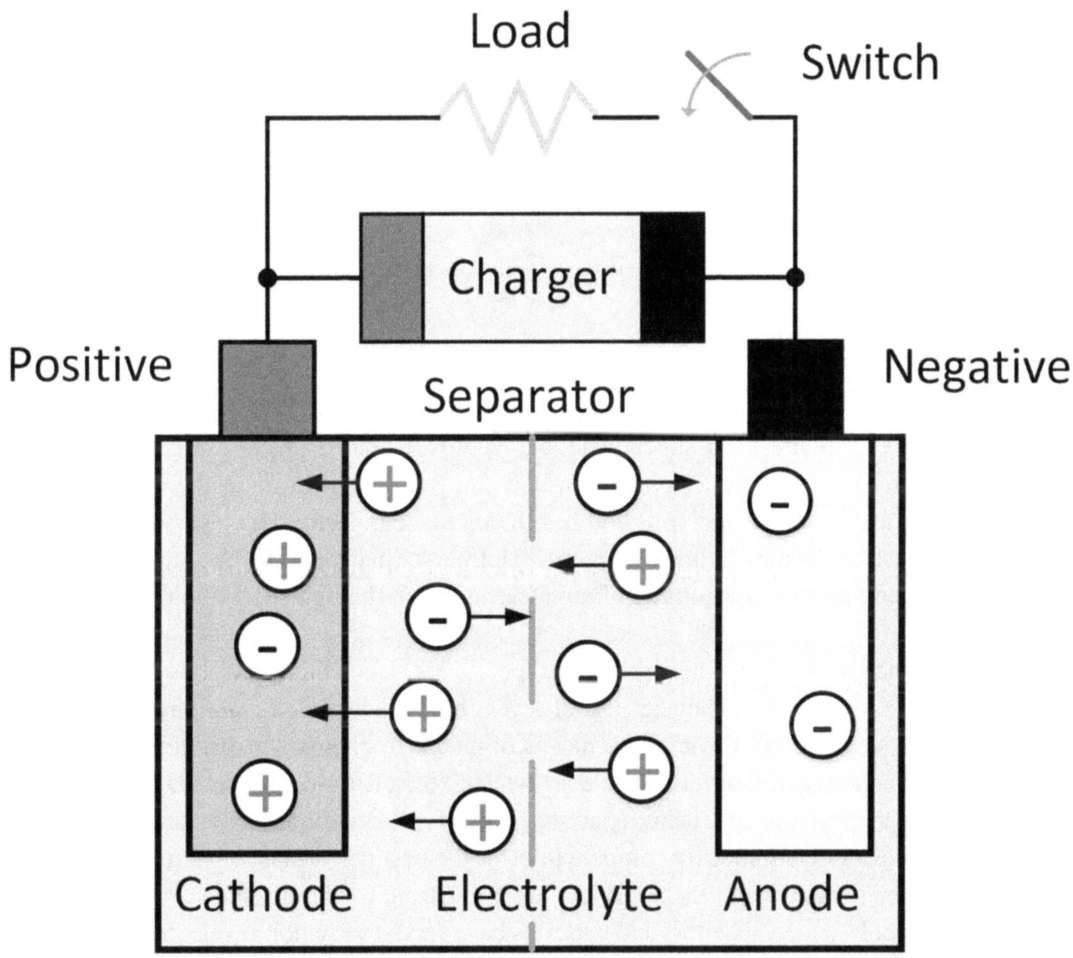

The insulator used in supercapacitors is not a solid dielectric as in the case of a conventional capacitor, but rather designed as a capacitor with electrostatic double-layer (EDLC) and/or electrochemical pseudocapacitor, as shown in Figure 2(b) and 2(c). In the case of an EDLC, carbon electrodes or their derivatives are normally using with the aim of achieving a capacitance value greater than their electrochemical counterpart, through a double Helmholtz layer, with a charge separation of a few angstroms, there is ac interface enclosed by the electrode surface and the electrolyte. In electrochemical pseudocapacitors, electrodes based on metal oxides or conductive polymers are using to carry out the charge-transfer by means of electrons based on the Faradaic principle through redox-type, intercalation, or electrosorption reactions. In addition, the electrolyte establishes a conductive ionic connection between the two electrodes and the electrolyte either asymmetrically or symmetrically through a potential applied during manufacturing.

Figure 2. Supercapacitor: (a) basic scheme, (b) capacitor with electrostatic double-layer, and (c) electrochemical pseudocapacitor

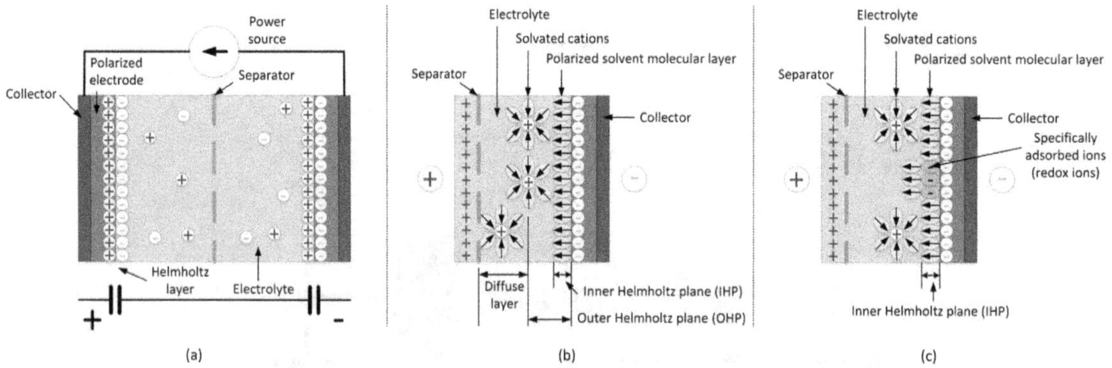

When a voltage is applying to a supercapacitor, electrical double-layers are generating on both electrodes. The first layer is on the electrode surface and the second layer in the electrolyte, which arises from the dissolved and solvated ions and has the opposite polarity. The layers are separating by a monolayer of solvent molecules, called the internal Helmholtz plane (IHP). The amount of charge on the electrode is matching by the magnitude of countercharges in the outer Helmholtz (OHP) plane, as depicted in Figure 2 (b).

Pseudo-capacitance occurs when adsorbed ions specifically leave the electrolyte and permeate the double-layer of the supercapacitor. It makes use of reversible Faradic redox reactions on the surface of proper electrodes to store electrical energy by means of an electrochemical capacitor. In addition, it is realizing through charge transfers amidst the electrolyte and the electrode, by means of desolvated and adsorbed ions and, whereby only one electron per unit charge is participating, as illustrated in Figure 2(c).

Energy storage devices continuously compete to offer the best performance parameters. In Table 1, the main performance parameters of these devices are comparing to identify the current state in which they operate. Commonly, supercapacitors and batteries have served as symbiotic electrochemical energy storage devices. In the case of supercapacitors, they easily transfer energy, while rechargeable batteries require a power bank to maintain their storage capacity. The power density in a supercapacitor is unprecedented; however, the rechargeable battery offers a higher energy density and breaking voltage. In addition, supercapacitors are lighter, have demanding operating limits, and a longer life expectancy.

The use of nanomaterials is revolutionizing innovation in the development of both devices. In the case of the supercapacitor, it is increasing its storage capacity and expanding its applications. In the case of rechargeable batteries, they are operating in extreme environments and work is being realizied to increase their energy transfer rate. The behavior of the electrical voltage and current of the metal-ion batteries and supercapacitors is illustrating in Figure 3.

Figure 3. Comparative graph of the electrical behavior of supercapacitors and metal-ion batteries

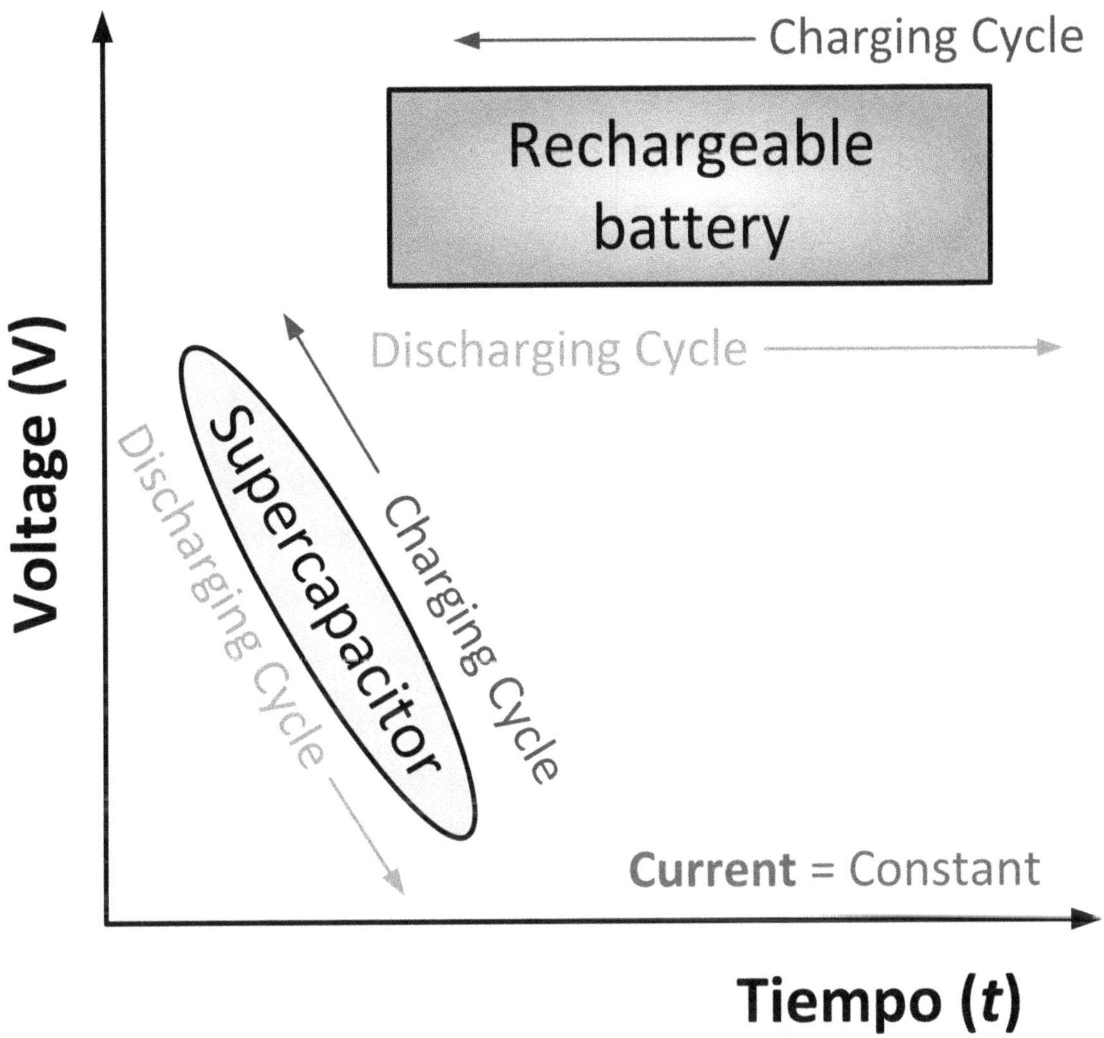

The search for solutions to implement electrochemical energy storage devices with high-energy efficiencies and environmentally friendly, metal-organic frameworks, represent an option that researchers around the world are developing to improve electrode performance, as is describing in the next section.

WHY USE METAL-ORGANIC FRAMEWORKS TO DESIGN ENERGY STORAGE DEVICES?

The main components of a metal-organic framework or MOFs are a metal ion, or a grouping of metal ions called a cluster, and an organic moiety known as a linker, as shown in Figure 4. Therefore, MOFs are classifying as inorganic-organic hybrid materials, and in addition, according to the number of organic units used as a bond, they are classified as mono-, di-, tri-, or tetravalent, so that their structures can be

Table 1. Main performance parameters of electrochemical energy storage devices

Parameter	Aluminum Electrolytic Capacitors			Supercapacitors: Electrostatic Double-Layer Supercapacitor			Supercapacitors: High Power Supercapacitor			Supercapacitors: Pseudocapacitor			Lithium-Ion Batteries			Unit
	Mininum value	Typical value	Maximum value	Mininum value	Typical value	Maximum value	Minimum value	Typical value	Maximum value	Minimum value	Typical value	Maximum value	Minimum value	Typical value	Maximum vale	
Maximum Voltage	4		630	1.2		3.3	2.2		3.3	2.2		3.8	2.5		4.2	Volts (V)
Recharge Cycles		unbounded		1×10^5		1×10^6	100		1×10^6	2×10^4		1×10^5	500		1×10^4	Cycles
Capacitance		≤ 2.7		0.1		470	100		12,000	300		3,300		--		Farads (F)
Specific energy	0.01		0.3	1.5		3.9	4		9	10		15	100		265	Watt-hour per kilogram (Wh/Kg)
Specific power		> 100		2		10	3		10	3		14	0.3		1.5	Watt per gram (W/g)
Operating Temperature range	-40		125	-40		70	-20		70	-20		70	-20		60	°C
Working life @ Room Temperature		> 20		5		10	5		10	5		10	3		5	Years (y)
Efficiency		99			95				95		90			90		%
Self-Discharge Time @ Room Temperature		Short-Time (days)			Middle-Time (weeks)			Middle-Time (weeks)			Long-Time (Month)			Long-Time (Month)		

one-dimensional, two-dimensional, or three-dimensional. The structure and properties of MOF depend on the choice of the metal or metals as well as the organic linker. The coordination number of the metal within the MOF establishes the shape and size of the pores since this determines the number of bonds and their orientation. Each MOF is making up of subunits called secondary building units (SBUs), which describe and organize their structures since there are different topologies that generate a multitude of different structures. Each SBU is linking to the main structure of the MOF through bridging bonds, which are basing on di- or tricarboxylic acids, and whose behavior is that of the rigid backbone. The main synthesis routes of the metal-organic frameworks are basing on solvothermal, mechanochemical, sonochemical, and microwave methods (Osman, 2019). Of these methods, the solvothermal approach is the most suitable for obtaining ordered symmetrical structures and tunable pore size. The strong chemical and thermal stability of MOFs is deriving from the carbon-carbon and metal-carbon chemical bonds. Metal-organic frameworks whose organic binders contain fluorine are preferred for applications with electrochemical activity such as rechargeable batteries and supercapacitors. The analysis of the synthesis and properties of these structures is studying by reticular chemistry, and it is not the purpose of this chapter, so the interested reader must read this topic specifically in many reviews found in the literature.

Figure 4. The basic structure of a metal-organic framework

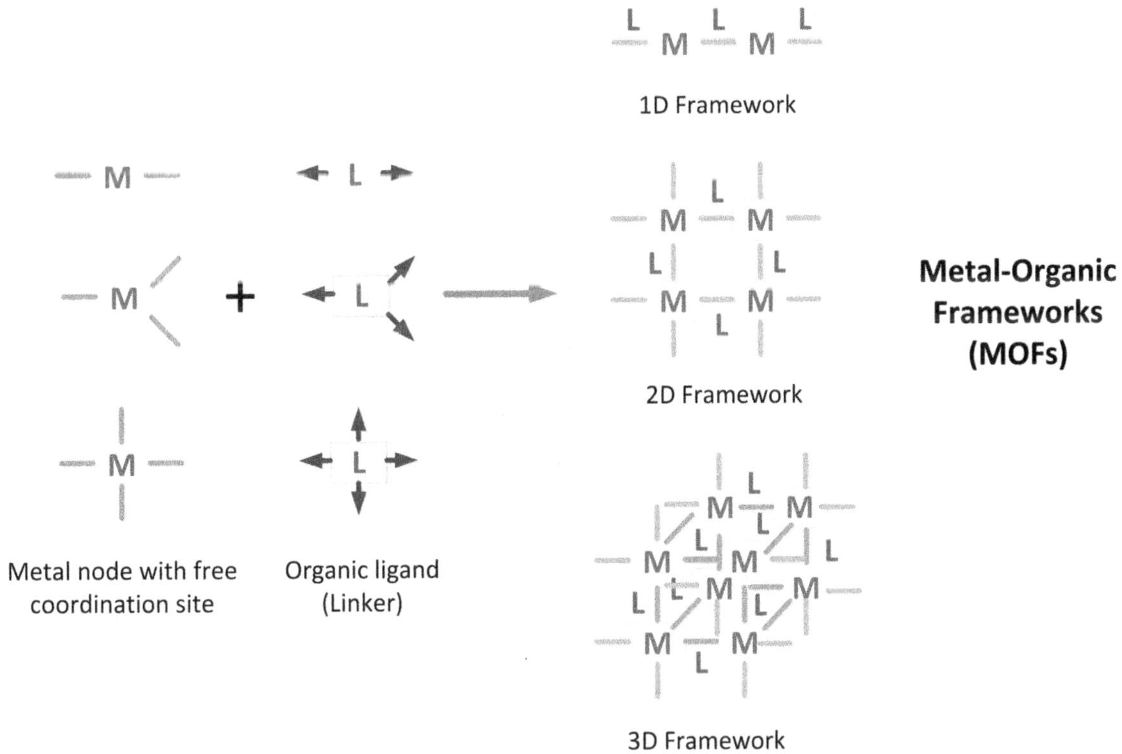

The metallic nodes or clusters are ions based on alkaline earth metals, transition metals or lanthanides (Xu, 2017). On the other hand, organic linkers are commonly multi-chain molecules with nitrogen or oxygen donating atoms such as pyridyl, carboxylates, polyamines, or others. Transition metals and lanthanides because they can have different valences, leads to different crystal structures when they are combining with the organic moieties in the metal-organic frameworks. The key advantage from a chemical point of view is that both metals and chemical linkers facilitate a catalytic activity that causes electrode materials to provide high energy density.

A MOF or metal-organic framework has interesting physical and chemical properties to design electrode materials for charge storage devices: rechargeable batteries and supercapacitors. Among its outstanding properties are tunable porous morphology, large specific surface area, metal sites, and ease of adding functional groups (Mehtab, 2019). Currently, the need for better electrode materials for storage devices has led researchers to consider the study of novel materials as an active and dynamic research area. For rechargeable batteries based on metal ions, the energy storage and conversion mechanism use the rocking chair mode to perform an ion exchange between the electrode materials. On the other hand, the ionic adsorption and/or desorption mechanism in an electrolyte is used in supercapacitors. Thus, the choice of material appropriate for the design of the electrode material should consider the next properties: 1) tunable pore size, 2) high electrical conductivity, and 3) large surface area. MOFs are assembling through a spacer node morphology to connect nanostructured materials by means of metal centers formed by clusters and/or ions as well as by organic linkers containing different organic moieties that build continuous and excessively porous crystalline structures. Among the advantages that can be attributed to metal-organic frameworks are highly controlled crystalline structural design, extraordinary synthetic versatility, excessively large surface area, chemical adaptability to different moieties, and matchless control of its porosity. The strategic advantage of metal-organic frameworks over other electrode materials is their ease of accommodating several different metal ions in their pores without causing considerable decomposition of their crystalline structure (Zhang, 2016). In this way, their tuning properties of both metal sites and pore size control can induce different densities of power and energy for the design of supercapacitors and rechargeable batteries. One of the strategies to raise the surface area of metal-organic frameworks involves increasing the length of the ligand, specifically when the structure has a high density.

An alternative to metal-organic frameworks (MOFs) are covalent organic frameworks (COFs), which use non-metallic elements such as hydrogen (H), boron (B), carbon (C), nitrogen (N) and oxygen (O) To produce extended structures, they are also being used to optimize the electrochemical performance of rechargeable batteries and supercapacitors (Zhang, 2016). The structures of the COFs as well as the structures of the MOFs provide electrode materials with a large number of cycles. The main disadvantage is that these structures do not achieve reversible redox activity.

Among the challenges that materials scientists are trying to solve to optimize the performance of metal-organic frameworks to be used in supercapacitors and rechargeable batteries are their surface area, structural stability, electrical conductivity, and flexibility (Baumann, 2019). In the next section, the state of the art of materials based on metal-organic frameworks for implementing supercapacitors and rechargeable batteries are discussing. Even though a large number of recent reviews have been publishing on the advances made in metal-organic frameworks aimed at electrochemical energy storage devices, the author considers that different points of view are necessary to disseminate the relevance of this type of material, as well as its contribution to the next innovations.

SOLUTIONS AND RECOMMENDATIONS

Because energy consumption and demand worldwide are increasing rapidly, materials researchers are developing alternatives for energy storage devices to store this energy in a cleaner and more sustainable way. Currently, rechargeable batteries are being developed with monovalent metal cations such as lithium (Li) (Yang, 2018, Zhong, 2019), sodium (Na) (Zhang, 2017, Zhong, 2019), and potassium (K) (Xiong, 2018), with divalent cations such as calcium (Ca), zinc (Zn), and magnesium (Mg), and with trivalent cations such as aluminum (Al) (Zhang, 2019). In the case of rechargeable batteries, researchers around the world have developed different proposals for electrode materials (Wang, 2019). The following examples have been identifying: Lithium-ion batteries (LIBs) (Yang, 2018; Zhong, 2019), sodium-ion batteries (NIBs) (Zhang, 2017; Zhong, 2019), potassium-ion batteries (PIBs) (Xiong, 2018), aluminum-ion batteries (AIBs) (Zhang, 2019), Li-O_2 batteries (Lang, 2019), Na-O_2 batteries (Liu, 2019), Li-CO_2 batteries (Li, 2018), Li-Se batteries (Jin, 2020), K-Se batteries (Huang, 2019), Li-S batteries (Chen, 2014), and Zinc-Air batteries (Wu, 2018). In the case of supercapacitors, different proposals have been developing to fulfill requirements either those based on pseudocapacitance or in electrostatic double-layer capacitance (Cao, 2015; Osman, 2019). Finally, hybrid battery-supercapacitor devices have also been presenting as physical deployment options (Wei, 2017; Zuo, 2017; Zhou, 2019; Tan, 2020).

Rechargeable Batteries

Numerous research papers on electrode materials have been proposing to improve the performance of rechargeable batteries. For example, electrodes based on metal oxides in the nanoscale or based on tin and silicon, which have offered high specific capacity, however, suffer from volumetric change, which reduces their stability, and therefore, the charge-discharge cycles are diminishing. With the introduction of MOFs, their self-assembly capabilities, along with their tunable porosity, large surface area, and effective redox activity, have been promoting the establishment of diffusion paths and storage spaces keeping the crystal structure stable, even when the energy storage device is subjected to exhaustive charge-and-discharge cycles (Shi, 2017). In the following paragraphs, some examples of proposals to implement the electrode materials for rechargeable batteries based on metal-organic frameworks are analyzing.

In lithium-ion battery (LIB), open structures of MOFs allow efficient delivery of electric charge as well as reversible ion insertion and/or extraction, and metal cations increase the active sites used for redox reactions (Mehtab, 2019). In addition, its structural adaptability does not modify its performance capabilities such as high lithium storage capacity and high current density. Given their chemical versatility to work with different conditions, MOFs have been proposing as anode materials with adequate but not excellent performance, so their efficiency should be improving in order to adapt them to reliable commercial developments.

Zeolitic imidazolate frameworks (Zn/Ni-ZIF-8) were synthesized by pyrolysis to design materials electrodes for lithium-ion batteries to produce a nitrogen-doped carbon-based composite material with zinc and nickel metal sites (Yang, 2018). The use of zinc and nickel provided the battery with excellent ratio capacity, a long life cycle, and cyclical reversibility so that the electrode material can be applying as an anode. The porous structure given by the MOF facilitates the diffusion process of lithium ions in charge-discharge processes.

TiO_2 structures similar to a porous cake can be obtained from MOFs (MIL-125) for sodium-ion batteries (Zhang, 2017). MOF gives the electrode material high storage capacity, cycling stability and an outstanding performance ratio. Proposals for porous metal oxides obtained from MOFs should be developing to increase the storage capacity of sodium ions in electrical energy storage devices.

Zeolitic imidazolate frameworks (ZIFs) can be using to synthesize by pyrolysis nitrogen-doped carbon nanotubes as the electrode material for a potassium-ion battery (Xiong, 2018). The structure obtained by the MOF facilitates and increases the intercalation rate of potassium-ions, which results in a very efficient electrochemical performance for electric charge and discharge processes.

MOFs have been using to obtain cathodes for rechargeable aluminum-ion batteries to design electrode materials based on metal oxide composites and porous carbon structures (Zhang, 2019). Among the proposed options are Co_3O_4, Fe_2O_3 or $CoFe_2O_4$ with reduced graphene oxide to provide improved electrochemical activity and stable charge-discharge processes with an efficiency above 99.6%. The Co_3O_4-based electrode material showed the best electrochemical behavior with respect to capacity and number of cycles than the other proposals.

The application of a manganese-based MOF to the air electrode catalyst in rechargeable lithium-oxygen batteries, synthesized using a solvothermal method, has been demonstrated (Lang, 2019). The advantage of using MOF as a catalyst in the gas electrode lies in its excellent reversibility and ability to withstand the oxidation process. The catalyst has an exceptional influence on the electrochemical performance of the battery, especially on the number of cycles and the specific energy delivered by the battery.

In search of improved electrocatalytic activity, Co_3O_4@$MnCo_2O_{4.5}$ nanocubes were synthesizing for rechargeable Na-O_2 batteries from MOFs (h-Co_3O_4@$MnCo_2O_{4.5}$ Ns) (Liu, 2019). The MOF, in addition to providing the structure to the electrode material, allows a more efficient flow of the O_2 through the pores as diffusion channels and by means of the metallic sites, the electro-catalytic activity is optimizing.

MOFs can be used as porous catalysts in CO_2 electrodes for rechargeable Li-CO_2 batteries, since they can capture CO_2 in large proportions and the active metal sites allow the decomposition of Li_2CO_3 (Li, 2018). The possible family of metal-organic frameworks is very varied in composition and structure, and in this study the behavior of the following frameworks Mn_2(dobdc), Mn(bdc), Mn($C_2H_2N_3$)$_2$, Mn(HCOO)$_2$, Co_2(dobdc), Ni_2(dobdc), Fe(bdc), and Cu(bdc) was performed. In this case, the Mn_2(dobdc) framework was the one with the highest discharge capacity, while the Mn(HCOOO)$_2$ framework retained the loading potential even at a reduced number of cycles.

A porous carbon-selenium-based composite material with a fish-like surface morphology was synthesizing by pyrolysis using a self-assembled MOF (PCNS/Se) loaded with selenium (Jin, 2020). The MOF establishes the porous support structure required for the electrode in the rechargeable lithium-selenium battery. This also prevents selenium from dissolving in the electrolyte during cycling, and therefore supports the mechanical stress from volumetric expansion during discharge.

A porous carbon-based composite doped with nitrogen and selenium was synthesizing through a MOF by carbonation (Huang, 2019). In this case, the role of the MOF is to provide the crystalline structure of the porous material and to confine the selenium ions to be applying in a rechargeable potassium-selenium battery. The precursor MOF used to obtain a nitrogen-doped porous carbon matrix was ZIF-67 with a pore size of approximately 250 nm.

A composite material consisting of a three-dimensional structure consisting of a sheet of graphene, sulfur and, a zeolitic imidazolate framework (ZIF-8) was synthesized through a MOF to be used as a cathode material in a rechargeable lithium-sulfur battery (Chen, 2014). The role of the MOF was to be a scaffold to produce a highly interconnected structure both structurally and electrically between the various materials noted above.

In the design of rechargeable zinc-air batteries, porous carbon polyhedra containing zinc can be obtained through cobalt-based zeolitic imidazolate frameworks, and doped with zinc (ZnCo-ZIFs) (Wu, 2018). The resulting composite material exhibits a synergistic effect between nitrogen-doped carbon and cobalt nanoparticles that increases electrical conductivity and surface area. The function of the MOF in this design is to act as a precursor or sacrificial template to prepare derivatives with the required crystalline structure such as that of nitrogen-doped porous polyhedron and with active sites of cobalt and zinc.

In Table 2, a comparison of electrode performance for batteries based on metal-organic frameworks is presenting. From the data reported in Table 2, it is evident that the current density is regularly low for most types of rechargeable batteries, but high for types based on metal ions. The number of cycles that rechargeable batteries offer is still low, because even some types do not exceed 100 cycles. The discharge capacity for some types of batteries is very high, while for others it is low.

Table 2. Performance of electrodes based on metal-organic frameworks for different rechargeable batteries

Storage Device Type	Current density (mA/g)	Number of cycles	Discharge capacity (mAh/g)	Reference
Lithium-ion battery (LIB)	5000	400	1105.2	Yang, 2018
Sodium-ion battery (SIB)	50	50	250	Zhang, 2017
Potassium-ion battery (PIB)	2000	500	102	Xiong, 2018
Aluminum-ion battery (AIB)	200	100	100	Zhang, 2019
Li-O_2 battery	100	35	500	Lang, 2019
Na-O_2 battery	100	135	8,400	Liu, 2019
Li-CO_2 battery	50	50	18,022	Li, 2018
Li-Se battery	--	200	311.8	Jin, 2020
K-Se battery	50	100	130	Huang, 2019
Li-S battery	168	120	561	Chen, 2014
Zinc-air battery	--	100	741	Wu, 2018

Supercapacitors

Metal-organic frameworks are being introduced as promising electrode materials for supercapacitors thanks to their large surface area, accessibility to active sites, tunable morphology, as well as their ability to produce pseudo-capacitance (Zhao, 2019). In the following paragraphs, some examples of proposals to implement the electrode materials for supercapacitors based on metal-organic frameworks are analyzing.

For the design of supercapacitors operating under the electrostatic double-layer mechanism, the active electrode materials are regularly basing on carbon materials such as carbon nanotubes, carbon fibers, graphene, and porous carbon (Osman, 2019). Among these materials, porous carbon has a large surface area, long-term stability in aqueous solutions, low electrochemical impedance, and good pore size distribution. Given the versatility that MOFs can perform as a self-template, it can be combining with carbon-based materials to provide excellent stability during charge-discharge processes in supercapacitors. In the case of supercapacitors, their efficiency can be controlled through the type of pore morphology, which can be either completely open, completely closed or partially closed.

Table 3. Performance of electrodes based on metal-organic frameworks for different supercapacitors

Electrode Material	Current density (mA/g)	Number of cycles	Specific capacitance (F/g)	Reference
Reduced graphene oxide-wrapped MoO_3 Composites	6,000	6,000	404 @ 0.5 A/g	Cao, 2015
Ni(4,4'-bpy)(tfbdc)$(H_2O)_2$, Ni-MOF	1,000	50	2548 @ 1 A/g	Shi, 2017
Zinc fluoride/Carbon Composite	50,000-100,000	100,000	322-360 @ 1 A/g	Osman, 2019
Mxene/Metal-Porphyrin Frameworks (MPFs) film	100	30,000	326.1 @ 0.1 A/g	Zhao, 2019

In the case of supercapacitors based on electrochemical pseudo-capacitance, electrode materials such as metal oxides have been using but show low electrical conductivity, low specific capacitance, and effective for few charge-discharge cycles (Cao, 2015). It is possible using MOFs to combine metal oxides with carbon materials such as graphene to increase the performance parameters of a supercapacitor. Graphene provides the supercapacitor with mechanical properties that guarantee the structural stability required during cycling processes, which guarantees a long-life cycle and a good ratio capacity, and electrical properties that improve the transfer of electric charge.

Active electrode materials based on two-dimensional materials such as nano-sheets of porphyrins with metals are being investigated, although the first results demonstrate low electrical conductivity and poor chemical stability (Zhao, 2019). Different strategies to solve the above problems have been suggesting without achieving success in all requirements, however, the use of two-dimensional materials promises to improve chemical stability as well as provide the electrical conductivity required to obtain an electrode material for a supercapacitor that satisfies the needs of a commercial device. Among the two-dimensional materials that have been advising to optimize the performance of the electrode materials are MXenes, a family of two-dimensional transition metal carbides, nitrides and carbonitrides. These materials have high electrical conductivity and a hydrophilic nature that makes them candidates for supercapacitor design, but their self-stacking reduces access to electrolyte ions. To solve this problem, it is necessary to intercalate active electrode materials to facilitate ion transport through the electrolyte, so metal-organic frameworks are required to achieve the required performance parameters in a supercapacitor.

Electrode materials from nickel-based metal-organic frameworks have been synthesized by hydrothermal approximation and grinding treatment to produce nanoparticles for supercapacitors (Shi, 2017). The unique characteristic structure of Ni-MOFs in supercapacitors increases electrochemical performance by establishing short pathways for diffusion of lithium ions through the electrolyte. These nanoparticles based on Ni-MOFs (Ni(4,4'-bpy)(tfbcd)$(H_2O)_2$) provided a high specific capacitance.

A porous amorphous structure was synthesized by a solvothermal approach from a zinc-based fluorinated MOF to be used as a cathode material for a supercapacitor (Osman, 2019). The results demonstrated that the resulting material has high electrochemical stability, excellent capacitance retention, and high current densities. The coral morphology obtained gives the electrode material exceptional surface adsorption-desorption properties related to the charge storage mechanism.

In Table 3, a comparison of electrode performance for supercapacitors based on metal-organic frameworks is summarizing. According to the values reported in Table 3, supercapacitors offer a high current density, a high number of charge-discharge cycles, and a high specific capacitance.

Electrochemical Storage Hybrid Devices

In this decade, the hybrid battery-supercapacitor device or BSH has been introducing (Zhou, 2019; Tan, 2020). This device so-called because one of the electrodes is a battery type and the other is a capacitor type electrode. This was a proposed idea to optimize power density, cycling life, and energy density. The energy density is achieving by means of the battery-type electrode, while the supercapacitor-type electrode obtains the power density (Zuo, 2017). The use of metal-organic frameworks reduces the possibility that during the insertion and/or disinsertion processes of the charge carrier ions, a volumetric expansion of the electrode material occurs; this destroys the structure at high cycling rates and reduces electrical conductivity (Wei, 2017). Unfortunately, its main limiting practice is a low energy density. So far, the main types of these devices are those based on lithium-sodium-ions, with redox electrolytes, acidic-alkaline compounds, and with pseudocapacitive electrode. In Table 4, the electrical behavior of different hybrid electrochemical storage devices based on metal-organic frameworks is summarizing.

Table 4. Performance of electrodes based on metal-organic frameworks for different electrochemical storage hybrid devices

Electrode Material	Current density (mA/g)	Number of cycles	Discharge capacity (mAh/g)	Energy density (Wh/kg)	Reference
Reduced graphene oxide nanosheets/ Porous carbon coated Fe_3C nanoparticles	1,000	5000	95.3	46.2 @ 1.2 kW/kg 28.3 @ 0.7 kW/kg	Tan, 2020
Porous $NiCo_2S_4$-N-doped carbon	1,000	5000	--	51.98	Zhou, 2019
Co_3O_4/Ni	1,000	3000	209	--	Zhang, 2018
Polyoxometalate/Reduced graphene oxide	2,000-3,000	100-400	1075	--	Wei, 2017

According to Table 4, the data reveal that the current density, number of operating cycles, discharge capacity and energy density offered by hybrid devices are intermediate between those reported for rechargeable batteries and supercapacitors. In this way, it can be concluded that a wide range of performance parameters are offered to users according to their requirements by using rechargeable batteries, supercapacitors, and battery-supercapacitor hybrid devices.

FUTURE RESEARCH DIRECTIONS

After analyzing the theoretical studies and experimental results of various research groups around the world, it can be concluding that the enormous advantages offered by metal-organic frameworks make them materials of interest for the development of electrode materials for storage devices of energy. Despite all the progress made so far, there are some challenges that must be overcome to achieve unprecedented commercial success and establish them as reliable materials for electrode development, as described below. Stability in any operating environment must be optimized to reduce the cost of production and increase its commercial viability. The electrical conductivity of metal-organic frameworks must be increasing to raise their charge storage capacity and delivery rate to be used more extensively in supercapacitors and rechargeable batteries. MOFs derived from carbon materials have not been fully exploiting despite the variety of types and morphologies that can be obtained, so it is suggesting developing more research and technological development in this regard. A refined emphasis on the use of metals, metal oxides and heteroatoms for the doped metal-organic frameworks must be working to achieve even a more interesting success in the electrochemical performance of energy storage devices. The catalytic activity should be optimized especially for metal-organic frameworks with two or more metal sites to fully exploiting their multifunctionality. Undoubtedly, one of the concerns that scientists around the world have is whether they really understand the type of chemical interaction between different functional groups, which would allow it to be working more widely to increase the performance parameters of supercapacitors, fuel cells, and rechargeable batteries. Proposals for porous metal oxides obtained from MOFs should be developing to increase the storage capacity of metal ions in electrical energy storage devices. In the case of BSH devices, the challenges to be overcome are in the development of aqueous electrolytes, new anode materials, and large work potentials.

CONCLUSION

The metal-organic frameworks represent a support material for the development of electrode materials, whose multifunctionality must be fully exploited by the innovative designs that have been and will be proposing by different groups of researchers worldwide. In the future, electrode material designs must continue using the properties of the pores, the multifunctionality of organic and inorganic components, tunable morphology, and the surface area, where the properties of nanomaterials are fully using. The key factors that govern the performance of energy storage devices such as supercapacitors and rechargeable batteries are the electrode materials and electrolyte compounds. The larger the surface area of the electrodes, the greater the capacity of electric charge storage of supercapacitors and rechargeable batteries. Therefore, a synergistic effect between the different materials used in the electrode material for supercapacitors and rechargeable batteries is required, to guarantee the optimal performance of energy storage devices in highly demanding portable applications. The research and technological development associated with electrode materials will continue to develop in this century with new proposals based on metal-organic frameworks.

ACKNOWLEDGMENT

The author thanks the access to scientific articles supplied by CONRICYT. It is economically sponsoring by CONACYT in Mexico. In addition, he wants to thank his wife and son for their support in time to conduct this research.

REFERENCES

Baumann, A. E., Burns, D. A., Liu, B., & Thoi, S. (2019). Metal-Organic Framework Functionalization and Design Strategies for Advanced Electrochemical Energy Storage Devices. *Communications Chemistry*, *2*(1), 86. doi:10.103842004-019-0184-6

Cao, X., Zheng, B., Shi, W., Yang, J., Fan, Z., Luo, Z., Rui, X., Chen, B., Yan, Q., & Zhang, H. (2015). Reduced Graphene Oxide-Wrapped MoO_3 Composites prepared by using Metal-Organic Framework as Precursor for All-Solid-State Flexible Supercapacitors. *Advanced Materials*, *27*(32), 4695–4701. doi:10.1002/adma.201501310 PMID:26178419

Chen, R., Zhao, T., Tian, T., Cao, S., Coxon, P. R., Xi, K., Fairen-Jimenez, D., Kumar, R. V., & Cheetham, A. K. (2014). Graphene-Wrapped Sulfur/Metal Organic Framework-derived Microporous Carbon Composite for Lithium Sulfur Batteries. *APL Materials*, *2*(12), 124109. doi:10.1063/1.4901751

Hua, Y., Li, X., Chen, C., & Pang, H. (2019). Cobalt based Metal-Organic Frameworks and Their Derivatives for Electrochemical Energy Conversion and Storage. *Chemical Engineering Journal*, *370*, 37–59. doi:10.1016/j.cej.2019.03.163

Huang, X., Xu, Q., Gao, W., Yang, T., Zhan, R., Deng, J., Guo, B., Tao, M., Liu, H., & Xu, M. (2019). Rechargeable K-Se Batteries based on Metal-Organic Frameworks-derived Porous Carbon Matrix confined Selenium as Cathode Materials. *Journal of Colloid and Interface Science*, *539*, 326–331. doi:10.1016/j.jcis.2018.12.083 PMID:30594007

Jin, W.-W., Li, H.-J., Zou, J.-Z., Iguva, S., Zhang, Q., Zeng, S.-Z., Xu, G.-Z., & Zeng, X.-R. (2020). Metal Organic Framework-derived Carbon Nanosheets with Fish-Scale Surface Morphology as Cathode Materials for Lithium-Selenium Batteries. *Journal of Alloys and Compounds*, *820*, 153084. doi:10.1016/j.jallcom.2019.153084

Kung, C.-W., Han, P.-C., Chuang, C.-H., & Wu, K. C.-W. (2019). Electronically Conductive Metal-Organic Framework-based Materials. *APL Materials*, *7*(11), 110902. doi:10.1063/1.5125487

Lang, X., Dong, C., Cai, K., Li, L., & Zhang, Q. (2019). Highly Active Cluster Structure Manganese-based Metal Organic Frameworks (Mn-MOFs) like Corals in the Sea Synthesized by Facile Solvothermal Method as Gas Electrode Catalysts for Lithium-Oxygen Batteries. *International Journal of Energy Research*, *44*(2), 1256–1263. doi:10.1002/er.4954

Li, S., Dong, Y., Zhou, J., Liu, Y., Wang, J., Gao, X., Han, Y., Qi, P., & Wang, B. (2018). Carbon Dioxide in the Cage: Manganese Metal-Organic Frameworks for High Performance CO_2 Electrodes in Li-CO_2 Batteries. *Energy & Environmental Science*, *11*(5), 1318–1325. doi:10.1039/C8EE00415C

Liu, Y., Chi, X., Han, Q., Du, Y., Huang, J., Lin, X., & Liu, Y. (2019). Metal-Organic Framework-derived Hierarchical Co3O4@MnCo2O4.5 Nanocubes with enhanced Electrocatalytic Activity for Na-O_2 Batteries. *Nanoscale*, *11*(12), 5285–5294. doi:10.1039/C8NR08818G PMID:30843015

Mehtab, T., Yasin, G., Arif, M., Shakeel, M., Korai, R. M., Nadeem, M., Muhammad, N., & Lu, X. (2019). Metal-Organic Frameworks for Energy Storage Devices: Batteries and Supercapacitors. *Journal of Energy Storage*, *21*, 632–646. doi:10.1016/j.est.2018.12.025

Osman, S., Senthil, R. A., Pan, J., & Sun, Y. (2019). A Novel Coral structured Porous-like Amorphous Carbon derived from Zinc-based Fluorinated Metal-Organic Framework as Superior Cathode Material for High Performance Supercapacitors. *Journal of Power Sources*, *414*, 401–411. doi:10.1016/j.jpowsour.2019.01.026

Rosseinsky, M. J. (2014). Perspective: Metal-Organic Frameworks – Opportunities and Challenges. *APL Materials*, *2*(12), 124001. doi:10.1063/1.4904880

Shi, C., Wang, X., Gao, Y., Rong, H., Song, Y., Liu, H.-J., & Liu, Q. (2017). Nickel Metal-Organic Framework Nanoparticles as Electrode Materials for Li-Ion Batteries and Supercapacitors. *Journal of Solid State Electrochemistry*, *21*(8), 2415–2423. doi:10.100710008-017-3591-6

Tan, Q., Chen, X., Wan, H., Zhang, B., Liu, X., Li, L., Wang, C., Gan, Y., Liang, P., Wang, Y., Zhang, J., Wang, H., Miao, L., Jiang, J., van Aken, P. A., & Wang, H. (2020). Metal-Organic Framework-derived High Conductivity Fe_3C with Porous Carbon on Graphene as Advanced Anode Materials for Aqueous Battery-Supercapacitor Hybrid Devices. *Journal of Power Sources*, *448*, 227403. doi:10.1016/j.jpowsour.2019.227403

Tang, B., Huang, S., Fang, Y., Hu, J., Malonzo, C., Truhlar, D. G., & Stein, A. (2016). Mechanism of Electrochemical Lithiation of a Metal-Organic Framework without Redox-Active Nodes. *The Journal of Chemical Physics*, *144*(19), 194702. doi:10.1063/1.4948706 PMID:27208960

Wang, L., Han, Y., Feng, X., Zhou, J., Qi, P., & Wang, B. (2018). Metal-Organic Frameworks for Energy Storage: Batteries and Supercapacitors. *Coordination Chemistry Reviews*, *307*, 361–381. doi:10.1016/j.ccr.2015.09.002

Wang, Z., Tao, H., & Yue, Y. (2019). Metal-Organic-Framework-based Cathodes for Enhancing the Electrochemical Performances of Batteries: A Review. *ChemElectroChem*, *6*(21), 5358–5374. doi:10.1002/celc.201900843

Wei, T., Zhang, M., Wu, P., Tang, Y.-J., Li, S.-L., Shen, F.-C., Wang, X.-L., Zhou, X.-P., & Lan, Y.-Q. (2017). POM-based Metal-Organic Framework/Reduced Graphene Oxide Nanocomposites with Hybrid Behavior of Battery-Supercapacitor for Superior Lithium Storage. *Nano Energy*, *34*, 205–214. doi:10.1016/j.nanoen.2017.02.028

Wu, X., Meng, G., Liu, W., Li, T., Yang, Q., Sun, X., & Liu, J. (2018). Metal-Organic Framework-derived, Zn-doped Porous Carbon Polyhedra with enhanced Activity as Bifunctional Catalysts for Rechargeable Zinc-Air Batteries. *Nano Research*, *11*(1), 163–173. doi:10.100712274-017-1615-2

Wu, Y., Qiu, X., Liang, F., Zhang, Q., Koo, A., Dai, Y., Lei, Y., & Sun, X. (2019). A Metal-Organic-Framework-derived Bifunctional Catalysts for Hybrid Sodium-Air Batteries. *Applied Catalysis B: Environmental*, *241*, 407–414. doi:10.1016/j.apcatb.2018.09.063

Xiong, P., Zhao, X., & Xu, Y. (2018). Nitrogen-doped Carbon Nanotubes derived from Metal-Organic Frameworks for Potassium-Ion battery Anodes. *ChemSusChem*, *11*(1), 202–208. doi:10.1002/cssc.201701759 PMID:28994244

Xu, G., Nie, P., Dou, H., Ding, B., Li, L., & Zhang, X. (2017). Exploring Metal Organic Frameworks for Energy Storage in Batteries and Supercapacitors. *Materials Today*, *20*(4), 191–209. doi:10.1016/j.mattod.2016.10.003

Xu, Y., Li, Q., Xue, H., & Pang, H. (2018). Metal-Organic Frameworks for Direct Electrochemical Applications. *Coordination Chemistry Reviews*, *376*, 292–318. doi:10.1016/j.ccr.2018.08.010

Yang, H., Cui, W., Han, Y., & Wang, B. (2018). Porous Nanocomposite derived from Zn, Ni-Bimetallic Metal-Organic Framework as an Anode Material for Lithium-Ion Batteries. *Chinese Chemical Letters*, *29*(6), 842–844. doi:10.1016/j.cclet.2017.09.024

Zeng, X., Yang, Z., Meng, J., Chen, L., Chen, H., & Qin, H. (2019). The Cube-like ZnO/C Composites derived from Metal Organic Framework-5 as Anodic Material with High Electrochemical Performance for Ni-Zn Rechargeable Battery. *Journal of Power Sources*, *438*, 226986. doi:10.1016/j.jpowsour.2019.226986

Zhang, K., Lee, T. H., Cha, J. H., Jang, H. W., Choi, J.-W., Mahmoudi, M., & Shokouhimehr, M. (2019). Metal-Organic Framework-derived Metal Oxide Nanoparticles@Reduced Graphene Oxide Composites as Cathode Materials for Rechargeable Aluminium-Ion Batteries. *Scientific Reports*, *9*(1), 13739. doi:10.103841598-019-50156-6 PMID:31551435

Zhang, L., Zhang, Y., Huang, S., Yuan, Y., Li, H., Jin, Z., Wu, J., Liao, Q., Hu, L., Lu, J., Ruan, S., & Zeng, Y.-J. (2018). Co_3O_4/Ni-based MOFs on Carbon Cloth for Flexible Alkaline Battery-Supercapacitor Hybrid Devices and near-infrared Photocatalytic Hydrogen Evolution. *Electrochimica Acta*, *281*, 189–197. doi:10.1016/j.electacta.2018.05.162

Zhang, X., Wang, M., Zhu, G., Li, D., Yan, D., Lu, T., & Pan, L. (2017). Porous Cake-like TiO_2 derived from Metal-Organic Frameworks as Superior Anode Material for Sodium Ion Batteries. *Ceramics International*, *43*(2), 2898–2402. doi:10.1016/j.ceramint.2016.11.028

Zhang, Z., & Awaga, K. (2016). Redox-Active Metal-Organic Frameworks as Electrode Materials for Batteries. *MRS Bulletin*, *41*(11), 883–889. doi:10.1557/mrs.2016.245

Zhao, W., Peng, J., Wang, W., Jin, B., Chen, T., Liu, S., Zhao, Q., & Huang, W. (2019). Interlayer Hydrogen-bonded Metal Porphyrin Frameworks/MXene Hybrid Film with High Capacitance for Flexible All-Solid-State Supercapacitors. *Small*, *15*(18), 1901351. doi:10.1002mll.201901351 PMID:30957989

Zhong, M., Kong, L., Li, N., Liu, Y.-Y., Zhu, J., & Bu, X.-H. (2019). Synthesis of MOF-derived Nanostructures and Their Applications as Anodes in Lithium and Sodium Ion Batteries. *Coordination Chemistry Reviews*, *388*, 172–201. doi:10.1016/j.ccr.2019.02.029

Zhou, M., Zhao, H., Ko, F., Servati, P., Bahi, A., Soltanian, S., Ge, F., Zhao, Y., & Cai, Z. (2019). Metal Organic Frameworks-derived Porous $NiCo_2S_4$ Nanorods and N-doped Carbon for High-Performance Battery-Supercapacitor Hybrid Device. *Journal of Power Sources*, *440*, 227146. doi:10.1016/j.jpowsour.2019.227146

Zuo, W., Li, R., Zhou, C., Li, Y., Xia, J., & Liu, J. (2017). Battery-Supercapacitor Hybrid Devices: Recent Progress and Future Prospects. *Advancement of Science*, *4*(7), 1600539. doi:10.1002/advs.201600539 PMID:28725528

ADDITIONAL READING

Chen, Y., Meng, Y., Zhang, C., Yang, H., Xue, Y., Yuan, A., Shen, X., & Xu, K. (2019). Yolk-shelled ZnO-NiO Microspheres derived from Tetracyanide-metallic-Frameworks as Bifunctional Electrodes for High-Performance Lithium-Ion Batteries and Supercapacitors. *Journal of Power Sources*, *421*, 41–49. doi:10.1016/j.jpowsour.2019.03.006

Dühnen, S., Nölle, R., Wrogemann, J., Winter, M., & Placke, T. (2019). Reversible Anion Storage in a Metal-Organic Framework for Dual-Ion Battery Systems. *Journal of the Electrochemical Society*, *166*(3), A5474–A5482. doi:10.1149/2.0681903jes

Jiao, G., Gu, Y., Wang, J., Wu, D., Tao, S., Chu, S., Liu, Y., Qian, B., & Chu, W. (2019). Porous CoP/C@ MCNTs Hybrid Composite derived from Metal-Organic Frameworks for High-Performance Lithium-Ion Batteries. *Journal of Materials Science*, *54*(4), 3273–3283. doi:10.100710853-018-3064-z

Lv, Y., Wu, Z., Fang, Y., Qian, X., Asiri, A. M., Tu, B., & Zhao, D. (2014). Hierarchical Mesoporous/ Microporous Carbon with Graphitized Frameworks for High-Performance Lithium-Ion Batteries. *APL Materials*, *2*(11), 113302. doi:10.1063/1.4897201

Mirzadeh, E., & Akhbari, K. (2016). Synthesis of Nanomaterials with Desirable Morphologies from Metal-Organic Frameworks for Various Applications. *CrystEngComm*, *18*(39), 7410–7424. doi:10.1039/ C6CE01076H

Nam, K. W., Park, S. S., dos Reis, R., Dravid, V. P., Kim, H., Mirkin, C. A., & Stoddart, J. F. (2019). Conductive 2D Metal-Organic Framework for High-Performance Cathodes in Aqueous Rechargeable Zinc Batteries. *Nature Communications*, *10*(1), 4948. doi:10.103841467-019-12857-4 PMID:31666515

Qi, J., Wang, P., Yan, Y., Zheng, X., Cao, J., & Feng, J. (2019). MCo_2O_4 (M= Co, Mn, Ni, Zn) Nanosheet Arrays constructed by Two-Dimension Metal-Organic Frameworks as Binder-Free Electrodes for Lithium-Ion Batteries. *Vacuum*, *169*, 108959. doi:10.1016/j.vacuum.2019.108959

Zhang, J., Zhang, D., Dai, G., Fang, Q., & Qiu, S. (2016). Nitrogen-doped Nanoporous Carbons through Direct Carbonization of a Metal-Biomolecule Framework for Supercapacitor. *Chinese Journal of Chemistry*, *34*(2), 203–209. doi:10.1002/cjoc.201500685

KEY TERMS AND DEFINITIONS

Coordination Polymer: Inorganic or organometallic polymer structure containing metal cation centers linked by ligands.

Covalent Organic Framework (COF): Two-dimensional and three-dimensional organic solids with extended structures in which building blocks are linked by strong covalent bonds based on light elements such as H, B, C, N, and O.

Electrode: Electrical conductor used to make contact with a nonmetallic part of a circuit.

Ligand: Ion or molecule (functional group) that binds to a central metal atom to form a coordination complex.

Metal-Organic Framework (MOF): Class of compounds consisting of coordinated metal ions or clusters to organic ligands to form one-, two-, or three-dimensional structures.

Rechargeable Battery: Long-term energy storage device that can be charging many times again after being discharging by applying DC current to its terminals.

Supercapacitor: An energy storage device that can store 10 to 100 times more energy per unit mass or volume than an electrolytic capacitor, with a simpler and faster-charging process and faster charge delivery.

Two-Dimensional Material: Crystalline material consisting of a single layer of atoms or a single layer of two or more covalently bonding elements.

Chapter 4
Applications of Quartz Crystal Microbalances Modified With Metal Organic Frameworks:
A Review

Fabian N. Murrieta-Rico
https://orcid.org/0000-0001-9829-3013
Universidad Nacional Autónoma de México, Mexico

Rosario I. Yocupicio-Gaxiola
Centro de Investigación Científica y de Educación Superior de Ensenada (CICESE), Mexico

Vitalii Petranovskii
Universidad Nacional Autónoma de México, Mexico

Donald H. Galván
Universidad Nacional Autónoma de México, Mexico

Joel Antunez-Garcia
https://orcid.org/0000-0003-3668-1701
Universidad Nacional Autónoma de México, Mexico

ABSTRACT

Metal-organic frameworks (MOFs) are versatile materials that are of interest due to their application and properties. MOFs are highly crystalline and porous materials; they are composed of organic bridging ligands, acting as linkers, and a three-dimensional (3D) network of metal ions that are secondary building units. Since the MOFs have a high surface area, high porosity, tunable topography, and their structures are quite diverse, these materials are used in process of separation/purification, gas/energy storage, drug delivery, catalysis, and chemical sensors. Since the MOFs can be modified to selectively adsorb chemical species, they can be used as sensitive layer for modification of sensors. This process allows the sensor to detect the target analyte. Quartz crystal microbalances (QCMs) are highly sensitive mass sensors. In this chapter, the authors review the literature related to QCMs modified with MOFs. In particular, the relationship between target analyte, class of MOF, and instrument used for measurement of frequency variations.

DOI: 10.4018/978-1-7998-4760-1.ch004

INTRODUCTION

The importance of a material comes from its functionality for solving a particular problem. The emergence of novel materials that can help to remedy certain issues requires processes, where it is necessary to have a good control of the final features of material, namely, textural, structural and chemical properties. For many applications as catalysis, sensing, and adsorption, porosity is a key point to development of good devices. In this sense, zeolites had been used as starting point for the development of new ideas and improvements. In this context, advancements starting from zeolites had been good, but such an approach presents some limitations due to its inorganic nature. The above, open the way for the study of a new kind of materials which combine of zeolitic features and the functionality brought from the inclusion of different organic species to the system. Therefore metal-organic frameworks are particularly suitable for that. Metal-organic frameworks (MOFs) stand for a new kind of crystalline porous materials also defined in a more general way as "coordination polymer", "coordination framework", "metallosupramolecular network" and "hybrid materials" (Batten et al., 2013; Kepert 2010). The MOFs structures are constructed by the linkage of metal ions or ion cluster through organic ligands to form different dimensional structures, namely, one-dimensional (1D), two-dimensional (2D) or three-dimensional (3D) architectures (Bataille et al., 2012; Xia et al., 2017; Chen et al., 2013). The use of MOFs is as ancient as the use of Prussian blue (more than 300 years), but the study of coordination polymers or MOFs as such is reported during the forties (Griffith, 1943) and the big contribution to the field was by the work of Robson and co-workers, who introduce the "node-and-spacer" approach (Hoskins et al., 1990), Kitagawa et al. (Kitagawa et al., 1991) and Yaghi et al. (Yaghi et al., 1995). The combination of metal ions or cluster ions with organic compounds resulting in crystalline structures opens a wide range of possibilities of features such as structure, size and shape of pore of final product. The MOFs, as well as zeolites, display very good features such as high adsorption capacities, big surface area, and pore volume. Although, the MOFs porosity is better than displayed by zeolites, since those are constructed from tethaedral units which can give rise to a finite number of Secondary building units (SBU), while MOFs can form SBUs from metal atom or cluster and one-, two- or three-dimensional extended inorganic substructures (Butova, 2016) which give rise to a wide range of possibilities with regard of topologies of the final material.

Nomenclature

As was already mentioned MOF is the abbreviation generally used to refer to this kind of crystalline materials. When the abbreviation MOF is followed by a number it represents a specific framework, as examples: MOF-74, MOF-177, MOF-253. When a structure has the same symmetry that other MOF but different composition is known as an isoreticular framework and is denoted adding the prefix IR to the MOF. In the same way, some MOFs are denoted with the letters of the place where they were discovered, as is the case of MIL, HKUST or UiO (Barthelet et al., 2002; Chui et al., 1999; Kandiah et al., 2010) which correspond to Materials of Institut Lavoisier, Hong Kong University of Science and Technology, and Universitetet i Oslo, respectively. In this regard, some efforts have been made to develop a more standardized nomenclature for all kind of synthesized MOF (O'Keeffe et al., 2008; Tranchemontagne et al., 2009) which could minimize the problem of give the correct name to new structures synthesized in the future.

Synthesis Methodologies

The synthesis of MOF materials include the conventional solvothermal methodology which is the most used, (Pachfule et al., 2011), sonochemichal (Son et al., 2008), solvothermal assisted by microwave (Bux et al., 2009), electrochemical (Van Assche et al., 2012) among others synthesis methodologies. The synthesis precursors of the metals generally are composed by inorganic or organic salts as sulfates, nitrates or acetates which are water-soluble. The organic linkers usually are composed by di, tri or tetravalent carboxylic acids or aromatics compounds with Nitrogen which often are dissolved in polar organic solvents like some alcohols, N,N-dimethylformamide, diethylformamide or water, among others. All components are mixed and are left to interact at temperature nearly to room temperature up to hydrothermal condition (near to 200°C). In this sense, there is a division between solvothermal and no-solvothermal methods. Solvothermal method works at the boiling temperature of solvent in closed and expensive reactors, while the no-solvothermal is performed below of the solvent boiling point in open systems.

Applications

It is a well-known fact that MOFs posses remarkable textural properties as result of their porosity, but in addition to that, MOFs can show a wide range of structures and functionalities depending on a particular array of metal or metal clusters with different organic linkers. All the above shows why MOF have been broadly linked and tested in different applications. In this regard, application in gas adsorption (Glover et al., 2011; Cao et al., 2019), catalysis (Feng et al., 2019; Liu et al., 2019 (a)), small size molecules separation (Jiang et al., 2019; Liu et al., 2018; Song et al., 2017) sensing (Wu et al., 2019; Xu et al., 2017; Li et al., 2019), drug delivery (Ahmad et al., 2019). Nonetheless, the accelerated discovery of new MOF with different topologies and the functionalization of their structures bring a huge amount of possibilities, not only the conventional utilities bring it by the porosity and textural features, but a bunch of new applications. In this sense, some studies of application as solar energy storage (D'Ans et al., 2019), supercapacitors (Yang et al., 2018), photosensitizer materials (Kaur et al., 2017), antibacterial nanostructure (Akbarzadeh, et al., 2020), bone repair(Liu et al., 2019 (b)), among others, have been recently carried out. Although, the widespread application of this kind of materials is due to its textural properties in combination with their adsorption capacity. In this regard, adsorption and exchange properties are very important features. Thus, a very interesting topic and with a lot of potential is the use and research of MOF in sensing applications. In this context, a lot of different molecules, ions and compounds of interest immersed in liquid (Ge et al., 2020; Gao, et al., 2020; Yu et al., 2020) or gas medium (Li et al., 2020; Kou et al.,2020; Bhanoth et al., 2019; Ma et al., 2020) have been studied. In the same way, the study and application of MOF directly supported on a quartz crystal microbalance sensor (QCM) is a very interesting and promising route to apply this kind of materials in sensing. In this regard, the widespread MOF on QCM application is in the humidity and CO_2 sensing (Benito et al., 2016; Si et al. 2011; Kosuru et al., 2016; Chappanda et al., 2018), but a lot of possibilities of sensing are available for different applications like environmental (Haghighi et al., 2020; Haghighi et al., 2019; Benito et al.,2016; Xu et al., 2018, Gu et al., 2017), health (Ma et al., 2020; Haghighi et al., 2019; Yamagiwa et al., 2014; Okada et al., 2017; Qiu et al., 2018), agricultural (Okur et al., 2020), energy (Streit, et al., 2012) among others. Therefore, the most widely extended study of MOF on QCM is directed to sensing and although a large proportion of these is on environmental and health, there is still a great opportunity niche in many fields.

As it is shown in Fig. 1a, interest in MOFs have increased in the last twenty years (Fig. 1a). Moreover, interest of sensors modified with MOF has grown since 2005 (Fig. 1b), and for QCM since 2008 (Fig. 1c). In general, from all the evaluated literature, the QCMs modified with MOF are a small portion of all the reports (Fig. 1d). In general, the QCMs modified with MOFs are a promising research topic that is gaining interest. For this reason, an early revision of these sensors is offered in this work. The aim of the present is to show how the QCMs are modified with MOFs, to contrast the kind of MOFs and the analytes of interest. In addition, since most of the reports regarding QCM show measurements of frequency shifts measurement, particular focus is made on the kind of frequency measurement instrument used in the reported experiments.

QUARTZ CRYSTAL MICROBALANCES MODIFIED WITH METAL-ORGANIC FRAMEWORKS

Fundaments of Quartz Crystal Microbalances

In 1959 Sauerbrey published his work entitled "Verwendung von Schwingquarzen zur Wägung dünner Schichten und zur Mikrowägung" (Saerbrey, 1959), where a deep characterization of the quartz crystal microbalances (QCM)s was carried on. These devices are sensors made of a thin quartz disk with metal electrodes in each side (Fig. 2a). A QCM has a proper frequency that shifts due to diverse physical effects such as deformation, temperature variations or mass loading on its surface. For application of a QCM, it needs to be connected into a circuit were a signal is generated. Therefore, the signal has an oscillation frequency defined by the QCM frequency; mentioned circuits are known as crystal oscillators, some examples include Coldpits or single-gate circuits (Frerking, 1978). Since the QCMs are very sensitive devices, their applications as mass sensor has been exhaustively researched. In some applications, the QCM surface is modified with a material that works as a sensitive layer; this means that such a layer interacts with specific chemical species. Due to their properties, porous materials are used as sensitive layer, this because adsorption of target analytes occur on the material surface.

Proper use of QCM as sensors has some restrictions. In case there is a uniform coating over QCM surface, the mass loading can be inferred from the direct measurement of the frequency shift in the generated signal, and the use of the Sauerbrey equation. Mentioned mathematical formalism is a linear relationship, which has been proven to work only when there are frequency shifts no more than 2% from the QCM original proper frequency value (Lu and Lewis, 1972). The QCM natural frequency can be increased by making thinner disks, and by increasing their diameter. This process makes more fragile disks, which are not suitable for applications outside laboratory. For using a QCM, a method to quantify the frequency shifts is required, there are different methods that include impedance analysis or direct frequency measurement. This last method is the fastest, but most of the frequency counters are designed for steady frequency values (Johansson, 2006), but it is important to note that recently a novel method has been proposed for enhancing the QCM applications (Murrieta-Rico, et al. 2015, Murrieta-Rico et al. 2019).

Figure 1. Historical report of literature regarding MOFs during the years report generated using Scopus (www.scopus.com). The query considered the terms MOF (a), MOF and sensors (b), and MOF and QCM (c). A comparison of the amount of reports in each query is presented in (d)

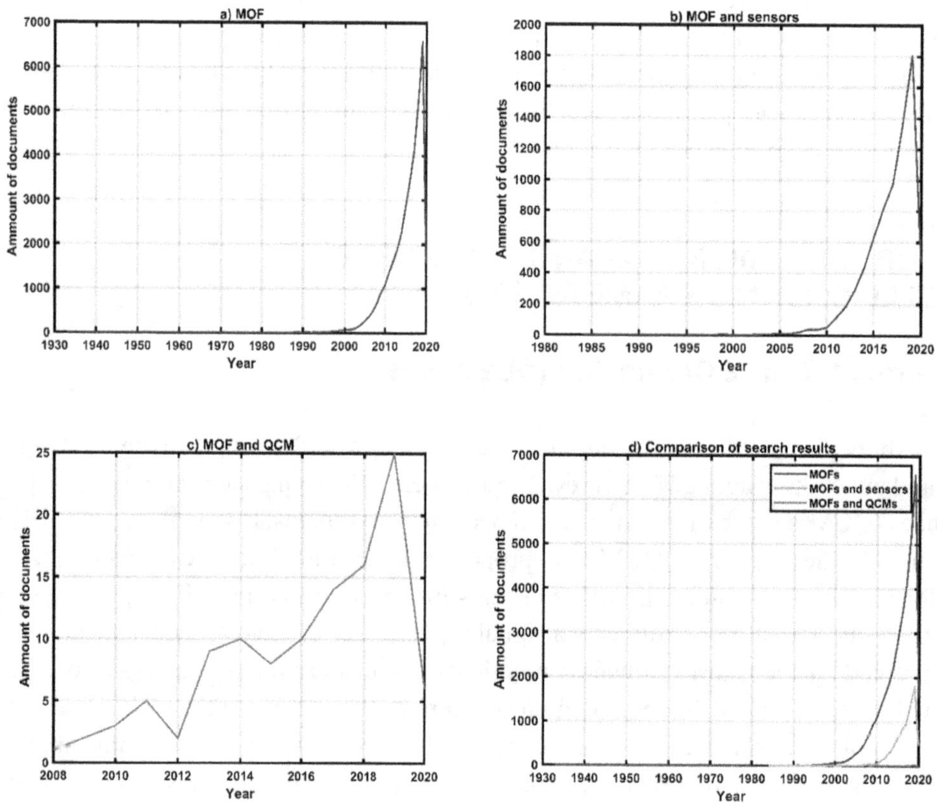

Since the QCMs were proposed by Sauerbrey as ultrasensitive weighing devices, quite diverse applications have been proposed. Some of them include which include dew-point measurement (Kwon et al. 2018), wax crystallization analysis (Cassiède et al. 2008), biological studies (Hayden et al. 2003; Lüthgens et al. 2003; Kurosawa et al. 2003; Yuan et al. 2011), viscosity analysis (Ash et al. 2003), vapor pressure measurement (Freedman et al. 2008), liquid-volume quantification (Liang et al. 2014), plasticizer monitoring (Hu et al. 2015) or sensors for detection of specific chemical species. As stated before, for this lastly mentioned task, the sensitive layer is selected according the target analyte, among the materials used to functionalize the QCM are zeolites (Osada et al. 1998; Sasaki et al. 2002; Huang et al. 2004; Ji et al. 2012; Zheng et al. 2012; Filippov et al. 2014), carbon nanotubes (Consales et al. 2006; Chen et al. 2005; Alvisi at al. 2020; Mermer et al. 2012; Lien et al. 2010; Yi et al. 2014). While mentioned materials have been exhaustively researched over the years, QCM modified with MOFs are beginning to be explored. In next section, applications regarding QCM modified with MOFs will be discussed in dept.

Figure 2. Elements of a QCM (a), single-gate oscillator circuit (b)

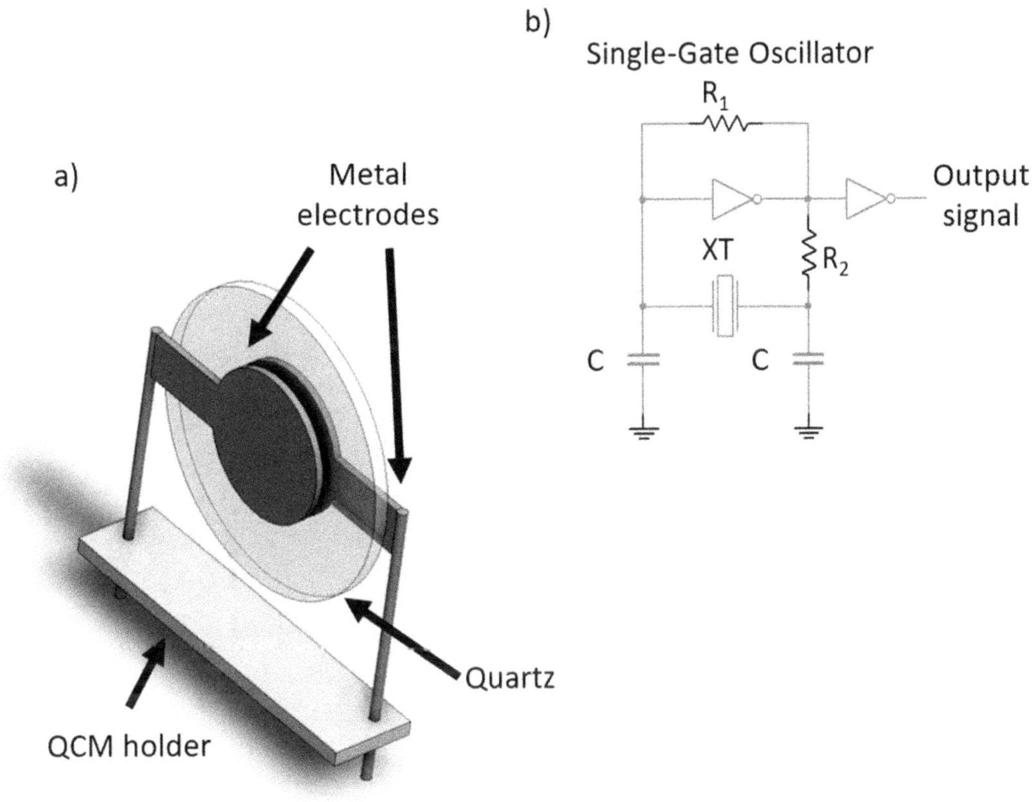

Applications of QCMs Modified With MOF

A sensor based on a QCM modified with amino functionalized Zr(IV)-MOF was reported by Ma et al (2020). The sensor was able to detect a colorectal cancer marker known as miR-221. The design of the proposed sensor is as follows. Streptavidin and UiO-66-NH_2 were used to create the complex SA@UiO-66-NH_2 (structure of UiO-66 is shown in Fig. 3).Then, the QCM surface was modified with a hairpin biotin modified probe DNA (PDNA), which generated a Hz. When target miRNA is added, the shape of PDNA changes, this process exposes the biotin site. After this process, where a 1uM of target miR-221 was added, a Hz was observed. As a result of the change in the PDNA shape, the SA@UiO-66-NH_2 complex can tie to the activated biotin PDNA. The sensitivity to the target miRNA is increased after an exponential amplification reaction (EXPAR) process. Considering miR-221, for this sensor the limit of detection was 6.9 fM, and the limit of quantity was 21 fM.

The marker detection was done in a QCM liquid cell, and authors state that Sauerbrey equation cannot be applied due to the "complexity" of the situation. A more detailed explanation is that mentioned equation only works, among other things, when the mass is loaded uniformly, and the natural frequency shifts no more than 2% from its original value. Of course, this is not valid for a QCM in a liquid media.

In addition, authors do not mention the kind of instrument used for measuring the shifts in the QCM frequency, and there is no information about a characterization of the QCM surface after its modification.

Figure 3. UiO-66

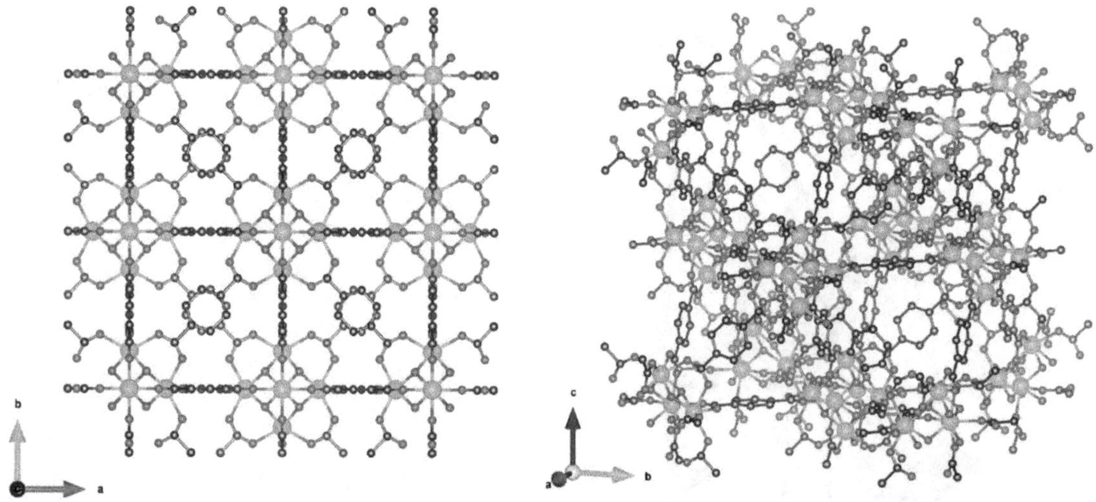

Figure 4. MOF-199

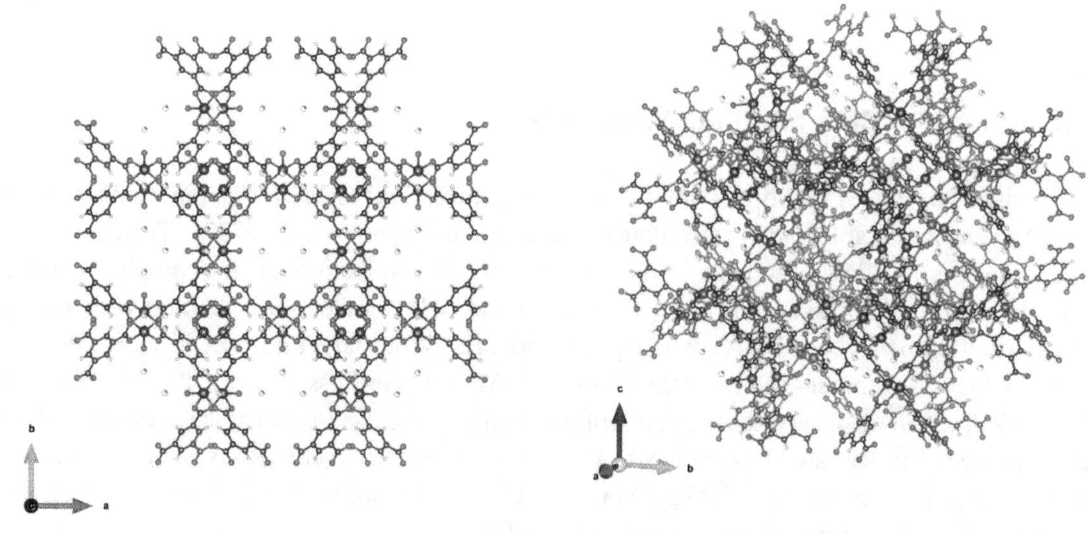

A sensor for alcohol detection was proposed by Aghakhani et al. (2019). The reported sensors were made QCMs coated with MOF-199 (Fig. 4) and m-MOF-199 (1H-benzotriazole (BTA) into the MOF-199 structure). The coatings were done using drop coating, spin coating, and electrospray. Each sensor was exposed to ethanol and the variations on each proper frequency were recorded The QCMs were connected into a Colpitts oscillator circuit and a frequency counter 8037 from Dagatron was used. A summary of maximum frequency variations is presented in Table 1. Sensors showed good sensitivity, selectivity, and a wide linear range for ethanol vapor measurement.

Methanol detection using a MOF modified QCM was reported by Si et al. (2011). The detector was made of a QCM coated with CAU-1 MOF (Fig. 5), and the coating interaction with methanol molecules was evaluated *in situ*. Different frequency shifts were observed under different concentrations of methanol. On one hand, at the lower concentration of 32 ppm, a negative frequency shift of 200 Hz was observed; on the other hand, at the highest concentration of 128 ppm a frequency shift of -800 Hz was quantified. Authors state that the QCM was provided by Maxtek, but there is no information about instrument used for measuring QCM frequency. For this reason, uncertainty of measurements is not available, as a consequence, the precision of the measurements obtained cannot be evaluated.

Figure 5. CAU-1

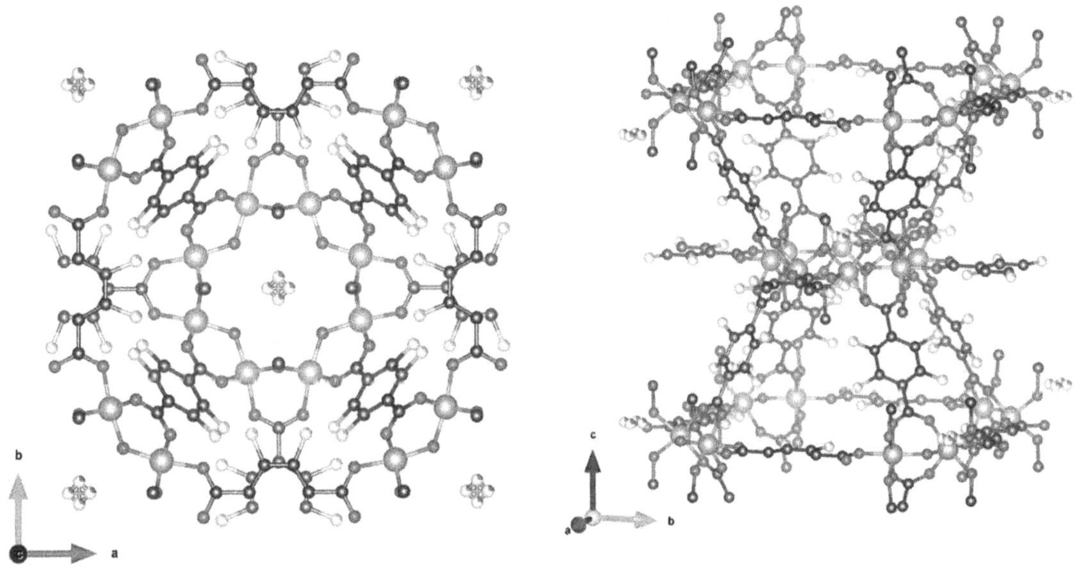

The epitaxial growth of SURMOF over a QCM is reported by Henke and Wöll (2013). The aim of this research was to assess adsorption and diffusion of ferrocene molecules into the MOF. This work shows no information of the measured frequency shifts, but instead, the adsorbed mass is presented. This approach will be feasible under the authors assumptions, but the original QCM frequency should be evaluated, this with the aim to define if the QCM operates in the Sauerbrey linear zone. Besides, authors do not state if there was done a temperature compensation for the quartz resonator (Frerking, 1978).

The modification of a QCM with a thin film of MOF type MIL-101 (Fig. 6) is reported by Benito et al. (2016). In this work, the QCM is modified using the Langmuir–Blodgett technique. In the experiment two QCMs were used, one modified with the MOF and the other one without coating. This last QCM served as reference, this allowed authors to compensate temperature and pressure effects. Both

QCMs had a proper frequency of 9 MHz. The modified QCM was exposed to a gas mixture flow of He and CO_2. At maximum pressure of 99.24 kPa, a frequency shift of -100 Hz was observed. No information is provide about the frequency measurement system, for this reason, uncertainty in measurements cannot be evaluated.

Figure 6. MIL-101

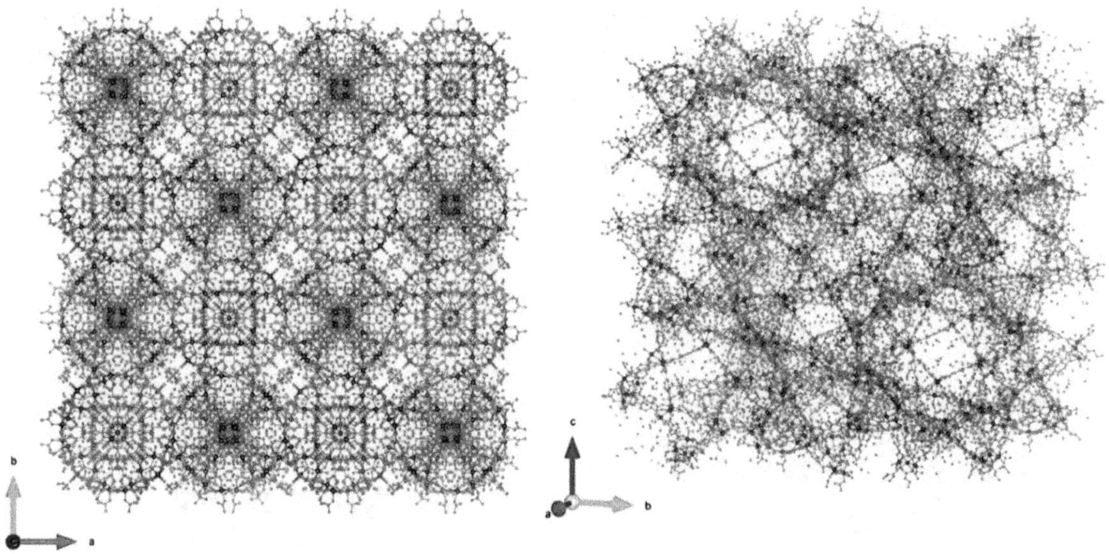

A QCM modified with NMOF (known as Al(OH)(1,4-NDC)) was reported as pyridine sensor (Xu et al., 2018). Structure of NMOF is presented in Fig. 7. It is reported that a QCM equipment from MAXTEK, INC Instruments (USA) was used in author's experiments. The QCM original frequency was in the range of 8.976 and 9.036 MHz. Prior to MOF coating, the QCM was cleaned using Piranha solvent and ultrasonicated, then it was rinsed with alcohol and double distilled water. Finally, the QCM was dried in air. The MOF was dispersed in an organic solvent, and from resulting solution, a separate volume was dispersed on the QCM surface. The QCM was dried in vacuum for 3 hours. During the sensing tests, the modified QCM was placed into a vacuum chamber were the sensor was exposed to different concentrations of pyridine. At 16 ppm, authors report a frequency shift of 1200 Hz. This result is interesting because a mass loading, the QCM decreases its proper frequency. Authors report that this frequency variation corresponds to the adsorption of pyridine by the NMOF.

Figure 7. NMOF named as Al(OH)(1,4-NDC)

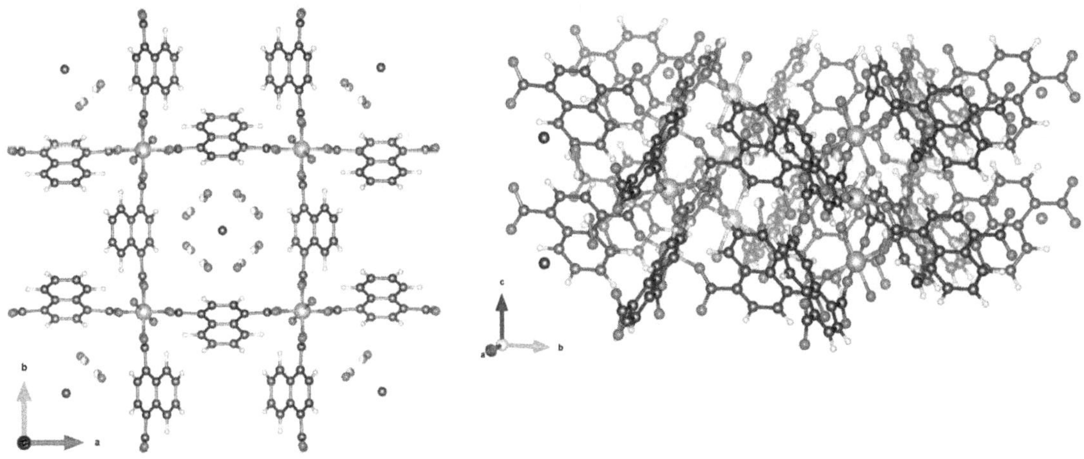

Table 1. Comparison of QCM modified with MOF

Target analyte	MOF	Approximated absolute measured frequency shift [Hz]	QCM proper frequency	Reference
Colorectal cancer marker (miR-221)	UiO-66-NH_2	120	5 MHz (AT-cut)	Ma et al. (2020)
Ethanol	MOF-199	140, 80, 170	6 MHz (AT-cut)	Aghakhani et al. (2019)
	m-MOF-199	140, 260, 200		
Methanol	CAU-1	800	9 MHz (AT-cut)	Si et al. (2011)

CONCLUSION

In this chapter, reports regarding quartz crystal microbalances modified with MOF were discussed. In most applications the MOF are deposited on a QCM, and then sensor is exposed to the analyte of interest. Since, the MOFs can be "tuned" for detecting a specific chemical specie, when a QCM is modified with MOF, small mass loading of target molecules on the MOF surface generates a variation in the QCM proper frequency. In this sense, the role of MOFs in this detection process is quite similar to other porous materials, but is important to remember that MOFs interact with organic molecules. This last effect is one of the main motivations for using MOFs as sensitive layer.

REFERENCES

Aghakhani, A., Mohamadi, F., & Ghadimi, J. (2019). Novel alcohol vapour sensor based on the mixed-ligand modified MOF-199 coated quartz crystal microbalance. *International Journal of Environmental Analytical Chemistry*, 1–18. doi:10.1080/03067319.2019.1689968

Akbarzadeh, F., Motaghi, M., Chauhan, N. P. S., & Sargazi, G. (2020). A novel synthesis of new antibacterial nanostructures based on Zn-MOF compound: Design, characterization and a high performance application. *Heliyon*, *6*(1), e03231. doi:10.1016/j.heliyon.2020.e03231 PMID:32021929

Alvisi, M., Aversa, P., Cassano, G., Serra, E., Tagliente, M. A., Schioppa, M., & Penza, M. (2012). Organic vapor detection by QCM sensors using CNT-composite films. In *Sensors and Microsystems* (pp. 79–85). Springer. doi:10.1007/978-1-4614-0935-9_14

Ash, D. C., Joyce, M. J., Barnes, C., Booth, C. J., & Jefferies, A. C. (2003). Viscosity measurement of industrial oils using the droplet quartz crystal microbalance. *Measurement Science & Technology*, *14*(11), 1955–1962. doi:10.1088/0957-0233/14/11/013

Barthelet, K., Marrot, J., Riou, D., & Férey, G. (2002). A breathing hybrid organic–inorganic solid with very large pores and high magnetic characteristics. *Angewandte Chemie International Edition*, *41*(2), 281–284. doi:10.1002/1521-3773(20020118)41:2<281::AID-ANIE281>3.0.CO;2-Y PMID:12491409

Bataille, T., Bracco, S., Comotti, A., Costantino, F., Guerri, A., Ienco, A., & Marmottini, F. (2012). Solvent dependent synthesis of micro-and nano-crystalline phosphinate based 1D tubular MOF: Structure and CO_2 adsorption selectivity. *CrystEngComm*, *14*(21), 7170–7173. doi:10.1039/c2ce26138c

Batten, S. R., Champness, N. R., Chen, X. M., Garcia-Martinez, J., Kitagawa, S., Öhrström, L., O'Keeffe, M., Suh, M. P., & Reedijk, J. (2013). Terminology of metal–organic frameworks and coordination polymers (IUPAC Recommendations 2013). *Pure and Applied Chemistry*, *85*(8), 1715–1724. doi:10.1351/PAC-REC-12-11-20

Bhanoth, S., Chappanda, K. N., Teja, V. M., Majhi, S. M., Surya, S. G., & More, Y. D. (2019). *A silver nanoparticle-anchored UiO-66 (Zr) metal-organic framework (MOF)-based capacitive H_2S gas sensor*. Academic Press.

Benito, J., Sorribas, S., Lucas, I., Coronas, J., & Gascon, I. (2016). Langmuir–Blodgett Films of the Metal–Organic Framework MIL-101 (Cr): Preparation, Characterization, and CO2 Adsorption Study Using a QCM-Based Setup. *ACS Applied Materials & Interfaces*, *8*(25), 16486–16492. doi:10.1021/acsami.6b04272 PMID:27268426

Butova, V. V. E., Soldatov, M. A., Guda, A. A., Lomachenko, K. A., & Lamberti, C. (2016). Metal-organic frameworks: Structure, properties, methods of synthesis and characterization. *Russian Chemical Reviews*, *85*(3), 280–307. doi:10.1070/RCR4554

Bux, H., Liang, F., Li, Y., Cravillon, J., Wiebcke, M., & Caro, J. (2009). Zeolitic imidazolate framework membrane with molecular sieving properties by microwave-assisted solvothermal synthesis. *Journal of the American Chemical Society*, *131*(44), 16000–16001. doi:10.1021/ja907359t PMID:19842668

Cao, Z., Chen, L., Li, S., Yu, M., Li, Z., Zhou, K., Liu, C., Jiang, F., & Hong, M. (2019). A Flexible Two-Fold Interpenetrated Indium MOF Exhibiting Dynamic Response to Gas Adsorption and High-Sensitivity Detection of Nitroaromatic Explosives. *Chemistry, an Asian Journal*, *14*(20), 3597–3602. doi:10.1002/asia.201900458 PMID:31069916

Cassiède, M., Pauly, J., Milhet, M., Rivaletto, M., Marrucho, I. M., Coutinho, J. A. P., & Daridon, J. L. (2008). A quartz crystal microbalance technique to study wax crystallization in the presence of gas. *Measurement Science & Technology*, *19*(6), 065704. doi:10.1088/0957-0233/19/6/065704

Chen, H. W., Wu, R. J., Chan, K. H., Sun, Y. L., & Su, P. G. (2005). The application of CNT/Nafion composite material to low humidity sensing measurement. *Sensors and Actuators. B, Chemical*, *104*(1), 80–84. doi:10.1016/j.snb.2004.04.105

Chappanda, K. N., Shekhah, O., Yassine, O., Patole, S. P., Eddaoudi, M., & Salama, K. N. (2018). The quest for highly sensitive QCM humidity sensors: The coating of CNT/MOF composite sensing films as case study. *Sensors and Actuators. B, Chemical*, *257*, 609–619. doi:10.1016/j.snb.2017.10.189

Chen, N., Li, M. X., Yang, P., He, X., Shao, M., & Zhu, S. R. (2013). Chiral coordination polymers with SHG-active and luminescence: An unusual homochiral 3D MOF constructed from achiral components. *Crystal Growth & Design*, *13*(6), 2650–2660. doi:10.1021/cg400426m

Chui, S. S. Y., Lo, S. M. F., Charmant, J. P., Orpen, A. G., & Williams, I. D. (1999). A chemically functionalizable nanoporous material [Cu3 (TMA) 2 (H2O) 3] n. *Science*, *283*(5405), 1148–1150. doi:10.1126cience.283.5405.1148 PMID:10024237

Consales, M., Campopiano, S., Cutolo, A., Penza, M., Aversa, P., Cassano, G., Giordano, M., & Cusano, A. (2006). Sensing properties of buffered and not buffered carbon nanotubes by fibre optic and acoustic sensors. *Measurement Science & Technology*, *17*(5), 1220–1228. doi:10.1088/0957-0233/17/5/S46

D'Ans, P., Courbon, E., Permyakova, A., Nouar, F., Simonnet-Jégat, C., Bourdreux, F., Malet, L., Serre, C., Frère, M., & Steunou, N. (2019). A new strontium bromide MOF composite with improved performance for solar energy storage application. *Journal of Energy Storage*, *25*, 100881. doi:10.1016/j.est.2019.100881

Feng, X., Song, Y., Li, Z., Kaufmann, M., Pi, Y., Chen, J. S., Xu, Z., Li, Z., Wang, C., & Lin, W. (2019). Metal–Organic Framework Stabilizes a Low-Coordinate Iridium Complex for Catalytic Methane Borylation. *Journal of the American Chemical Society*, *141*(28), 11196–11203. doi:10.1021/jacs.9b04285 PMID:31265280

Filippov, A. P., Strizhak, P. E., Vlasenko, N. V., Kochkin, Y. N., & Serebrii, T. G. (2014). Adsorption—Desorption Dynamics of Alcohols on H-Beta and H-CMK Zeolites Nanocrystallites Studied by Quartz Crystal Microbalance Method. *Adsorption Science and Technology*, *32*(10), 807–820. doi:10.1260/0263-6174.32.10.807

Freedman, A., Kebabian, P. L., Li, Z., Robinson, W. A., & Wormhoudt, J. C. (2008). Apparatus for determination of vapor pressures at ambient temperatures employing a Knudsen effusion cell and quartz crystal microbalance. *Measurement Science & Technology*, *19*(12), 125102. doi:10.1088/0957-0233/19/12/125102

Frerking, M. E. (1978). Temperature Compensation. In *Crystal Oscillator Design and Temperature Compensation* (pp. 130–176). Springer. doi:10.1007/978-94-011-6056-8_10

Gao, J., He, P., Yang, T., Wang, X., Zhou, L., He, Q., Jia, L., Deng, H., Zhang, H., Jia, B., & He, X. (2020). Short rod-like Ni-MOF anchored on graphene oxide nanosheets: A promising voltammetric platform for highly sensitive determination of p-chloronitrobenzene. *Journal of Electroanalytical Chemistry (Lausanne, Switzerland)*, *861*, 113954. doi:10.1016/j.jelechem.2020.113954

Ge, K. M., Wang, D., Xu, Z. J., & Chu, R. Q. (2020). A luminescent Eu (III)-MOF for selective sensing of Ag^+ in aqueous solution. *Journal of Molecular Structure*, *1208*, 127862. doi:10.1016/j.molstruc.2020.127862

Glover, T. G., Peterson, G. W., Schindler, B. J., Britt, D., & Yaghi, O. (2011). MOF-74 building unit has a direct impact on toxic gas adsorption. *Chemical Engineering Science*, *66*(2), 163–170. doi:10.1016/j.ces.2010.10.002

Griffith, R. L. (1943). The crystal structure of silver oxalate. *The Journal of Chemical Physics*, *11*(11), 499–505. doi:10.1063/1.1723789

Gu, Z. G., Fu, W. Q., Liu, M., & Zhang, J. (2017). Surface-mounted MOF templated fabrication of homochiral polymer thin film for enantioselective adsorption of drugs. *Chemical Communications*, *53*(9), 1470–1473. doi:10.1039/C6CC08342K PMID:28093585

Haghighi, E., & Zeinali, S. (2019). Nanoporous MIL-101 (Cr) as a sensing layer coated on a quartz crystal microbalance (QCM) nanosensor to detect volatile organic compounds (VOCs). *RSC Advances*, *9*(42), 24460–24470. doi:10.1039/C9RA04152D

Haghighi, E., & Zeinali, S. (2020). Formaldehyde detection using quartz crystal microbalance (QCM) nanosensor coated by nanoporous MIL-101 (Cr) film. *Microporous and Mesoporous Materials*, *300*, 110065. doi:10.1016/j.micromeso.2020.110065

Hayden, O., Bindeus, R., & Dickert, F. L. (2003). Combining atomic force microscope and quartz crystal microbalance studies for cell detection. *Measurement Science & Technology*, *14*(11), 1876–1881. doi:10.1088/0957-0233/14/11/004

Hoskins, B. F., & Robson, R. (1990). Design and construction of a new class of scaffolding-like materials comprising infinite polymeric frameworks of 3D-linked molecular rods. A reappraisal of the zinc cyanide and cadmium cyanide structures and the synthesis and structure of the diamond-related frameworks $[N(CH_3)_4][Cu^IZn^{II}(CN)_4]$ and Cu^I[4, 4′, 4′′, 4′′′-tetracyanotetraphenylmethane] $BF_4.xC_6H_5NO_2$. *Journal of the American Chemical Society*, *112*(4), 1546–1554. doi:10.1021/ja00160a038

Hu, R., Zhang, K., Fan, G., Luo, Z., & Li, G. (2015). Development of a high-sensitivity plasticizer sensor based on a quartz crystal microbalance modified with a nanostructured nickel hydroxide film. *Measurement Science & Technology*, *26*(5), 055102. doi:10.1088/0957-0233/26/5/055102

Huang, H., Zhou, J., Chen, S., Zeng, L., & Huang, Y. (2004). A highly sensitive QCM sensor coated with Ag^+-ZSM-5 film for medical diagnosis. *Sensors and Actuators. B, Chemical*, *101*(3), 316–321. doi:10.1016/j.snb.2004.04.001

Ji, X., Yao, W., Peng, J., Ren, N., Zhou, J., & Huang, Y. (2012). Evaluation of Cu-ZSM-5 zeolites as QCM sensor coatings for DMMP detection. *Sensors and Actuators. B, Chemical*, *166*, 50–55. doi:10.1016/j.snb.2011.12.014

Jiang, K., Zhang, L., Xia, T., Yang, Y., Li, B., Cui, Y., & Qian, G. (2019). A water-stable fcu-MOF material with exposed amino groups for the multi-functional separation of small molecules. *Science China Materials*, *62*(9), 1315–1322. doi:10.100740843-019-9427-5

Johansson, S. (2006, March). New frequency counting principle improves resolution. In *Proceedings of the 20th European Frequency and Time Forum* (pp. 139-146). IEEE.

Kaur, R., Kim, K. H., & Deep, A. (2017). A convenient electrolytic assembly of graphene-MOF composite thin film and its photoanodic application. *Applied Surface Science*, *396*, 1303–1309. doi:10.1016/j.apsusc.2016.11.150

Kandiah, M., Nilsen, M. H., Usseglio, S., Jakobsen, S., Olsbye, U., Tilset, M., Larabi, C., Quadrelli, E. A., Bonino, F., & Lillerud, K. P. (2010). Synthesis and stability of tagged UiO-66 Zr-MOFs. *Chemistry of Materials*, *22*(24), 6632–6640. doi:10.1021/cm102601v

Kosuru, L., Bouchaala, A., Jaber, N., & Younis, M. I. (2016). Humidity detection using metal organic framework coated on QCM. *Journal of Sensors*, *2016*, 2016. doi:10.1155/2016/4902790

Kou, D., Ma, W., Zhang, S., Li, R., & Zhang, Y. (2020). BTEX vapors detection with flexible MOF and functional polymer by means of composite photonic crystal. *ACS Applied Materials & Interfaces*, *12*(10), 11955–11964. doi:10.1021/acsami.9b22033

Kepert, C. J. (2010). Metal-Organic Framework Materials. In D. W. Bruce, D. O'Hare, R. I. Walton, D. W. Bruce, D. O'Hare, & R. I. Walton (Eds.), *Porous Materials*. doi:10.1002/9780470711385.ch1

Kitagawa, S., Matsuyama, S., Munakata, M., & Emori, T. (1991). Synthesis and crystal structures of novel one-dimensional polymers,[{M(bpen)X}$_\infty$][M= Cu^I, X= [{Cu (bpen)(CO)(CH3 CN)(PF6)}∞]. *Journal of the Chemical Society, Dalton Transactions: Inorganic Chemistry*, (11), 2869–2874. doi:10.1039/DT9910002869

Kwon, S. Y., Kim, J. C., & Choi, B. I. (2008). Accurate dew-point measurement over a wide temperature range using a quartz crystal microbalance dew-point sensor. *Measurement Science & Technology*, *19*(11), 115206. doi:10.1088/0957-0233/19/11/115206

Kurosawa, S., Aizawa, H., Tozuka, M., Nakamura, M., & Park, J. W. (2003). Immunosensors using a quartz crystal microbalance. *Measurement Science & Technology*, *14*(11), 1882–1887. doi:10.1088/0957-0233/14/11/005

Li, D., Zhang, S., Feng, X., Yang, H., Nie, F., & Zhang, W. (2019). A novel peroxidase mimetic Co-MOF enhanced luminol chemiluminescence and its application in glucose sensing. *Sensors and Actuators. B, Chemical*, *296*, 126631. doi:10.1016/j.snb.2019.126631

Li, S., Xie, L., He, M., Hu, X., Luo, G., Chen, C., & Zhu, Z. (2020). Metal-Organic Frameworks-Derived Bamboo-Like CuO/In2O3 Heterostructure for High-Performance H2S Gas Sensor with Low Operating Temperature. *Sensors and Actuators. B, Chemical*, *310*, 127828. doi:10.1016/j.snb.2020.127828

Liang, D., Zhang, J., Govindaiah, M. T., Tanguy, L., Ernst, A., Zengerle, R., & Koltay, P. (2014). Liquid volume measurement method for the picoliter to nanoliter volume range based on quartz crystal microbalance technology. *Measurement Science & Technology*, *25*(9), 095302. doi:10.1088/0957-0233/25/9/095302

Lien, T. T. N., Dai Lam, T., An, V. T. H., Hoang, T. V., Quang, D. T., Khieu, D. Q., ... Kim, J. S. (2010). Multi-wall carbon nanotubes (MWCNTs)-doped polypyrrole DNA biosensor for label-free detection of genetically modified organisms by QCM and EIS. *Talanta*, *80*(3), 1164–1169. doi:10.1016/j.talanta.2009.09.002 PMID:20006069

Liu, G., Chernikova, V., Liu, Y., Zhang, K., Belmabkhout, Y., Shekhah, O., Zhang, C., Yi, S., Eddaoudi, M., & Koros, W. J. (2018). Mixed matrix formulations with MOF molecular sieving for key energy-intensive separations. *Nature Materials*, *17*(3), 283–289. doi:10.103841563-017-0013-1 PMID:29434309

Liu, P., Redekop, E., Gao, X., Liu, W. C., Olsbye, U., & Somorjai, G. A. (2019). Oligomerization of Light Olefins Catalyzed by Brønsted-Acidic Metal–Organic Framework-808. *Journal of the American Chemical Society*, *141*(29), 11557–11564. doi:10.1021/jacs.9b03867 PMID:31264857

Liu, W., Yan, Z., Zhang, Z., Zhang, Y., Cai, G., & Li, Z. (2019). Bioactive and anti-corrosive bio-MOF-1 coating on magnesium alloy for bone repair application. *Journal of Alloys and Compounds*, *788*, 705–711. doi:10.1016/j.jallcom.2019.02.281

Lu, C. S., & Lewis, O. (1972). Investigation of film-thickness determination by oscillating quartz resonators with large mass load. *Journal of Applied Physics*, *43*(11), 4385–4390. doi:10.1063/1.1660931

Lüthgens, E., Herrig, A., Kastl, K., Steinem, C., Reiss, B., Wegener, J., Pignataro, B., & Janshoff, A. (2003). Adhesion of liposomes: A quartz crystal microbalance study. *Measurement Science & Technology*, *14*(11), 1865–1875. doi:10.1088/0957-0233/14/11/003

Ma, D., Zheng, S. R., Fan, J., Cai, S. L., Dai, Z., Zou, X. Y., Teng, S.-H., & Zhang, W. G. (2020). A new QCM signal enhancement strategy based on streptavidin@ metal-organic framework complex for miRNA detection. *Analytica Chimica Acta*, *1095*, 212–218. doi:10.1016/j.aca.2019.10.034 PMID:31864625

Ma, D., Zheng, S. R., Fan, J., Cai, S., Dai, Z., Zou, X., Teng, S.-H., & Zhang, W. G. (2020). A new amplification strategy for quartz crystal microbalance miRNA sensor based on selective interactions between Metal-organic framework and miRNA. *New Journal of Chemistry*, *44*(5), 1684–1688. doi:10.1039/C9NJ04600C

Ma, Z., Yuan, T., Fan, Y., Wang, L., Duan, Z., Du, W., Zhang, D., & Xu, J. (2020). A benzene vapor sensor based on a metal-organic framework-modified quartz crystal microbalance. *Sensors and Actuators. B, Chemical*, *311*, 127365. doi:10.1016/j.snb.2019.127365

Mermer, Ö., Okur, S., Sümer, F., Özbek, C., Sayın, S., & Yılmaz, M. (2012). *Gas sensing properties of carbon nanotubes modified with calixarene molecules measured by QCM techniques*. Academic Press.

Murrieta-Rico, F. N., Mercorelli, P., Sergiyenko, O. Y., Petranovskii, V., Hernandez-Balbuena, D., & Tyrsa, V. (2015). Mathematical modelling of molecular adsorption in zeolite coated frequency domain sensors. *IFAC-PapersOnLine*, *48*(1), 41–46. doi:10.1016/j.ifacol.2015.05.060

Murrieta-Rico, F. N., Petranovskii, V., Sergiyenko, O., Mercorelli, P., Antúnez-Garcia, J., de Dios Sanchez-Lopez, J., & Yocupicio-Gaxiola, R. I. (2019, October). Experimental analysis of measurement process for a QCM using the pulse coincidence method. In *IECON 2019-45th Annual Conference of the IEEE Industrial Electronics Society* (Vol. 1, pp. 4657-4662). IEEE. 10.1109/IECON.2019.8927337

O'Keeffe, M., Peskov, M. A., Ramsden, S. J., & Yaghi, O. M. (2008). The reticular chemistry structure resource (RCSR) database of, and symbols for, crystal nets. *Accounts of Chemical Research*, *41*(12), 1782–1789. doi:10.1021/ar800124u PMID:18834152

Qiu, X., Xu, X. Y., Chen, X., Wu, Y., & Guo, H. (2018). Preparation of a molecularly imprinted sensor based on quartz crystal microbalance for specific recognition of sialic acid in human urine. *Analytical and Bioanalytical Chemistry*, *410*(18), 4387–4395. doi:10.100700216-018-1094-7 PMID:29736700

Okada, K., Sawai, S., Ikigaki, K., Tokudome, Y., Falcaro, P., & Takahashi, M. (2017). Electrochemical sensing and catalysis using Cu 3 (BTC) 2 coating electrodes from Cu (OH) 2 films. *CrystEngComm*, *19*(29), 4194–4200. doi:10.1039/C7CE00416H

Okur, S., Zhang, Z., Sarheed, M., Nick, P., Lemmer, U., & Heinke, L. (2020). Towards a MOF e-Nose: A SURMOF sensor array for detection and discrimination of plant oil scents and their mixtures. *Sensors and Actuators. B, Chemical*, *306*, 127502. doi:10.1016/j.snb.2019.127502

Osada, M., Sasaki, I., Nishioka, M., Sadakata, M., & Okubo, T. (1998). Synthesis of a faujasite thin layer and its application for SO2 sensing at elevated temperatures. *Microporous and Mesoporous Materials*, *23*(5-6), 287–294. doi:10.1016/S1387-1811(98)00125-5

Pachfule, P., Das, R., Poddar, P., & Banerjee, R. (2011). Solvothermal synthesis, structure, and properties of metal organic framework isomers derived from a partially fluorinated link. *Crystal Growth & Design*, *11*(4), 1215–1222. doi:10.1021/cg101414x

Sasaki, I., Tsuchiya, H., Nishioka, M., Sadakata, M., & Okubo, T. (2002). Gas sensing with zeolite-coated quartz crystal microbalances—Principal component analysis approach. *Sensors and Actuators. B, Chemical*, *86*(1), 26–33. doi:10.1016/S0925-4005(02)00132-6

Sauerbrey, G. (1959). Verwendung von Schwingquarzen zur Wägung dünner Schichten und zur Mikrowägung. *Zeitschrift für Physik*, *155*(2), 206–222. doi:10.1007/BF01337937

Si, X., Jiao, C., Li, F., Zhang, J., Wang, S., Liu, S., Li, Z., Sun, L., Xu, F., Gabelica, Z., & Schick, C. (2011). High and selective CO_2 uptake, H_2 storage and methanol sensing on the amine-decorated 12-connected MOF CAU-1. *Energy & Environmental Science*, *4*(11), 4522–4527. doi:10.1039/c1ee01380g

Son, W. J., Kim, J., Kim, J., & Ahn, W. S. (2008). Sonochemical synthesis of MOF-5. *Chemical Communications*, (47), 6336–6338. doi:10.1039/b814740j PMID:19048147

Song, Z., Qiu, F., Zaia, E. W., Wang, Z., Kunz, M., Guo, J., Brady, M., Mi, B., & Urban, J. J. (2017). Dual-channel, molecular-sieving core/shell ZIF@ MOF architectures as engineered fillers in hybrid membranes for highly selective CO2 separation. *Nano Letters*, *17*(11), 6752–6758. doi:10.1021/acs.nanolett.7b02910 PMID:29072837

Streit, H. C., Adlung, M., Shekhah, O., Stammer, X., Arslan, H. K., Zybaylo, O., Ladnorg, T., Gliemann, H., Franzreb, M., Wöll, C., & Wickleder, C. (2012). Surface-Anchored MOF-Based Photonic Antennae. *ChemPhysChem*, *13*(11), 2699–2702. doi:10.1002/cphc.201200262 PMID:22700363

Tranchemontagne, D. J., Mendoza-Cortés, J. L., O'Keeffe, M., & Yaghi, O. M. (2009). Secondary building units, nets and bonding in the chemistry of metal–organic frameworks. *Chemical Society Reviews*, *38*(5), 1257–1283. doi:10.1039/b817735j PMID:19384437

Van Assche, T. R., Desmet, G., Ameloot, R., De Vos, D. E., Terryn, H., & Denayer, J. F. (2012). Electrochemical synthesis of thin HKUST-1 layers on copper mesh. *Microporous and Mesoporous Materials*, *158*, 209–213. doi:10.1016/j.micromeso.2012.03.029

Wu, D., Liu, J., Jin, J., Cheng, J., Wang, M., Yang, G., & Wang, Y. Y. (2019). New Doubly Interpenetrated MOF with [Zn4O] Clusters and Its Doped Isomorphic MOF: Sensing, Dye, and Gas Adsorption Capacity. *Crystal Growth & Design*, *19*(11), 6774–6783. doi:10.1021/acs.cgd.9b01193

Xia, H., Zhang, J., Yang, Z., Guo, S., Guo, S., & Xu, Q. (2017). 2D MOF nanoflake-assembled spherical microstructures for enhanced supercapacitor and electrocatalysis performances. *Nano-Micro Letters*, *9*(4), 43. doi:10.100740820-017-0144-6 PMID:30393738

Xu, F., Sun, L., Huang, P., Sun, Y., Zheng, Q., Zou, Y., Chu, H., Yan, E., Zhang, H., Wang, J., & Du, Y. (2018). A pyridine vapor sensor based on metal-organic framework-modified quartz crystal microbalance. *Sensors and Actuators. B, Chemical*, *254*, 872–877. doi:10.1016/j.snb.2017.07.026

Xu, G. W., Wu, Y. P., Dong, W. W., Zhao, J., Wu, X. Q., Li, D. S., & Zhang, Q. (2017). A multifunctional Tb-MOF for highly discriminative sensing of Eu3+/Dy3+ and as a catalyst support of Ag nanoparticles. *Small*, *13*(22), 1602996. doi:10.1002mll.201602996 PMID:28418186

Yaghi, O. M., & Li, H. (1995). Hydrothermal synthesis of a metal-organic framework containing large rectangular channels. *Journal of the American Chemical Society*, *117*(41), 10401–10402. doi:10.1021/ja00146a033

Yamagiwa, H., Sato, S., Fukawa, T., Ikehara, T., Maeda, R., Mihara, T., & Kimura, M. (2014). Detection of volatile organic compounds by weight-detectable sensors coated with metal-organic frameworks. *Scientific Reports*, *4*(1), 1–6. PMID:25175808

Yang, F., Li, W., & Tang, B. (2018). Facile synthesis of amorphous UiO-66 (Zr-MOF) for supercapacitor application. *Journal of Alloys and Compounds*, *733*, 8–14. doi:10.1016/j.jallcom.2017.10.129

Yi, P., & Chen, K. L. (2014). Release kinetics of multiwalled carbon nanotubes deposited on silica surfaces: Quartz crystal microbalance with dissipation (QCM-D) measurements and modeling. *Environmental Science & Technology*, *48*(8), 4406–4413. doi:10.1021/es405471u PMID:24654566

Yu, C. X., Hu, F. L., Song, J. G., Zhang, J. L., Liu, S. S., Wang, B. X., Meng, H., Liu, L.-L., & Ma, L. F. (2020). Ultrathin two-dimensional metal-organic framework nanosheets decorated with tetra-pyridyl calix [4] arene: Design, synthesis and application in pesticide detection. *Sensors and Actuators. B, Chemical*, *310*, 127819. doi:10.1016/j.snb.2020.127819

Yuan, Y. J., Van der Werff, M. J., & Xu, W. L. (2011). A field-programmable-gate-array-based high-speed transceiver for a quartz crystal microbalance induced bond-rupture sensor. *Measurement Science & Technology*, *22*(4), 045201. doi:10.1088/0957-0233/22/4/045201

Zheng, Y., Li, X., & Dutta, P. K. (2012). Exploitation of unique properties of zeolites in the development of gas sensors. *Sensors (Basel)*, *12*(4), 5170–5194. doi:10.3390120405170 PMID:22666081

ADDITIONAL READING

García, H., & Navalón, S. (Eds.). (2018). *Metal-Organic Frameworks: Applications in Separations and Catalysis*. John Wiley & Sons. doi:10.1002/9783527809097

Gautschi, G. (2002). Piezoelectric sensors. In *Piezoelectric Sensorics* (pp. 73–91). Springer. doi:10.1007/978-3-662-04732-3_5

Johannsmann, D. (2015). *The quartz crystal microbalance in soft matter research. Fundamentals and modeling*. Springer International Publishing.

Liu, P., & Chen, G. F. (2014). *Porous materials: processing and applications*. Elsevier.

Tressler, J. F., Alkoy, S., & Newnham, R. E. (1998). Piezoelectric sensors and sensor materials. *Journal of Electroceramics*, *2*(4), 257–272. doi:10.1023/A:1009926623551

Steinem, C., & Janshoff, A. (Eds.). (2007). Piezoelectric sensors (Vol. 5). Springer Science & Business Media.

KEY TERMS AND DEFINITIONS

Analyte: A chemical substance that is of interest to detect.

Frequency Shift: This is the variation of the frequency of a signal. Usually, this term is used when a sensor with frequency output has a frequency variation after it is stimulated.

MOF: Acronym of Metal-Organic Framework. This is a crystalline material that is made of a metal center with an organic linker.

QCM: Acronym of Quartz Crystal Microbalance. This is a sensor made of a thin quartz disk with a metal electrode on each side.

Sensitive Layer: This is the name of a thin layer of a material that is intended to interact with a specific chemical specie. Usually, this layer is grown or synthesized in a sensor, this process gives the sensor the ability to interact with a particular analyte.

Chapter 5
Application of Metal-Organic Framework Adsorbents for Water Defluoridation

Wilfrida Nyanduko Nyairo
Maseno University, Kenya

Emily Ng'eno
Masinde Muliro University of Science and Technology, Kenya

Victor Odhiambo Shikuku
Kaimosi Friends University College, Kenya

Patrick Ssebugere
Makerere University, Uganda

ABSTRACT

Defluoridation of drinking water is a subject of great concern due to the side effects of exposure to fluorine such as fluorosis, damage to vital organs such as the brain, liver, and kidney, among others. Fluorine finds itself in water resources both by geological and anthropogenic processes. Adsorption still remains the popular method of defluoridation and water treatment in general due to its inherent advantages such as simplicity and low-cost. This chapter discusses at length the up-to-date data on the preparation and adsorption capacities of metal-organic framework (MOF) materials as emerging adsorbents for sequestration of fluoride from water. Their synthesis, efficiency, and suggested adsorption mechanisms are examined. It is demonstrated that MOFs are low-cost, efficient, and environmentally benign alternative adsorbents for the removal of fluoride from aqueous solution.

DOI: 10.4018/978-1-7998-4760-1.ch005

INTRODUCTION

Though an essential element in living systems, such as in bone and teeth development, fluorine has been recognized as a water contaminant for many decades (Rehman et al., 2015). At concentrations above the stipulated limit of 1.5 mg/L, excess intake of fluoride has been associated with several disorders such as fluorosis, fetal cerebral function, and neurotransmitters among others (Tang and Zhang, 2016). Therefore, much work has been published on the methods for removal of fluorine from drinking water. Such defluoridation methods include precipitation (Budyanto et al., 2015), electro-coagulation (Vasudevan et al., 2009), membrane filtration (Shen et al., 2015) and adsorption (Meenakshi et al., 2008). Among these methods, adsorption still remains the widely used method for the removal of fluoride from water due to its efficiency and inexpensiveness. Additionally, adsorption process is an eco-friendly process since it does not generate toxic by-products. Several adsorbents, derived from a plethora of precursor materials, have been examined for the sequestration of fluoride. These include; alumina (Rojas-Mayorga et al., 2015), mixed oxides (Ghosh et al., 2014), activated carbon (Du et al., 2015), clay minerals (Meenakshi et al., 2008), and ion-exchange resins (Viswanathan and Meenakshi, 2009) among others. Among an emerging class of adsorbents are metal organic frameworks (MOFs). These have gained considerable attention as alternative materials for the uptake of fluorides due to the tunability of their structures and textural properties (surface areas) providing new surfaces and functionality (Wang et al. 2019). MOFs have been exploited for various applications such as hydrogen storage and CO_2 adsorption among others. This chapter discusses the synthesis methods for preparation of MOFs and their application as adsorbents for the removal of fluorides from aqueous solutions. The adsorption capacities and mechanisms have also been presented.

Occurrence of Fluoride in Water

Elemental fluorine does not occur in nature due to its high reactivity. It however occurs as fluoride (F^-) in minerals such as cryolite, fluorspar and flourapatite among others. It is approximated to be 0.3 g/kg of the lithosphere (B. Kumar, 2015). It is also present as fluoride in manufactured compounds such as sodium fluoride (NaF), hydrogen fluoride (HF), aluminum fluoride (AlF_3), calcium fluoride (CaF_2), magnesium fluoride (MgF_2) and flourosilicic acid (H_2SiF_6). These compounds are used in industry in a variety of ways such as production of aluminum, steel, glass fiber, bricks, tiles and ceramics. Fluorides are also used in water fluoridation since it is an essential mineral for healthy bones and teeth. The combination of natural concentrations and anthropogenic emissions increases the amount of fluorides in drinking water beyond the maximum allowable limit of 1.5 mg/L stipulated by the World Health Organization (Karmakar et al., 2016; Wang et al., 2019). Exposure to fluorides vary depending on the geographical location and weather conditions but groundwater sources can surpass 35 mg/L (Tripathy et al., 2006). Generally, a concentration between 0.5 mg/L-1.5 mg/L is essential for the healthy growth of skeletal tissues, that is, teeth and bones. It prevents dental carries especially in children (B. Kumar, 2015). Beyond the aforementioned concentrations, it is detrimental to human health. Higher concentrations ranging around 1.5 mg/L to 3 mg/L in drinking water are most likely to cause dental fluorosis especially in warmer areas where water consumption is higher (He et al., 2016). Elevated intake of drinking water containing between 3 to 10 mg/L of fluoride causes skeletal fluorosis which may eventually lead to crippling skeletal fluorosis (Karmakar et al., 2018). It is also thought to affect fetal cerebral function and neurotransmitters (He et al., 2016). It may also cause reduced IQ and lasting adverse effects to the brain, liver, kidney and thyroid functions of young children (Karmakar et al., 2016; Massoudinejad et al., 2018).

These effects caused by increased amounts of fluorides have led to several developments in analytical techniques aimed at real-time monitoring of the fluoride ion as well as quantitative determination in different environmental matrices (Hinterholzinger et al., 2013). More innovations have gone into defluoridation of water sources and especially drinking water. These include adsorption, precipitation, ion exchange, electrocoagulation, electrodialysis, use of membranes and reverse osmosis (Lin et al., 2016; Wang et al., 2019). These technological advancements have been efficient but the cost associated with some of these methods for example the electrochemical methods are inhibitive. Adsorption and membrane use are two widely used methods of fluoride removal (He et al., 2016) since they are cost effective, scalable with easy technical use (Lin et al., 2016). Adsorption, especially using carbonaceous adsorbents has captivated a vast majority of researchers dealing with removal of contaminants from various environmental matrices. This can be justified by the fact that most of these carbonaceous materials are non-toxic to the human health. These carbon materials have been prepared using different flexible and simple methodologies. Among them include directly carbonizing them from their original organic materials by subjecting them to controlled heat thereby obtaining nanoporous biochars (Ngeno et al., 2016). These adsorbents have their own shortcomings comprising: diminished, disorganized structures and varied sizes which restraints their utilization. These drawbacks have led to the exploitation of other materials that could be used for defluoridation including mixed oxides, clay like materials, activated carbon, alumina and alumina based adsorbents, bio-polymer based composites, aluminum hydroxide impregnated limestone and ion-exchange resins/fibers (He et al., 2016). Most of these adsorbents worked effectively in laboratory experiments but in practical applications, separating them from water has proved to be a challenge (He et al., 2016). Their surface area was also found to be low and thus the material gets saturated too soon (Karmakar et al., 2016). Additionally, these adsorbents do not work effectively on low pH range, and there is also interference from competing anions such as nitrates, phosphates, sulfates and bicarbonates (Massoudinejad et al., 2016). These disadvantages have been pivotal in scientists looking for more porous and efficient adsorbents.

While advancing material science, chemists and material scientists have always had an interest in the preparation of hollow structures for industrial applications. The challenge, however, has always been the preparation of these cage-structured nanoporous complex substances where the structures and composition are designed to fit a certain target sorbate. Initially, inorganic and organic adsorbents were prepared independently but each had strengths and weaknesses. The organic sorbents had increased surface areas with the drawback being irregular non-uniform pores. Conversely, the inorganic adsorbents had lesser surface areas but with a more structured pore sizes with elevated thermal and mechanical stability (Massoudinejad et al., 2016). To obtain the various benefits, it was vital to merge the two diverse materials. These merged materials were pioneered by the end of the 1990s by Prof. Omar Yaghi of University of California, Berkeley (Rosi et al., 2002; Yaghi et al., 1998). These hollow structures were named metal-organic frameworks (MOFs), also known as supramolecular structures, porous coordination polymers (PCPs), or metal-organic polymers (Czaja et al., 2009). This shifted the removal of pollutants using carbonaceous materials to the use of this new innovation.

So what are MOFs? MOFs, depicted in Figure 1, are organic-inorganic composite, crystalline and porous substances made of a three-dimensional structured arrangement of positively charged metal ions attached to multidentate organic molecules acting as linkers forming a repeating, cage-like structure (Leus et al., 2010). The cages forms a hollow structure with an exceptionally enormous internal surface area (more than 7000 m^2g^{-1} in most MOF materials) (Godfrey et al., 2018) and a high micropore volume. This hybrid material allows distinctive structural flexibility in terms of topology, geometry and dimensions as it can be designed based on what the target sorbate is unlike other porous adsorbents. MOF synthesis provides an avenue for material scientists to design an adsorbent with constant pore measurements, defined pore shapes, tunable porosity, atomic-level structural uniformity, increased active sites due to more metal ions and versatile chemical functionalities (Czaja et al., 2009; Lin et al., 2016). Its structure and composition gives it a number of properties including increased surface-to-volume ratio; lesser density; enhanced thermal and mechanical stability; microreactor environment; higher loading capacities; total lack of non-accessible bulk volume; and lower transmission lengths of mass and charge (Czaja et al., 2009).

In most instances, the synthesis of MOFs uses soluble metal salts such as sulfates, nitrates or acetates while the organic linkers, which majorly are mono-, di-, tri- and tetracarboxylic acids, are delivered via amines or amides, mostly triethylamine, diethylformamide, or dimethylformamide (Czaja et al., 2009). MOF structures are self-assembled when these components are combined while stirring beginning at room temperature and gradually increasing the temperatures to 200 ^{o}C at solvothermal conditions in a couple of hours (Czaja et al., 2009). On large scale production, a salt-free electrochemical method may be used (Ameloot et al., 2009). After synthesis, MOFs are usually characterized to determine the crystallinity and phase purity using X-ray diffraction (XRD). Fourier transform infrared spectroscopy (FTIR) spectra is used to determine functional groups in the adsorbent. Field emission scanning electron microscopy (FE-SEM) is also usually performed to give the images and the nanoscale diameters of the synthesized crystals while Energy dispersive x-ray spectroscopy (EDAX) is done to determine the chemical composition of the crystals. Brunauer-Emmett-Teller (BET) characterization is carried out to ascertain the surface area and total pore volumes (Massoudinejad et al., 2016).

So far, myriads of different MOFs have been synthesized, characterized and used extensively in gas detection and storage (Drisdell et al., 2013), liquid separation and purification, electrochemical energy storage, heavy metal and organic pollutants removal, catalysis (Leus et al., 2010), and sensing (Chernikova et al., 2018; McDonald et al., 2015). Additionally, MOFs have been used indirectly as base materials for making composites of metals and carbon based compounds. Considering their extraordinary physico-chemical properties, they have also been used in refrigeration; storage and delivery of pharmaceuticals in the body (Sun et al., 2013). Furthermore, they have found use in monitoring body organs thus controlling animal conduct, divulging the mechanism of biological processes, monitoring health conditions, and treating ailments (Czaja et al., 2009). More extensively, MOFs have greatly been used as adsorbents for contaminants since they work well at lower temperatures and can be severally regenerated and reused (P. Kumar et al., 2019). MOFs also offer selective adsorption since their pore dimensions and geometry can easily be designed to enable the uptake of targeted sorbates. Synthesis therefore uses different kinds of metal ions and organic ligands depending on what the researchers are interested in adsorbing selectively in the tailor-made cages (Chernikova et al., 2018). This flexibility in shape and size surmounts the performance of all other adsorbents. Heavy metal ions, fluorides, toxic dyes, pharmaceuticals, compounds of nitrogen and sulfur are some of the contaminants where MOFs have been used (Karmakar et al., 2016). The next section of this chapter focuses on the synthesis of MOFs and their application in the removal of fluorides from water.

Figure 1. Skeletal structure of a MOF
(Adopted from https://www.nanowerk.com)

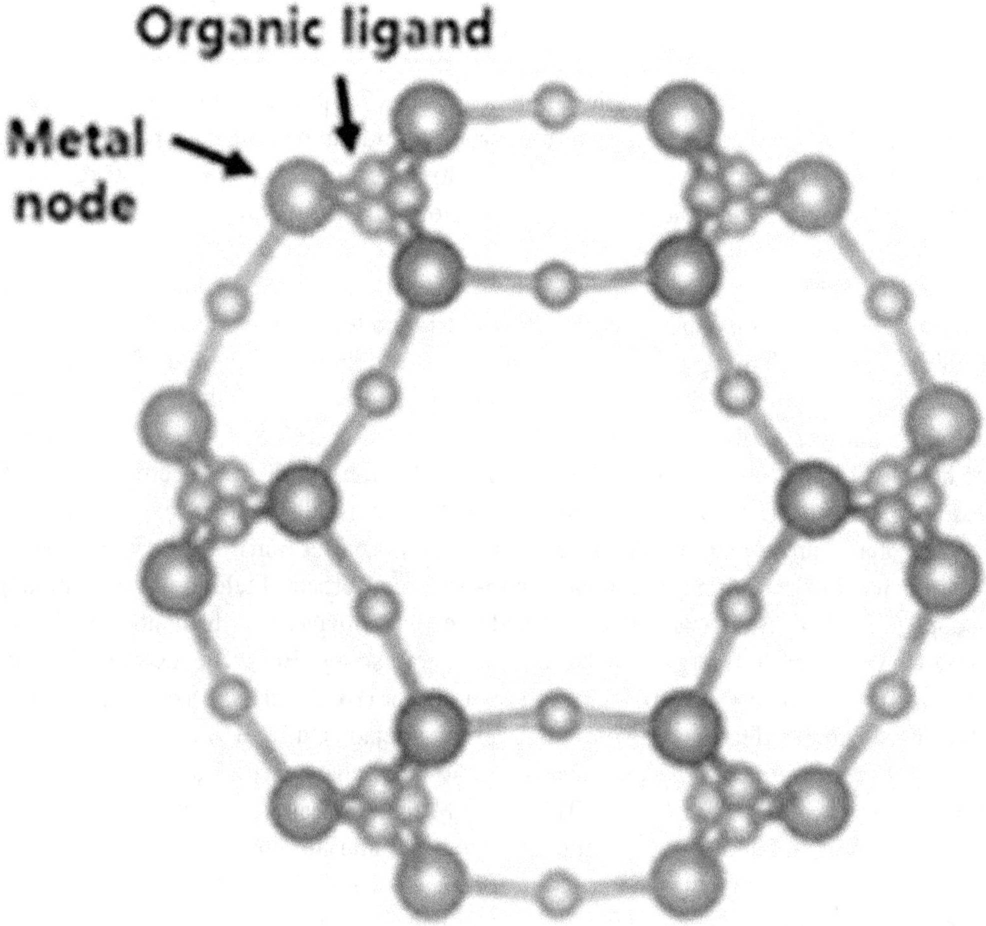

Synthesis of MOFs

The metal organic frameworks (MOFs) are synthesized through the conventional solvothermal/hydrothermal method (Karmakar et al., 2016; Ma et al., 2017) and non-solvothermal methods such as: sonochemical (Son et al., 2008), mechanochemical (Klimakow et al., 2010), microwave assisted (Cho et al., 2012), electrochemical (Muller et al., 2006; Hartmann et al., 2008) and other post-synthesis modification methods as described in literature (Stock and Biswas, 2012; Lee and Ahn 2013). Archetypical MOFs such as HKUST-1, AL-MIL-53, MOF-5, IRMOF-3, Cr-MIL-101, UiO-66, ZIF-8 and others have been synthesis using these methods.

Conventional solvothermal (CS) method involves mixing of the metal ions and organic linkers in a solvent which is then heated in Teflon lined autoclaves or in glass vials (Hartmann et al., 2008; Ma et al., 2017). When the solvent used for synthesis is water then the method is referred to as hydrothermal (Zhang et al., 2014). The specific structure of the MOFs can be tuned by varying the reaction conditions such as temperature, pressure, solvent composition and the concentration of the reactants (Kim et al., 2012). Some of the setbacks associated to this method are the long reaction time and use of organic solvents some of which are not environmentally friendly. The reported reaction time of MOFs through solvothermal synthesis ranges from hours to days (Chui et al., 1999; Kim et al., 2012)

Sonochemical (SC) method is a procedure that uses ultrasound energy (20 kHz-10 MHz) for synthesis. The high ultrasonic radiation provides energy for dispersion of the agglomerated organic particles. Li et al. (2009) synthesized microporous HKUST-1 which is copper (II) benzenetricarboxylate, $Cu_3(BTC)_2$ metal-organic frameworks using ultrasonic energy and the group reported that the method was efficient and environmentally friendly. Mesoporous MOF-5 synthesized using a sonicator with an adjustable maximum power output (200 W at 20 kHz) was reported by Son et al., (2008). The crystals were obtained by reacting $Zn(NO_3)_2.6H_2O$ and H_2BDC in 1-methyl-2-pyrrolidene (NMP) solvent under a nitrogen atmosphere within 30 min.

Mechanochemical (MC) method involves grinding of the metal salt and organic linkers mixture by use of a ball mill or pestle and mortar in the absence of a solvent. The grinding facilitates breaking of the bonds holding the molecules together which then undergo a chemical change. The water or volatile molecules produced during this reaction are removed by gentle heating of the mixture (Pichon and James, 2008; Klimakow et al., 2010). However, small amounts of solvents may be used in the mechanochemical technique; in this case the method is referred to as liquid assisted grinding (LAG). The presence of a solvent enhances the properties of the MOFs as indicated in a study by Pichon and James (2008). In the study, the authors reported that the addition of small amount of 100 μL methanol to the reaction between copper acetate and benzenetricarboxylic acid using mechanochemical process produced HKUST-1 with a higher BET surface area of 1364 m^2/g than the HKUST-1 synthesized in the absence of a solvent with a BET surface area of 1084 m^2/g. Nevertheless, this process is also considered to be environmentally friendly since it uses very little amounts of solvents or none at all to produce high quality yields.

Microwave (MW) assisted method employs microwave radiation to provide energy for the reaction between metal ions and organic ligands. The materials produced through this process is mostly microporous, the process also offers fast crystallization, phase selectivity, morphological control as well as narrow size distribution (Cho et al., 2012).

Electrochemical (EC) method produces MOFs that use a metal such as Zn, Cu, Mg and Co as cathode and organic linkers like 1,2,3- benzenetricarboxylic acid (1,2,3-H_3BTC), 1,3,5-H_3BTC, H_2BDC (benzenedicarboxylic acid), 2,5- dihydroxy-1,4-benzenedicarboxylic acid ($H_2BDC-(OH)_2$ among others. The anode goes into solution to provide the metal ions which then reacts with the dissolved organic linkers in a conducting salt and hydrogen is preferentially discharged at the cathode instead of the metal ions (Muller et al., 2006; Hartmann et al., 2008; Joaristi et al., 2012). Table 1 is a summary of some MOFs and method of synthesis.

Table 1. Some common MOFs and the method of synthesis

MOF	Metal	Ligand	Solvent	Synthesis	References
HKUST-1	$Cu(NO_3)_2.3H_2O$	H_3BTC	H_2O/EtOH	CS	Chui et al., 1999
	$CuOAc.2H_2O$	H_3BTC	H_2O/EtOH/ DMF	SC	Li et al., 2009
MOF-5	$Zn(NO_3)_2.6H_2O$	H_2BDC	NMP	SC	Son et al., 2008
MIL-53(Al)	$Al(NO_3)_3.9H_2O$	H_2BDC	H_2O/DMF	EC	Joaristi et al., 2012
UiO-66	$ZrCl_4$	H_2BDC	DMF	CS	Abdi etal., 2012
Cr-MIL-101	$Cr(NO_3)_2.9H_2O$	H_2BDC	H_2O/EtOH	MW	Jhung et al., 2007
Co-MOF-74	$Co(NO_3)_2.6H_2O$	H_4DHBDC	H_2O/EtOH	MW	Cho et al., 2012

Removal of Fluorides Using Metal Organic Framework Adsorbents

The most commonly used metal in MOFs tailored for fluoride removal is aluminum (Stock, 2011). This is because it is abundant and inexpensive. It is also forms a strong bond with the fluoride and that is why activated alumina is widely recognized for fluoride removal. Fumarate based MOFs have also been explored extensively because fumaric acid is a biological molecule that is mostly used as an organic linker. Studies that have synthesized MOFs utilizing aluminum include that of Wang et al. (2019). These scientists synthesized MIL-96(Al) to be used for fluoride removal from drinking water. MIL-96(Al) was generated by reacting $Al(NO_3)_3{\cdot}9H_2O$ with a 1,3,5-benzene-trycarboxylate (BTC) ligand in hydrothermal conditions. MIL-96(Al) was characterized and found to be thermally stable with a surface area of approximately $220m^2\ g^{-1}$. After varying parameters such as initial concentration, pH, adsorbent dosage and rival anions, the study established the optimum pH to be in the range of 6-9. Further, the matrix was spiked with rival anions and they were found to have minor influences on the efficiency of the adsorbent. After varying the initial concentration, the adsorption isotherm was best described by the Langmuir model implying a monolayer type of adsorption with an adsorption capacity of 42.19 mg Fg^{-1}. Experiments at diverse temperatures confirmed the reaction to be endothermic with the highest adsorption being achieved at room temperature, that is, 298 K. While optimizing the time required for maximum adsorption, the kinetics of the study indicated a fit into the pseudo second-order kinetics. Regeneration of the adsorbent was achievable using two methods: The first method involved heating the spent adsorbent in the oven to achieve desorption while the second involved the use of sodium hydroxide as a solvent to achieve desorption. The regenerated adsorbent was found to perform up to 61.8% of the adsorption capacity even after a dozen regenerations. The study confirmed MIL-96(Al) as an efficient adsorbent for elimination of fluorides from drinking water.

Karmakar et al. (2016) investigated how aluminum fumarate (AlFu) MOF, shown in Figure 2, can be used to remove fluorides from aqueous solutions. The adsorbent was fabricated using aluminium sulphate 18 hydrate ($Al_2(SO_4)_3{\cdot}18H_2O$) and fumaric acid under hydrothermal conditions. After characterization using BET analysis, AlFu MOF's surface area was established to be 1156 m^2/g and the mean pore size was 17 Å. The adsorbent demonstrated stability both in water and in heat with the adsorbent degrading only in temperatures beyond 718 K. From FTIR analysis, the study established that the hydroxyl ions in the AlFu MOF were substituted by the fluoride ions. After varying and optimizing different parameters, the investigation concluded that 0.75 g/L AlFu MOF was able to completely remove fluoride concentration of 30 mg/L in aqueous media at a pH of 7. The optimum adsorption for fluoride was 600 mg/g at 293 K while the Freundlich adsorption isotherm fitted well confirming a heterogenous type of adsorption. The adsorption process was also a spontaneous reaction which followed second order kinetics, suggesting that the rate limiting step was a chemisorption kind of reaction. Further, it was established that AlFu MOF could be regenerated using sodium hydroxide.

Figure 2. AlFu MOF building block and a section of the cage-like structure
(Karmakar et al., 2016).

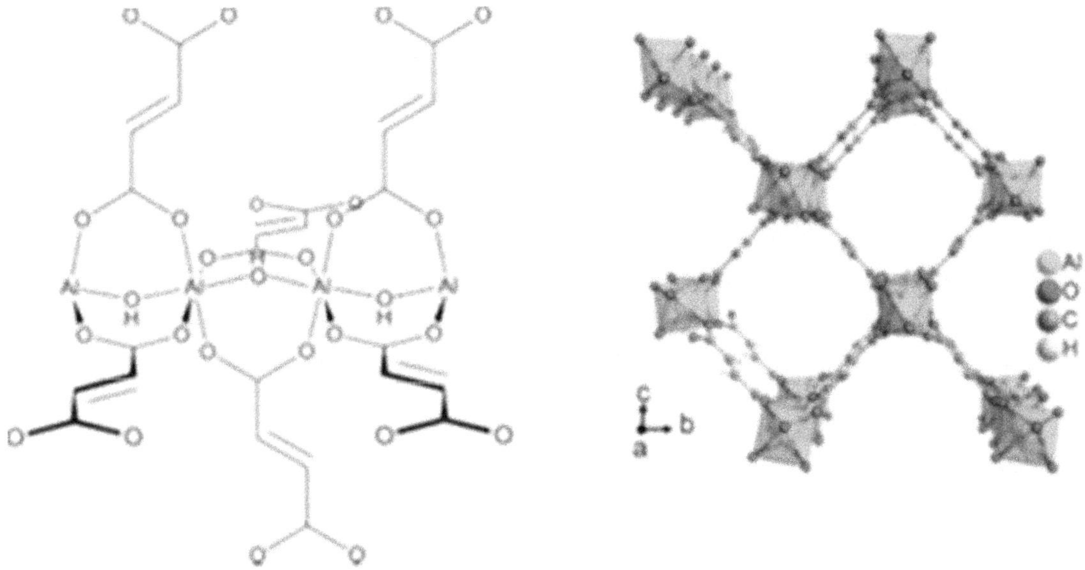

Karmakar et al. (2018) further carried out a study on how AlFu MOF incorporated with polyacrylonitrile (PAN) hollow fiber membranes can be used in fluoride removal from groundwater. PAN is a chemically and thermally stable molecule with excellent solvent resistance and that is why it was chosen as a copolymer. The hollow fiber membranes were prepared by first forming a matrix through dissolving PAN in dimethylformamide (DMF) under mechanical stirring in 60 °C for 4-5 hours. Different weight percentages of AlFu MOF were added to the matrix until a homogenous paste was formed. To ensure proper mixing, sonication was carried out. Membranes were then spun out and kept in a gelation bath before use. The fabricated membranes were found to be permeable and hydrophilic thus favoring fluoride removal. The enhanced membranes had an area of 0.026m^2 and were able to adsorb up to 12 mg/L at 35 kpa with a flow rate of 30 l/h. The optimum pH was found to be 7 and co-ions had no impact on the flux capacity of the membrane. The membranes were found to work successfully in actual samples of ground water and regeneration was further possible on the membranes using sodium hydroxide.

Zirconium element has also been used extensively for the synthesis of MOFs due to its thermal and chemical stability (Ke et al., 2018). Zhu et al. (2018) focused on fluoride adsorption using zirconium fumarate ($Zr_6O_4(OH)_4$ (fumarate)$_6$·xH_2O) framework (MOF-801), fabricated using $ZrCl_4$ and fumaric acid, in aqueous solution . MOF-801 was stable in water and very efficient at pH 2–10. High fluoride concentration in the feed solution had no influence at the adsorption capacity of the adsorbent neither did other present ions like Cl^-, NO_3^- and SO_4^{2-}. The kinetics of the batch experiments fitted well to a pseudo-second-order model implying that the rate determining step was a chemical reaction. The data also suited the Langmuir equation suggesting a monolayer kind of adsorption. MOF-801's fluoride adsorption was found to be highest at a temperature of 303 K with a capacity of 40 mg g^{-1} at, though temperature change was found to be least effective. Adsorption of fluoride was possible due to the substitution of fluoride ions in the feed solution with the hydroxyl ions in the MOF-801 adsorbent. The study concluded that MOF-801 was a potential adsorbent which can solve fluoride contamination in drinking water.

Another zirconium-based MOF was synthesized by Lin et al. (2016) and used for deflouridation. This type of MOF had a primary amine group on its ligand and was named UiO-66-NH_2. The ammine functional group was meant to make the fluoride removal an electrostatic process. Characterizing UiO-66-NH_2 MOF gave a mean length of approximately 100 nm with an octahedron shape. The BET surface area was 905 m^2 g^{-1} while the total pore volume was 0.43 cm^3 g^{-1}. The adsorbent was confirmed to be highly stable in aqueous solution and did not disintegrate or lose its crystallinity in the course of the sorption process. During the batch adsorption studies, the kinetics best fitted the pseudo-second order rate law and the thermodynamic studies displayed a spontaneous exothermic reaction with the enthalpy indicating a general physical adsorption with an adsorption capacity reaching 55 mg g^{-1} at 20 °C. Comparing UiO-66 and UiO-66-NH_2, the adsorption capacity of the latter was higher and can be accounted by the electrostatic interaction between F^- and the metal site. It was also noted that the adsorption capacity decreased with decreasing pH. Similar ions like chlorides and bromides did not alter the adsorption efficiency unlike ionic surfactants. The adsorbent was further found to work well as a packed column adsorbing up to 40 mg g-1. This study confirmed this MOF to be an excellent adsorbent.

He et al. (2016) also developed zirconium based MOF adsorbent and membrane for fluoride removal. Characterizing the adsorbents gave the surface area of the Zr-MOF powder adsorbent as 740.28 m^2/g and the Zr-MOF membrane measured 733.07 m^2/g while the pore volume was 6 Å. Batch experiments were done to determine the fluoride adsorption onto Zr-MOF adsorbent. The isotherm experiments were best described by the Langmuir isotherm model suggesting a monolayer kind of adsorption with the highest adsorption capacity being 102.40 mg/g at pH 7.0 when the initial fluoride concentration was 200 mg/L. The kinetics data confirmed a pseudo-second order model. Characterizing the adsorbent before and after use revealed that for successful fluoride removal, both surface hydroxyl groups and Zr(IV) active sites played significant roles. After adsorption, further desorption studies confirmed that the adsorbent could be regenerated using sodium hydroxide solution. The fabricated Zr-MOF membrane, depicted in Figure 3, was reinforced by an alumina substrate and used for filtration. Flow rate and initial fluoride concentration were two significant parameters which affected the membranes' capacity. With a thickness of 20 μm, the removal capacities were 5510, 5173, and 4664 L/m^2 for 5, 8 and 10 mg/L fluoride concentrations respectively. Both the Zr- MOF adsorbent and membrane were found to be viable for defluoridation.

Figure 3. Zr-MOF membrane reinforced by an alumina base
(Source: He et al., 2016).

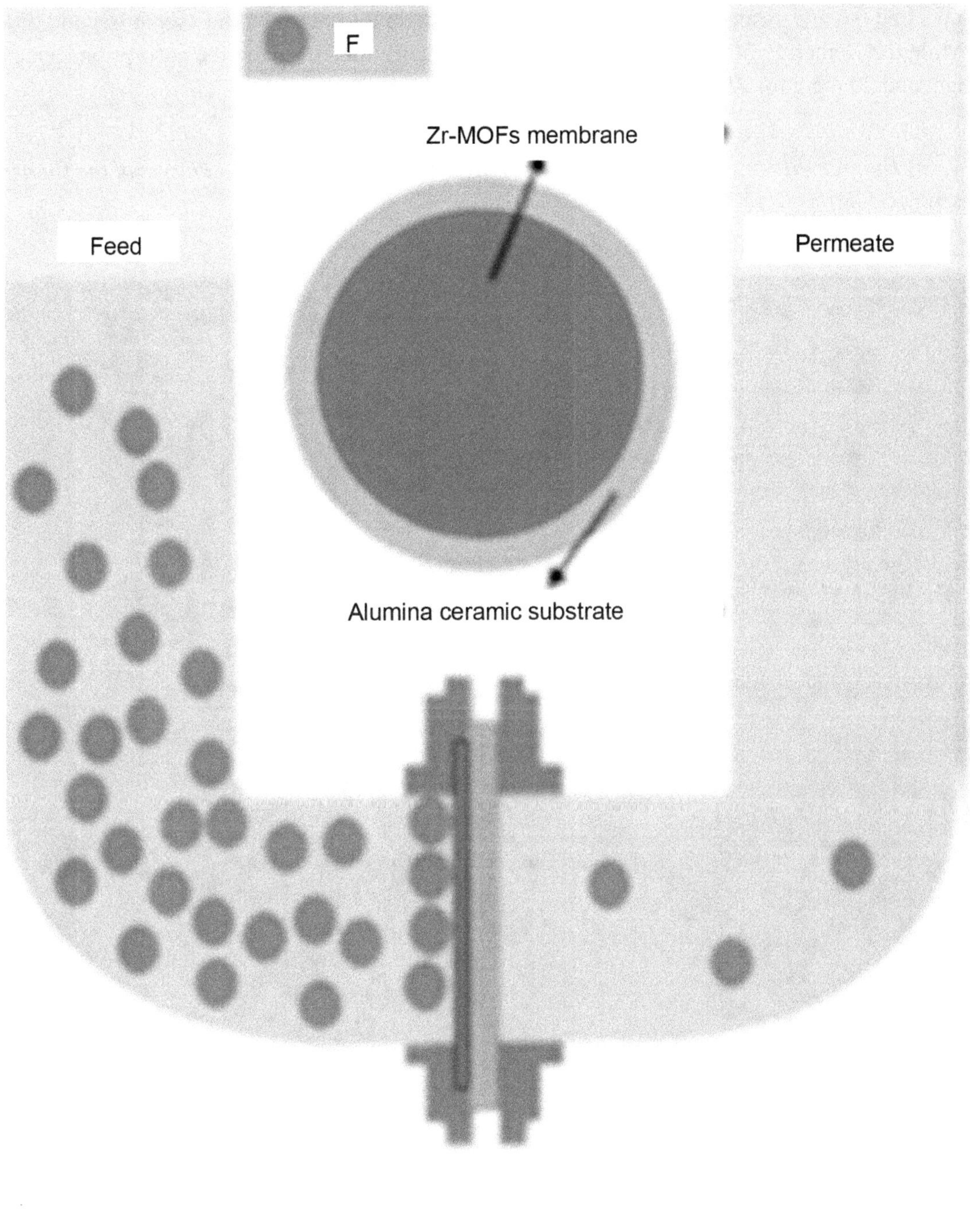

Kamarehie et al. (2018) also investigated two MOFs- ZIF-8 and Uio-66 for fluoride adsorption from water. The two MOFs were characterized and ZIF-8 had 1050 m^2/g as the surface area, a pore volume of $0.57 cm^3/g$ and a mean pore diameter of 4.5nm. Uio-66's surface area was 800 m^2/g while the total pore volume gave 0.45 cm^3/g and the average pore diameter was 3.2 nm. A range of parameters were varied and optimized. Neutral conditions were established to be appropriate for maximum adsorption. Increasing the adsorbent dosage increased the adsorption and ZIF-8 and Uio-66 adsorbed optimally at dosages 0.2 and 0.6 g/L, respectively. The kinetic data was best illustrated by the pseudo-second order model while the isotherms fitted well to the Langmuir isotherm model giving the adsorption capacity as 25 mg/g and 20 mg/g for ZIF-8 and Uio-66, respectively.

Figure 4. (a) Diagram illustrating the synthesis of MOF-801, (b) proposed mechanisms of the fluoride removal by MOF-801
(Source: Ke et al., 2018)

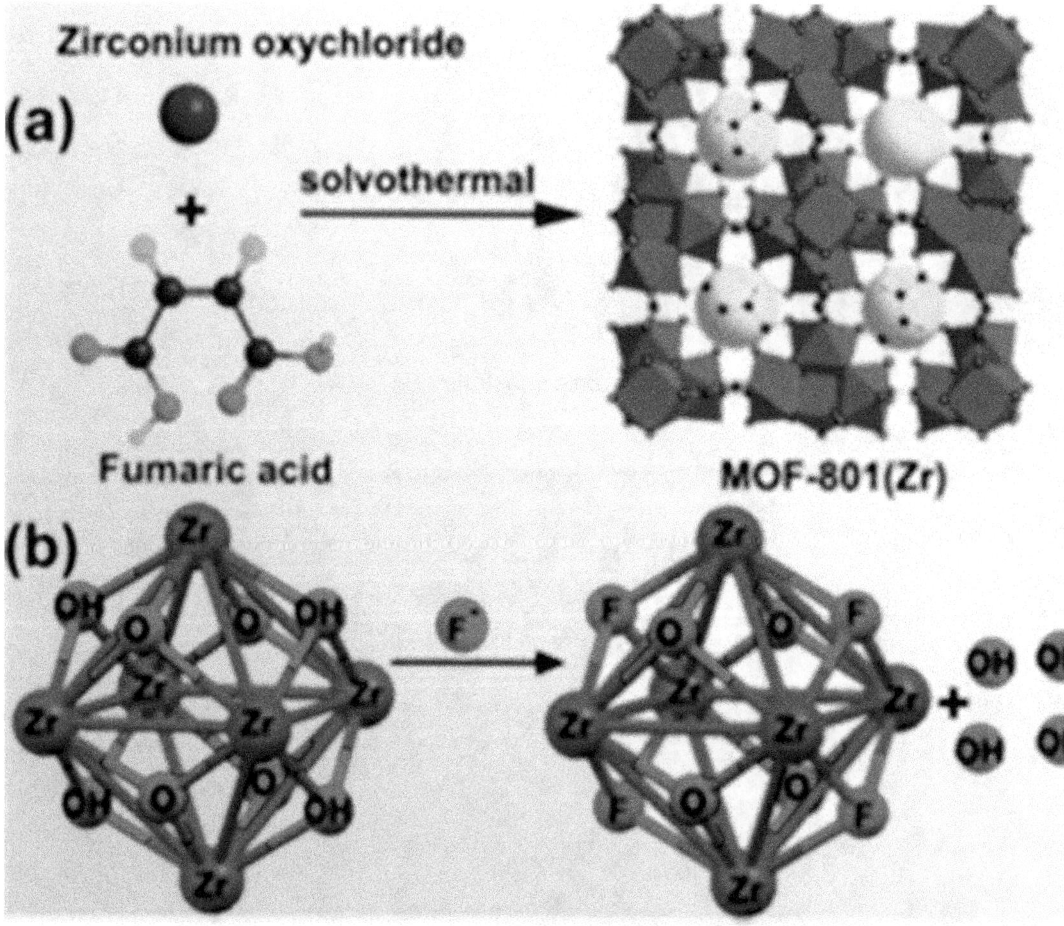

Massoudinejad et al. (2016) performed a study with a similarity to that done by Lin et al. (2016) earlier mentioned. The research synthesized another Zr(IV) based MOF called Uio-66 with amine functional groups for fluoride removal from environmental matrices. The research group varied and optimized different parameters used in batch adsorption using a statistical and mathematical software called the central composite design, hence saving on time. pH was confirmed to be the most dominant parameter in the adsorption process. Initial fluoride concentration was also directly proportional to the amount of fluoride adsorbed. 41.5 mg/L maximum adsorption, which accounted for 88.14% efficiency, was achieved with the following optimized conditions: pH 7, initial ion concentration of 12.18 mg/L, adsorbent dose of 0.4 g and contact time of 32.4 minutes. Pseudo-second order and Langmuir isotherm models were established to best fit the adsorption data.

In a related study, Massoudinejad et al. (2018) used Uio-66 without the ammine for fluoride removal. The findings showed that the highest fluoride adsorption capacity of Uio-66 was 31.09 mg g^{-1}. The experimental data were best illustrated by the pseudo-second order model and the Langmuir model. The findings also confirmed that Uio-66 adsorbent had a high capacity for fluoride adsorption and could be functionalized to enhance the adsorption capacity.

Two other fumarate-based MOFs (MOF-801 and calcium fumarate (CaFu) MOF) were prepared for the targeted elimination of fluoride from brick tea infusion (Ke et al., 2018). MOF-801 belongs to the Zirconium (IV)-based MOFs with a chemical formula $Zr_6O_4(OH)_4(fumarate)_6$ and its synthesis has been summarized in Figure 4 below. The CaFu MOF was prepared from $Ca(CH_3COO)_2$ and fumaric acid. The SEM and TEM images revealed that the MOF-801 were spherical nanoparticles with average diameters of 150 nm. After varying and optimizing parameters, 80% fluoride elimination was accomplished by MOF-801 in the first 5 minutes at room temperature, while 166.11 mg g^{-1} adsorption capacity was obtained for calcium CaFu MOF at 373 K. It was observed that the hydroxyl group in MOFs played a significant role where they were substituted by fluoride ions. By using the tea bag model, the MOFs worked efficiently and selectively. This work demonstrated that fumarate-based MOFs are potential adsorbents for the targeted defluoridation not only from aqueous solutions but also from brick tea infusion.

Aluminium based MOFs such as MIL-96 are considered ideal for defluoridation because they contain hydroxyl ions that aid in the uptake of fluoride ions from water. Zhang et al 2014 reported hydrothermal synthesis of MIL-96 which was obtained by reacting $Al(NO_3)_3.9H_2O$ and H_3BTC in the presence of water in an autoclave at 200°C for 24 h. The batch adsorption studies done by Zhang and workers reported that the performance of MIL-96 was independent of pH and coexisting anions. The F^- adsorption capacity of MIL-96 was found to be 21.8 mg/g and the Al in the residual solutions treated with adsorbent were observed to be within the WHO safe limit (< 0.2 mg/L). Aluminium fumarate metal-organic framework (AlFu MOF) prepared through hydrothermal synthesis by heating of a mixture of hydrated aluminium sulphate $[Al_2(SO_4)_3.18H_2O]$, fumaric acid and water at 60°C has been reported (Leung et al., 2012). The highly water stable adsorbent has been applied in the removal of fluoride from water by

A study conducted by Zhao et al. (2014) shows the defluoridation capacity of UiO-66 (Zr) to be 41.36 mg/g. It was observed that the initial pH of the fluoride solutions remained constant in the neutral range 6.0-9-0 but had a negative effect at pH>9. The group suggested that the defluoridation performance of the MOFs could be enhanced by increasing the –OH groups in their structures. The co-existing ions were observed to have an influence on the fluoride ion adsorption capacity of the MOF.

Ma et al. (2017) studied the adsorption capacity for fluoride ions from water using lanthanide based adsorbent prepared through solvothermal method. The hydrated cerium nitrate $[Ce(NO_3)_2.6H_2O]$ and 2,5-di(3′,5′-dicarboxylphenyl)benzene mixture was dissolved in dimethylformamide (DMF) in the presence of nitric acid. The crystals obtained by heating the mixture in a capped vial in an oven for 75°C for 72 h was found to have an adsorption capacity for fluoride ions of 103 mg/g.

CONCLUSION

MOFs have got a high potential for a wider use in adsorption of fluoride because of its exceptional features including tailored controlled structure, tunable pore dimensions, large internal surface area and numerous framework functionalities. If well structured, they are also highly reversible and can undergo adsorption and desorption with no loss of uptake capacity or crystallinity even after various cycles. These strengths make them incomparable to frequently used microporous inorganic materials like zeolites and carbonaceous materials.

From the adsorption experiments performed by varied scientists, it is evident that the efficiencies of most synthesized MOFs were above average, even beyond 80% in some studies. It is also noted that most studies established neutral pH and room temperature as optimum conditions for the best performance of most MOFs. The adsorption kinetic data in most of the experiments also fitted well to pseudo-second order model. This implies that the rate determining step in majority of the studies was chemisorption. From their enthalpies however, the overall reaction, suggested physisorption. Apart from one noted study, the rest of the mentioned studies confirmed the isotherm data to fit well into Langmuir model indicating a monolayer kind of adsorption. The outcome of the experiments also illustrated that MOFs could be regenerated and reused more than once.

Even with all their strengths, MOFs have not been fully explored. Most researchers have shied away from using them because if they are poorly designed, they become unstable and may show degradation on exposure or desorption during use.

REFERENCES

Abdi, H. R., Tian, H., Ang, H. M., Tade, M. O., Buckley, C. E., & Wang, S. (2012). Nanosize Zr-metal-organic framework (UiO-66) for hydrogen and carbon dioxide storage. *Chemical Engineering Journal*, *187*, 415–420. doi:10.1016/j.cej.2012.01.104

Ameloot, R., Stappers, L., Fransaer, J., Alaerts, L., Sels, B. F., & De Vos, D. E. (2009). Patterned growth of metal-organic framework coatings by electrochemical synthesis. *Chemistry of Materials*, *21*(13), 2580–2582. doi:10.1021/cm900069f

Budyanto, S., Kuo, Y. L., & Liu, J. C. (2015). Adsorption and precipitation of fluoride on calcite nanoparticles: A spectroscopic study. *Separation and Purification Technology*, *150*, 325–331. doi:10.1016/j.seppur.2015.07.016

Chernikova, V., Yassine, O., Shekhah, O., Eddaoudi, M., & Salama, K. N. (2018). Highly sensitive and selective SO2 MOF sensor: The integration of MFM-300 MOF as a sensitive layer on a capacitive interdigitated electrode. *Journal of Materials Chemistry. A, Materials for Energy and Sustainability*, *6*(14), 5550–5554. doi:10.1039/C7TA10538J

Cho, H. Y., Yang, D. A., Kim, J., Jeong, S. Y., & Ahn, W. S. (2012). CO_2 adsorption and catalytic application of Co-MOF-74 synthesized by microwave heating. *Catalysis Today*, *185*(1), 35–40. doi:10.1016/j.cattod.2011.08.019

Chui, S. S. Y., Lo, S. M. F., Charmant, J. P. H., Orpen, A. G., & Williams, I. D. (1999). A chemically functionalizeable nanoporous material $[Cu_3(TMA)_2.(H_2O)_3]_n$. *Science*, *283*(5405), 1148–1150. doi:10.1126cience.283.5405.1148 PMID:10024237

Czaja, A. U., Trukhan, N., & Müller, U. (2009). Industrial applications of metal–organic frameworks. *Chemical Society Reviews*, *38*(5), 1284–1293. doi:10.1039/b804680h PMID:19384438

Drisdell, W. S., Poloni, R., McDonald, T. M., Long, J. R., Smit, B., Neaton, J. B., & Kortright, J. B. (2013). Probing Adsorption Interactions in Metal–Organic Frameworks using X-ray Spectroscopy. *Journal of the American Chemical Society*, *135*(48), 18183–18190. doi:10.1021/ja408972f PMID:24224556

Du, Z., Deng, S., Chen, Y., Wang, B., Huang, J., Wang, Y., & Yu, G. (2015). Removal of perfluorinated carboxylates from washing wastewater of perfluorooctanesulfonyl fluoride using activated carbons and resins. *Journal of Hazardous Materials*, *286*, 136–143. doi:10.1016/j.jhazmat.2014.12.037 PMID:25585266

Ghosh, A., Chakrabarti, S., Biswas, K., & Ghosh, U. C. (2014). Agglomerated nanoparticles of hydrous Ce(IV)+Zr(IV) mixed oxide: Preparation, characterization and physicochemical aspects on fluoride adsorption. *Applied Surface Science*, *307*, 665–676. doi:10.1016/j.apsusc.2014.04.095

Godfrey, H. G. W., da Silva, I., Briggs, L., Carter, J. H., Morris, C. G., Savage, M., & Schröder, M. (2018). Ammonia Storage by Reversible Host–Guest Site Exchange in a Robust Metal–Organic Framework. *Angewandte Chemie International Edition*, *57*(45), 14778–14781. doi:10.1002/anie.201808316 PMID:30098090

Hartmann, M., Kunz, S., Himsl, D., Tangermann, O., Ernst, S., & Wagener, A. (2008). Adsorptive separation of isobutene and isobutene on $Cu_3(BTC)_2$. *Langmuir*, *24*(16), 8634–8642. doi:10.1021/la8008656 PMID:18611044

He, J., Cai, X., Chen, K., Li, Y., Zhang, K., Jin, Z., & Kong, L. (2016). Performance of a novelly-defined zirconium metal-organic frameworks adsorption membrane in fluoride removal. *Journal of Colloid and Interface Science*, *484*, 162–172. doi:10.1016/j.jcis.2016.08.074 PMID:27610471

Hinterholzinger, F. M., Rühle, B., Wuttke, S., Karaghiosoff, K., & Bein, T. (2013). Highly sensitive and selective fluoride detection in water through fluorophore release from a metal-organic framework. *Scientific Reports*, *3*(1), 1–7. doi:10.1038rep02562 PMID:24008779

Jhung, S. H., Lee, J. H., Yoon, J. W., Serre, C., Ferey, G., & Chang, J. S. (2007). Microwave synthesis of chromium terephthalate MIL-101 its benzene sorption. *Advanced Materials*, *19*(1), 121–124. doi:10.1002/adma.200601604

Joaristi, A. M., Juan-Alcanpiz, J., Serra-Crespo, P., Kapteijin, F., & Gascon, J. (2012). Electrochemical synthesis of some Zn^{2+}, Cu^{2+} and Al^{3+} metal-organic frameworks. *Crystal Growth & Design*, *12*(7), 3489–3498. doi:10.1021/cg300552w

Kamarehie, B., Noraee, Z., Jafari, A., Ghaderpoori, M., Karami, M. A., & Ghaderpoury, A. (2018). Data on the fluoride adsorption from aqueous solutions by metal-organic frameworks (ZIF-8 and Uio-66). *Data in Brief*, *20*, 799–804. doi:10.1016/j.dib.2018.08.159 PMID:30211278

Karmakar, S., Bhattacharjee, S., & De, S. (2018). Aluminium fumarate metal organic framework incorporated polyacrylonitrile hollow fiber membranes: Spinning, characterization and application in fluoride removal from groundwater. *Chemical Engineering Journal*, *334*, 41–53. doi:10.1016/j.cej.2017.10.021

Karmakar, S., Dechanik, J., Janiak, C., & De, S. (2016). Aluminium fumarate metal-organic framework; A super adsorbent for fluoride removal from water. *Journal of Hazardous Materials*, *303*, 10–20. doi:10.1016/j.jhazmat.2015.10.030 PMID:26513559

Karmakar, S., Dechnik, J., Janiak, C., & De, S. (2016). Aluminium fumarate metal-organic framework: A super adsorbent for fluoride from water. *Journal of Hazardous Materials*, *303*, 10–20. doi:10.1016/j.jhazmat.2015.10.030 PMID:26513559

Ke, F., Peng, C., Zhang, T., Zhang, M., Zhou, C., Cai, H., Zhu, J., & Wan, X. (2018). Fumarate-based metal-organic frameworks as a new platform for highly selective removal of fluoride from brick tea. *Scientific Reports*, *8*(1), 1–11. doi:10.103841598-018-19277-2 PMID:29343778

Kim, J., Kim, S. H., Yang, S. T., & Ahn, W. S. (2012). Bench-scale preparation of $Cu_3(BTC)_2$ by ethanol reflux: Synthesis optimization and adsorption/catalytic application. *Microporous and Mesoporous Materials*, *161*, 48–55. doi:10.1016/j.micromeso.2012.05.021

Klimakow, M., Klobes, P., Thunemann, A. F., Rademann, K., & Emmerling, F. (2010). Mechanochemical synthesis of metal-organic frameworks: A fast and facile approach towards quantitative yields and high specific surface areas. *Chemistry of Materials*, *22*(18), 5216–5221. doi:10.1021/cm1012119

Kumar, B. (2015). *Fluoride in Groundwater and its Health Hazards: A Review in Indian Scenario*. Academic Press.

Kumar, P., Anand, B., Tsang, Y. F., Kim, K.-H., Khullar, S., & Wang, B. (2019). Regeneration, degradation, and toxicity effect of MOFs: Opportunities and challenges. *Environmental Research*, *176*, 108488. doi:10.1016/j.envres.2019.05.019 PMID:31295665

Lee, Y. R., & Ahn, W. S. (2013). Synthesis of metal-organic frameworks: A mini review. *Korean Journal of Chemical Engineering*, *30*(9), 1667–1680. doi:10.100711814-013-0140-6

Leung, E., Muller, U., Trukhan, N., Mattenheimer, H., Cox, G., & Blei, S. (2012). *Process preparing porous metal organic frameworks based on aluminium fumarate*. US patent (05.04.12) 2012/0082864A1

Leus, K., Muylaert, I., Van Speybroeck, V., Marin, G. B., & Van Der Voort, P. (2010). *A coordinative saturated vanadium containing metal organic framework that shows a remarkable catalytic activity*. doi:10.1016/S0167-2991(10)75053-9

Li, Z. Q., Qiu, L. G., Su, T., Wu, Y., Wang, W., Wu, Z. Y., & Jiang, X. (2009). Ultrasound synthesis of microporous metal-organic framework $Cu_3(BTC)_2$ at ambient temperature and pressure: An efficient and environmentally friendly method. *Materials Letters*, *63*(1), 78–80. doi:10.1016/j.matlet.2008.09.010

Lin, K.-Y. A., Liu, Y.-T., & Chen, S.-Y. (2016). Adsorption of fluoride to UiO-66-NH2 in water: Stability, kinetic, isotherm and thermodynamic studies. *Journal of Colloid and Interface Science*, *461*, 79–87. doi:10.1016/j.jcis.2015.08.061 PMID:26397913

Lin, K.Y.A., Liu, Y.T., & Chen, S.Y. (2016). Adsorption of fluoride to UiO-66-NH_2 in water: stability, kinetic, isotherm and thermodynamic studies. *Journal of Colloid and Interfaces Science, 461*, 79-87.

Ma, A., Ke, F., Jiang, J., Yuan, Q., Luo, Z., Liu, J., & Kumar, A. (2017). Two lanthanide based metal-organic frameworks for highly efficient adsorption and removal of fluoride ion from water. *CrystEng-Comm, 19*(16), 2172–2177. doi:10.1039/C7CE00291B

Massoudinejad, M., Ghaderpoori, M., Shahsavani, A., & Amini, M. M. (2016). Adsorption of fluoride over a metal organic framework Uio-66 functionalized with amine groups and optimization with response surface methodology. *Journal of Molecular Liquids, 221*, 279–286. doi:10.1016/j.molliq.2016.05.087

Massoudinejad, M., Shahsavani, A., Kamarehie, B., Jafari, A., Ghaderpoori, M., Amini, M. M., & Gha-derpoury, A. (2018). Highly efficient adsorption of fluoride from aqueous solutions by metal organic frameworks: Modeling, isotherms, and kinetics. *Fluoride, 51*(4), 355–365.

McDonald, T. M., Mason, J. A., Kong, X., Bloch, E. D., Gygi, D., Dani, A., Crocellà, V., Giordanino, F., Odoh, S. O., Drisdell, W. S., Vlaisavljevich, B., Dzubak, A. L., Poloni, R., Schnell, S. K., Planas, N., Lee, K., Pascal, T., Wan, L. F., Prendergast, D., ... Long, J. R. (2015). Cooperative insertion of CO2 in diamine-appended metal-organic frameworks. *Nature, 519*(7543), 303–308. doi:10.1038/nature14327 PMID:25762144

Meenakshi, S., Sundaram, C. S., & Sukumar, R. (2008). Enhanced fluoride sorption by mechano-chemically activated kaolinites. *Journal of Hazardous Materials, 153*(1-2), 164–172. doi:10.1016/j.jhazmat.2007.08.031 PMID:17897780

Muller, U., Schubert, M., Teich, F., & Puetter, H. (2006). Metal-organic frameworks- prospective industrial applications. *Journal of Materials Chemistry, 16*(7), 626–636. doi:10.1039/B511962F

Ngeno, E. C., Orata, F. O., Baraza, L. D., Shikuku, V. O., & Kimosop, S. J. (2016). Adsorption of Caffeine and Ciprofloxacin onto Pyrolitically Derived Water Hyacinth Biochar: Isothermal, Kinetic and Thermodynamic Studies. *Journal of Chemistry and Chemical Engineering, 10*(4). Advance online publication. doi:10.17265/1934-7375/2016.04.006

Pichon, A., & James, S. L. (2008). An array-based study of reactivity under solvent-free mechanochemical conditions-insights and trends. *CrystEngComm, 10*(12), 1839–1847. doi:10.1039/b810857a

Rehman, M. A., Yusoff, I., & Alias, Y. (2015). Fluoride adsorption by doped and un-doped magnetic ferrites CuCeFeO: Preparation, characterization, optimization and modeling for effectual remediation technologies. *Journal of Hazardous Materials, 299*, 316–324. doi:10.1016/j.jhazmat.2015.06.030 PMID:26143194

Rojas-Mayorga, C. K., Bonilla-Petriciolet, A., Sánchez-Ruiz, F. J., Moreno-Pérez, J., Reynel-Ávila, H. E., Aguayo-Villarreal, I. A., & Mendoza-Castillo, D. I. (2015). Breakthrough curve modeling of liquid-phase adsorption of fluoride ions on aluminum doped bone char using micro-columns: Effectiveness of data fitting approaches. *Journal of Molecular Liquids, 208*, 114–121. doi:10.1016/j.molliq.2015.04.045

Rosi, N. L., Eddaoudi, M., Kim, J., O'Keeffe, M., & Yaghi, O. M. (2002). Advances in the chemistry of metal-organic frameworks. *CrystEngComm, 4*(68), 401–404. doi:10.1039/B203193K

Shen, J., Mkongo, G., Abbt-Braun, G., Ceppi, S. L., Richards, B. S., & Schäfer, A. I. (2015). Renewable energy powered membrane technology: Fluoride removal in a rural community in northern Tanzania. *Separation and Purification Technology*, *149*, 349–361. doi:10.1016/j.seppur.2015.05.027

Son, W. J., Kim, J., Kim, J., & Ahn, W. S. (2008). Sonochemical synthesis of MOF-5. *ChemComm.*, *2008*, 6336–6338. PMID:19048147

Stock, N. (2011). Metal-organic frameworks: aluminium-based frameworks. Encyclopedia of Inorganic and Bioinorganic Chemistry, 1-16.

Stock, N., & Biswas, S. (2012). Synthesis of metal-organic frameworks: Routes to various topologies, morphologies and composites. *Chemical Reviews*, *112*(2), 933–969. doi:10.1021/cr200304e PMID:22098087

Sun, C.-Y., Qin, C., Wang, X.-L., & Su, Z.-M. (2013). Metal-organic frameworks as potential drug delivery systems. *Expert Opinion on Drug Delivery*, *10*(1), 89–101. doi:10.1517/17425247.2013.741583 PMID:23140545

Tang, D., & Zhang, G. (2016). Efficient removal of fluoride by hierarchical Ce–Fe bimetal oxides adsorbent: Thermodynamics, kinetics and mechanism. *Chemical Engineering Journal*, *283*, 721–729. doi:10.1016/j.cej.2015.08.019

Tripathy, S. S., Bersillon, J.-L., & Gopal, K. (2006). Removal of fluoride from drinking water by adsorption onto alum-impregnated activated alumina. *Separation and Purification Technology*, *50*(3), 310–317. doi:10.1016/j.seppur.2005.11.036

Vasudevan, S., Lakshmi, J., & Sozhan, G. (2009). Studies on a Mg-Al-Zn alloy as an anode for the removal of fluoride from drinking water in an electrocoagulation process. *Clean-Soil Air Water*, *37*(4-5), 372–378. doi:10.1002/clen.200900031

Viswanathan, N., & Meenakshi, S. (2009). Role of metal ion incorporation in ion exchange resin on the selectivity of fluoride. *Journal of Hazardous Materials*, *162*(2-3), 920–930. doi:10.1016/j.jhazmat.2008.05.118 PMID:18619734

Wang, X., Zhu, H., Sun, T., Liu, Y., Han, T., Lu, J., Dai, H., & Zhai, L. (2019). Synthesis and Study of an Efficient Metal-Organic Framework Adsorbent (MIL-96 (Al)) for Fluoride Removal from Water. *Journal of Nanomaterials*, *2019*, 2019. doi:10.1155/2019/3128179

Yaghi, O. M., Li, H., Davis, C., Richardson, D., & Groy, T. L. (1998). Synthetic Strategies, Structure Patterns, and Emerging Properties in the Chemistry of Modular Porous Solids. *Accounts of Chemical Research*, *31*(8), 474–484. doi:10.1021/ar970151f

Zhang, N., Yang, X., Yu, X., Jia, Y., Wang, J., Kong, L., Jim, Z., Sun, B., Luo, T., & Liu, J. (2014). Al-1,2,5-benzenetricarboxylic metal-organic frameworks: A promising adsorbent for defluoridation of water with pH insensitivity and low aluminium residual. *Chemical Engineering Journal*, *252*, 220–229. doi:10.1016/j.cej.2014.04.090

Zhao, X., Liu, D., Huang, H., Zhang, W., Yang, Q., & Zhong, C. (2014). The stability and defluoridation performance of MOFs in fluoride solutions. *Microporous and Mesoporous Materials*, *185*, 72–78. doi:10.1016/j.micromeso.2013.11.002

Zhu, X.-H., Yang, C.-X., & Yan, X.-P. (2018). Metal-organic framework-801 for efficient removal of fluoride from water. *Microporous and Mesoporous Materials*, *259*, 163–170. doi:10.1016/j.micromeso.2017.10.001

Chapter 6
Metal-Organic Frameworks for Heavy Metal Removal From Water

Mintu Maan Dutta
Arya Vidyapeeth College, India

Paran Jyoti Borpatra
Gauhati University, India

ABSTRACT

Clean and safe drinking water is one of the most important and essential components on earth for human survival. However, due to the various anthropogenic activities, the surface and groundwater have been contaminated with heavy metals and radionuclides, and it has become a serious problem of concern globally. These non-biodegradable heavy metal ions from water tends to accumulate in the soil and living organisms and causes serious health hazards in humans. Thus, the finding of recyclable, water-stable MOFs, and multi-functional MOFs (hybrid MOFs) with superior adsorption capability have emerged as a novel material for the removal of heavy metal species (such as Cr(III)/Cr(VI), Cu(II), Hg(II), Cd(II), As(III)/As(V), Pb(II), radionuclides U(VI), Se(IV)/(VI), Cs(I) from aqueous solution. This chapter deals with the recent development of MOFs and their application in heavy metal removal from water.

INTRODUCTION

Water is an essential source of life on which the life cycle depends. But from the last few decades, the presence of heavy metals in surface and groundwater is posing a serious universal concern. (Bakker, 2012; Vörösmarty *et al.*, 2010) It affects both the environment as well as human health. Industries such as electroplating, battery manufacturing, metallurgy, tannery and metal finishing are the primary sources for the generation of these types of pollutants and are found commonly in water. (Fu and Wang, 2011; Azizian, 2015) Unlike other organic pollutants, heavy metals are naturally non-biodegradable and have long biological half-life, which easily enters the food chain through marine animals and creates countless

DOI: 10.4018/978-1-7998-4760-1.ch006

health problems. (Babel and Kurniawan, 2004; Wan Ngah and Hanafiah, 2008) Generally, when density exceed 5 g per cubic centimetre, then these heavy metals are considered toxic. (Barakat, 2011) They cause numerous health risks, which includes growth and development reduction, organ damage, cancer, nervous system damage and finally causes death. For example, exposure of lead causes the expansion of autoimmunity in which the immune system may attack its own cells, creates various new diseases such as rheumatoid arthritis, kidney diseases, problem in circulatory system, and finally damages the brain. (Odum, 2000) Nowadays, adsorption technique has achieved a great interest in removal of heavy metals due to its effectiveness, economic nature, simple operating conditions and wide pH range. (Pan *et al.*, 2009) The metal-organic frameworks (MFOs) are considered as a new class of adsorbent (mesoporous and microporous) materials. They are of great potential with various range of applications such as catalysis, gas storage, sensing and separations. (O'Keeffe and Yaghi, 2012)

OCCURRENCES OF HEAVY METALS

Sources of Heavy Metal in Soil

Heavy metal elements are found within our environment like in the water reservoirs, in excess of soil and atmosphere, due to numerous anthropogenic activities. The natural sources of heavy metal contamination are mainly through the interactions of various types of rocks (i.e., igneous, sedimentary and metamorphic) with the surrounding environment. (Siegel, 2002) In rock-forming minerals, the heavy metals are usually found. For example, nickel, cobalt, manganese, zinc, copper, and vanadium have most easily leached due to mineral weathering. In addition to the metals such as scandium, yttrium and other rare earth metal elements have intermediate stability and uranium or hafnium have weathering resistivity. When hot hydrothermal fluids pervade the rocks, all these metals are then concentrated and induces various chemical reactions which causes precipitation of minerals and formation of ores. (Bradl, 2005) Within sedimentary rocks, such kind of deposits are usually found because of their porous structures and high permeability.

An alternate to store heavy metals in soil include its storage as relocated rock fragments, insoluble minerals and organic matter (i.e., the solid phase), as water and air trapped within the soil (i.e., the liquid phase). These two phases (i.e., solid phase and liquid phase) interact with each other with different ions passing through the system. (Alloway, 1995) Different types of rock that the soil originated from, defies the concentration and uniqueness of the heavy metals. Most of the heavy metals are originated in the third layer of the soil. Elements of this layer are dissolved within the upper layer and then these dissolved elements move downwards or sideways to the lower layer, where these elements are deposited. (Drever, 1997) The cationic elements of the heavy metals may get attracted by the third layer due to the high concentration of iron oxyhydroxides and clay. (Bradl, 2005)

Sources of Heavy Metal in Water

Heavy metals are carried by surface water (such as streams, rivers, lakes, springs and ponds) over a large distance and the chemical composition depends on the geological topography over where they flow. The presence of heavy metals in surface water influences the chemical, biological and physical properties of water. It affects the living organisms, pH of the water, redox potential, adsorption and desorption properties, evaporation-dilution method and lastly temperature. (Bradl, 2005) However, few heavy metals are very important for biological systems such as zinc, copper etc., but continuous and extreme intake can cause poison.

Groundwater is very important to human life. It is the direct source of drinking water. The groundwater is mainly contaminated through various organic and inorganic pollutants. Inorganic pollutants such as various heavy metals, salts etc. are released into the environment from agricultural areas, urban areas, industrial sites, landfills, mining disposal sites, dredged sediments, hazardous waste sites, leaking storage tanks, line sources (motorways, railways, sewerage, systems, etc.). (Bradl, 2005) This contamination may affect both the aquatic as well as terrestrial animals, and finally human health. (Bradl, 2005)

Human-Made Sources of Heavy Metal

Industrial activities have played an important role to heavy metal contamination in the environment. Industrial contamination sources include mainly coal combustion, mining, fossil fuel combustion, effluent streams and the disposal of the waste product. Literature reveals that there are various human-made sources of heavy metal pollutants and their toxicity to plants and animals are listed below. (Bradl, 2005; Adriano, 2001)

Arsenic: Key anthropogenic source of arsenic includes its use as additive in animal feed, algaecides, herbicides, insecticides, fungicides, pesticides, rodenticides, sheep dip, tanning and textile industries, pigments, veterinary medicine, ceramics, special glasses, electrical generation (coal and geothermal), fireworks etc. Excess of As causes Phyto toxicity- to plants, arsenicosis (i.e., blackfoot disease- BFD) & keratosis in animals.

Cadmium: Main anthropogenic source of cadmium includes its use in neutron absorbers (within nuclear reactors), nickel-cadmium batteries, anti-corrosive metal coatings, alloys, plastic stabilizers, coal combustion. Excess of Cd causes *itai-itai* disease.

Chromium: Basic anthropogenic source of chromium includes its use in plating, ferro-alloys manufacturing, textiles and leather tanning industries, wood treatment etc. Excess of Cr^{3+} is detrimental to mammals and Cr^{6+} is carcinogenic.

Lead: Alloys, ceramics, plastics, glassware, lead-acid batteries, cable sheathings, sheets, pipes and tubing, ordinance, antiknock agents are the main anthropogenic source of lead. Excess of lead causes lead poisoning and sometimes fatal.

Mercury: Key anthropogenic source of chromium includes its use in amalgamation (the process of metal extraction), electrical and measuring apparatus, catalyst, dental fillings, Hg vapor lamps, X-ray tubes, pharmaceuticals, scientific instruments, electrodes, chloralkali cell's mobile cathode. Excess of Hg causes Minamata disease in humans and causes Biomagnification in aquatic environment.

Copper: Key anthropogenic source of copper includes its use water pipes, chemicals and pharmaceutical equipment, kitchenware, roofing, alloys. Compared to As, Cd, Cr, Copper is less toxic.

Zinc: Key anthropogenic source of zinc includes its use as Zn alloys, PVC stabilizers, in chemicals and medicines, anti-corrosion coating, rubber industries, welding and soldering fluxes etc. Excess of Zn may affect cholesterol metabolism in humans.

IMPACTS OF HEAVY METALS

Effect on Plants

Heavy metals such as Cd, As, Pb, Hg and Se does not play any role for plants' growth or physiological function. Other heavy metals like Co, Fe, Cu, Mo, Mn, Zn and Ni are very important for the metabolism of plants as well as plants growth. However, when the concentration exceeds, the same become toxic in nature. (Rascio and Izzo, 2011) The soil productivity decreases with the increase of Pb in soil surface and causes different plants problems like photosynthesis problem, reduction of mitosis and water adsorption ability, brown short roots, stunted foliage and wilting of older leaves. (Bhattacharyya *et al.*, 2008) Heavy metals are generally toxic in nature and causes weak plant growth, chlorosis, reduction of nutrient uptake, reduction in the ability of nitrogen fixation, yield depression, delay in seed germination and plant metabolism disorder. (Guala *et al.*, 2010)

Effect on Human Health

Through food chain, heavy metals enter the human body through consumption of plants having heavy metal in the soil at very high concentration. Contamination of food crops is the main reason of human exposure to heavy metals. When these heavy metals are not metabolized by the body, then, it becomes very toxic and accumulates in soft tissues. (Sobha *et al.*, 2007)

Cadmium (Cd) is a very harmful toxic heavy metal whose specific gravity 8.65 times higher than water. It directly attacks the kidney, liver, lungs, placenta, brain and bones. (Sobha *et al.*, 2007) It shows different symptoms like nausea, abdominal cramps, vomiting, dyspnea and muscular weakness depending on the rate of exposure. In case of severe exposure, it may cause pulmonary odema and death. (Duruibe *et al.*, 2007) After the *Itai-Itai* disease in Japan, Cd poisoning got the attention in the entire world. Cd may also be responsible for various clinical conditions in lesser or greater extent including cardiac failure, cancers, anosmia, osteoporosis, cerebrovascular infarction, emphysema and proteinuria contract formation in the eyes. (Lalor, 2008)

Copper (Cu) is a very important element to mammalian nutrition. Cu acts as an electron acceptor or donor in metalloenzymes proteins. However, high-level exposure of Cu may show various adverse health issues. Consumption of food and drinking water is the primary source of Cu exposure to human. High intake of Cu causes genetic predisposition. (Stern *et al.*, 2007) Excessive intake of Cu also causes widespread capillary damage, hepatic and renal damage, mucosal irritation and corrosion. It also leads to the gastrointestinal irritation and possible necrotic changes in liver and kidney.

Lead (Pb) is very harmful to human health. Pb affects both physiological and neurological system of human body. Pb poisoning may bring about kidney disfunction, reproduction system problems, liver and brain disease and finally death. (Odum, 2000) Even at very low concentration, Pb may affect human body. (Kazemipour *et al.*, 2008) Pb poisoning also causes teratogenic effect, haemoglobin synthesis inhibition, damage of central nervous system (CNS) and peripheral nervous system (PNS). It is also responsible for various diseases such as gastrointestinal problems, anemia, anoxia, high blood pressure, fatigue, pregnancy difficulties, muscle and joint pain. (Odum, 2000)

When **Zinc (Zn)** is orally taken, it is considered to be relatively non-toxic. However, excess amount of exposure results in the loss of growth and reproduction. Zn poisoning is also responsible for various diseases like diarrhea, vomiting, bloody urine, liver failure, kidney failure and anemia. (Duruibe *et al.*, 2007)

Mercury (Hg) is very toxic element. No biochemical and physiological function is known in human body till now. Poisoning by inorganic Hg causes gastrointestinal disorders and congenital malformation. Organic form of Hg which include monomethyl and dimethylmercury causes acrodynia (Pink disease), gingivitis, stomatitis, erethism (sensitivity of an organ or body part to stimulation or an abnormal irritation), neurological disorders, total damage of the brain and damage of central nervous system (CNS). (Duruibe *et al.*, 2007)

Different Types of MOF Used

High surface area and high porosity of the MOF makes it more advantageous for its use as adsorbent materials. (Zhang *et al.*, 2017) Various reports explains the use of Metal-organic frameworks can be used as adsorbents for the removal of wide range of heavy metal species and radionuclides such as arsenic, lead, mercury, cadmium, copper, chromium, uranium, selenium etc. (Feng *et al.*, 2018)

Zirconium Phosphonate Based Metal-Organic Framework (MOF)

Zhou *et al* reported the synthesis of Zirconium phosphonate based metal-organic framework material with a formula of $[C_4mpyr]_2[Zr_{1.5}(Tppm)_{0.5}F_4.6H_2O]$ (SZ-2). (Zhou *et al.*, 2018) Strong electrostatic attractions between metal cations and negatively charged MOF atoms helps in the adsorption of heavy metal cations from water. The nano porous nature of the material makes it suitable for the removal of hazardous heavy metal cations (such as Cd^{2+}, Hg^{2+} and Pb^{2+}) from water with an absorption capacity > 95.

Fe-BTC/PDA MOF: (1,3,5-benzenetricaboxylate/Polydopamine)

Queen *et al* synthesized cheap, water stable Fe-BTC/PDA, 1,3,5-benzenetricaboxylate/Polydopamine composite and used it as a novel MOF-polymer for the selective removal of heavy metal cations (i.e. Pb^{2+} and Hg^{2+}) from water. (Queen *et al.* 2018) During the process, Fe-BTC (also called as MIL-100) was treated with dopamine and a rapid polymerization to polydopamine (PDA) takes place within its pores via the metal sites. Removal capacities of different metal ions were obtained to be more than 99.8% i.e. 1634 mg of Hg^{2+} and 394 mg of Pb^{2+} can be removed using 1 gram of composite, which is almost 10 times and 2.6 times the capacity that observed with the unmodified Fe-BTC.

TEPA-MIL-101(Cr): (tetraethylenepentamine)

Zhang and his coworkers reported the functionalization of amino group into the metal-organic frameworks material using tetraethylenepentamine (TEPA). (Zhang *et al.*, 2018) Introduction of amino group considerably increases the absorption property towards uranyl ion U(VI). Adsorption kinetics and adsorption isotherms are best represented by Freundlich model and Langmuir adsorption isotherm. This material attained a maximum adsorption capacity of 350 mg/g for uranium (VI) at pH=4.5.

Manganese Based MOF ($Mn(BDC)(H_2O)_2$): 1, 4-Benzenedicarboxylic Acid

MOF based on manganese with general formula ($Mn(BDC)(H_2O)_2$) was successfully synthesized by Salam *et al* via. microwave assisted solvothermal method. (Abd El Salam and Zaki, 2019) The product obtained were highly pure and crystalline in nature. They showed high adsorption capacities towards the removal of Cd(II), Hg(II) and Pb(II) ions from aqueous medium and maximum removal efficiency of the heavy metal ions were obtained at pH=5. Thus, the experimental data fits well with the Langmuir adsorption isotherm model and the kinetics data fits well with the pseudo-second order kinetic equation.

MOF-808-EDTA: Ethylenediaminetetraacetic Acid

Zhong *et al* synthesized MOF-808-EDTA by incorporating a non-specific chelating group i.e. ethylenediaminetetraacetic acid (EDTA) with six binding sites which helps in binding of heavy metal species. (Zhong *et al.*, 2018) This robust MOF was successful for the adsorption of 22 heavy metal ions comprising of hard (La^{3+}, Nd^{3+}, Pr^{3+}, Ce^{3+}, Eu^{3+}, Gd^{3+}, Zr^{4+}, Fe^{3+}, Mn^{2+}); soft (Hg^{2+}, Pd^{2+}, Rh^{3+}, Ru^{2+}, Cd^{2+}, Pt^{2+}) and borderline (Pb^{2+}, Sn^{2+}, Sb^{3+}, Zn^{2+}, Cu^{2+}, Co^{2+}, Ni^{2+}) Lewis metal ions with the removal efficiency of >99%.

Zr-based Stable Metal-Organic Frameworks DUT-67 (Zr)

Shu *et al* successfully synthesized metal-organic framework based on Zirconium with the composition of $[Zr_6O_6(OH)_2(tdc)_4(CH_3COO)_2]$ commonly referred to as DUT-67 (Zr) where DUT: (Dresden University of Technology). (Shu *et al.*, 2018) DUT-67 (Zr) was synthesized using 2, 5-thiophenedicarboxylic acid (H_2tdc) and Zirconium chloride as the precursors. The MOF obtained possess high surface area and is highly stable with water and acid. DUT-67 (Zr) MOF was used as adsorbent for the removal of dissolved mercury (Hg^{2+}) and methyl mercury (CH_3Hg^+) ions form aqueous solutions. The maximum removal efficiency of the ions were obtained to be ~90%.

Copper-Based Metal Organic Framework (Cu-MOF)

Copper-based metal organic framework (Cu-MOF) synthesized via solvothermal method was fabricated by Yusuff *et al.* (Yusuf *et al.*, 2019) During the synthesis process they have used copper nitrate trihydrate $[Cu(NO_3)_2{\bullet}3H_2O]$ and benzene 1,3,5-tricarboxylic acid, $[C_6H_3\text{-}1,3,5\text{-}(COOH)_3$ (H_3BTC) as the precursor. The synthesized product Cu-MOF was used as adsorbent for the removal of cadmium ions (Cd^{2+}) from aqueous solution. Removal efficiency increased with an increase in contact time, initial concentration and adsorbent dose. Thus, the experimental data fits well with the Langmuir adsorption isotherm model and the kinetics data fits well with the pseudo-second order kinetic equation.

Ni(II) Based Phosphonate Metal Organic Framework

Visa *et al* reported the synthesis of Ni(II) based phosphonate metal organic framework using nickel acetate $Ni(CH_3COO)_2 \cdot 4H_2O$, phosphonoacetic (CP), *N,N*-bis(phosphonomethyl)glycine (Gly) or vinylphosphonic (VP) under hydrothermal conditions. (Visa *et al.*, 2017) The synthesized MOF was used as adsorbent for the removal of chromium ions [Cr(VI)] from aqueous solution and the removal efficiency of Cr(VI) ions using various Ni(II) based MOF follows the order Ni-CP < Ni-Gly ≤ Ni-VP. Thus, the adsorption data fits well with the Langmuir adsorption isotherm model and the kinetics data fits well with the pseudo-second order kinetic equation.

MOF Incorporated Polymeric Membranes

Das *et al* reported the synthesis of MOF-5 (Zinc based MOF) embedded membranes with three different polymers namely Cellulose acetate (CA), Polyethersulfone (PES), and

Polyvinylidene fluoride (PVDF) via. simple Solvothermal method. (Das *et al.*, 2018) Thereafter the MOF was employed for the removal of hazardous heavy metal ions (Co^{2+}, Cu^{2+}) from aqueous solution. From the experimental data, it is evident that the performance and hydrophobicity of the MOF incorporated polymeric materials increases several fold compared to the simple MOF-5.

Mesoporous Metal-Organic Framework-5 (MOF-5)

Zepeda *et al* synthesized Mesoporous metal-organic framework-5 (MOF-5) in a Teflon-lined stirrer using terephthalic acid and zinc nitrate hexahydrate as the main precursors to yield the desired $Zn_4O(BDC)_3$ where BDC: Benzene dicarboxylate. MOF-5 obtained was used as adsorbent for the removal of lead ions Pb(II) from aqueous solutions. (Zepeda *et al.*, 2016) Thus, the experimental adsorption data fits well with the Langmuir and BET isotherm models and the kinetics data fits well with the pseudo-second order kinetic equation. Maximum adsorption capacity of MOF-5 was obtained to be 658.5mg/g at pH=5 and 45°C.

iNP@MOF-Beads

iNP@MOF-Beads were synthesized by Maspoch *et al* by incorporating inorganic nanoparticles (iNPs) (such as CeO_2, Fe_3O_4) into MOFs (such as Zr-MOFs, UiO-66 or UiO-66-$(SH)_2$ via a one-step continuous-flow spray-drying process to yield the desired CeO_2@UiO-66, CeO_2@UiO-66-$(SH)_2$, CeO_2/Fe_3O_4@UiO-66-$(SH)_2$ composite microbeads. (Zepeda *et al.*, 2020) The composite microbeads were used as adsorbent for the removal of multiple heavy metals (such as As(III and V), Cd(II), Cr(III and VI), Cu(II), Pb(II) and Hg(II)) from water. The use of magnetic microbeads CeO_2/Fe_3O_4@UiO-66-$(SH)_2$ for the similar metal-adsorption capacity was also studied and the results are similar to the original non-magnetic microbeads. The main advantages of magnetic microbeads is that it can be easily removed from treated water by an external magnet.

TMU-56

TMU-56 was developed by Morsali *et al* by incorporating N-donor ligand (N^1, N^2- di(pyridine-4-yl) oxalamide) into the cavities of a Zn-MOF. (Morsali *et al.*, 2020) Large pore size and high densities of strong metal chelating sites of TMU-56 MOF makes it an effective adsorbent for the removal of carcinogenic lead ions Pb(II) from aqueous solutions. Thus, the experimental adsorption data fits well with the Langmuir and BET isotherm models and the kinetics data fits well with the pseudo-second order kinetic model. Maximum adsorption capacity of lead ions was found to be 1130 mg/g with a removal efficiency >97% in 20 sec.

Advancements in the Chemistry of MOF for the Removal of Various Metal Ions

Table 1.

MOFs	Synthesis method	Adsorption isotherm and Kinetics	Adsorption capacity (mg/g)		Reference
			As(III)	As(V)	
Fe-BTC (1,3,5-Benzenetricarboxylic)	Solvothermal	Langmuir isotherm, pseudo-second-order kinetic equation		12.29	(Zhu *et al.*, 2012)
MIL-53(Fe)	HF free-solvothermal method.	Langmuir isotherm, pseudo-second-order kinetic equation		21.27	(Vu *et al.*, 2015)
MOF-808 $[Zr_6O_4(OH)_4(BTC)_2(HCOO)_6]$	Microwave irradiation method	pseudo-second-order kinetic equation		24.83	(Li *et al.*, 2015)
UiO-66	--	Langmuir isotherm and Freundlich isotherm		303.34	(Wang *et al.*, 2015)
ZIF-8		Langmuir isotherm, pseudo-second-order kinetic equation	49.49	60.03	(Jian *et al.*, 2015)
β-MnO_2@ZIF-8		Langmuir isotherm, pseudo-second-order kinetic equation	140.27	--	(Jian *et al.*, 2016)
$CoFe_2O_4$@MIL-100(Fe)		Langmuir isotherm, pseudo-second-order kinetic equation	143.6	114.8	(Yang and Yin, 2017)
Fe_3O_4@MIL-101(Cr)		Langmuir isotherm,	121.5	80.0	(Folens *et al.*, 2016)
Fe_3O_4@ZIF-8		Langmuir isotherm, pseudo-second-order kinetic equation	100		(Huo *et al.*, 2018)

Recently, the synthesis and use of water stable MOFs have shown significant results in water treatment. Recently reports have been found where other composites such as

metal nanoparticles, metal oxides, graphene, quantum dots and carbon nanotubes have been incorporated with MOFs, which have shown significantly better results than MOFs alone. (Chen *et al.*, 2017; Falcaro *et al.*, 2016). In addition, Few MOF-composites have also the potential as green and recyclable materials. ((Zepeda *et al.*, 2020; Feng *et al.*, 2018)

Table 2. Removal of Cr(III) and Cr(VI) Species from Aqueous Solution

MOFs	Adsorption isotherm and Kinetics	Adsorption capacity (mg/g)		Reference
		Cr(III)	Cr(VI)	
Ag-triazolate MOF	Langmuir isotherm		38.6	(Li *et al.*, 2017)
IL-MIL-100(Fe) IL: Ionic Liquid	Langmuir isotherm, pseudo-second-order kinetic model		285.7	(Nasrollahpour and Moradi, 2017)
TMU-30 Isonicotinate N-oxide	Langmuir isotherm, pseudo-second-order kinetic model		145	(Aboutorabi *et al.*, 2016)
UiO-66-NH_2	Langmuir isotherm, pseudo-second-order kinetic model		32.36	(Wu *et al.*, 2017)
UiO-66-NHC(S)NHMe	Langmuir isotherm	117		(Saleem *et al.*, 2016)
ZJU-101, Zirconium cluster based MOF	pseudo-second-order kinetic model		245	(Zhang *et al.*, 2015)
MP@ZIF-8 Magnetic (Fe_3O_4) Polydopamine @ Zeolitic Idazolate Frameworks-8	Langmuir isotherm, pseudo-second-order kinetic model		136.56	(Zhu *et al.*, 2017)
Cu-BTC MOFs; BTC: Benzenetricarboxylates	Langmuir isotherm model, pseudo-second-order kinetic model		48	Maleki *et al.*, 2015
Fe_3O_4@ MIL-100(Fe) magnetic microspheres	--		13	Yang *et al.*, 2016
Chitosan-MOF Composite	Pseudo-second-order kinetic model		93.6	Wang *et al.*, 2016

Table 3. Removal of Pb(II) Species from Aqueous Solution

MOFs	Adsorption isotherm and Kinetics	Adsorption capacity (mg/g)	Reference
		Pb(II)	
Cu-MOFs/Fe_3O_4	Pseudo-second-order kinetic model, Freundlich adsorption isotherm model, Dubinin-Radushkevich (D-R) isothermal adsorption model	219.00	(Shi *et al.*, 2018)
Fe_3O_4@TAR, TAR: 4- (thiazolylazo) resorcinol (TAR)	Box-Behnken design (BBD) was applied for fitting a second-order response surface	185	(Ghorbani-Kalhor, 2016)
(Fe_3O_4-ED)/MIL-101(Fe)	Box-Behnken design (BBD) was applied for fitting a second-order response surface	198	(Babazadeh *et al.*, 2015)
IOMN@NH_2-MIL53(Al) Magnetic nanocomposite material based on an aluminium MOF	Freundlich isotherm model	492.4	(Ricco *et al.*, 2015)
UiO-66-NHC(S)NHMe	Langmuir isotherm	232	(Saleem *et al.*, 2016)
NH_2-Zr-MOFs	Pseudo-second-order kinetic model	166.74	(Wang *et al.*, 2017)
MOF-808-EDTA	Langmuir isotherm, Pseudo-second-order kinetic model	313	(Peng *et al.*, 2018)
Melamine-MOFs	Pseudo-second-order kinetic model	122.0	(Yin *et al.*, 2018)
Nanofibrous MOF	Langmuir isotherm, Pseudo-second-order kinetic model	170.74	Efome *et al*, 2018
Cu- and Co-MOFs based on based on Copper- and cobalt-1,2,4,5-tetrabenzenecarboxylates	Freundlich isotherm model, Second-order kinetic model	Cu-MOFs: 76.50% Co-MOFs: 70.01%	David *et al.*, 2015
HKUST-1-MW [Cu_3(1,3,5-benzene-tri-carboxilicacid)$_2$] and HKUST-1-MW@ $H_3PW_{12}O_{40}$	Pseudo-second-order equation	98.18	Zou *et al.*, 2014
HS-mSi@MOF-5	Langmuir isotherm, Pseudo-second-order kinetic model	312	Zhang *et al.*, 2016

Table 4. Removal of Hg(II) Species from Aqueous Solution

MOFs	Adsorption isotherm and Kinetics	Adsorption capacity (mg/g)	Reference
		Hg(II)	
Fe_3O_4-2,5-dimercapto-1,3,4-thiadiazole/ metal-organic framework	Optimized by Box–Behnken design	124	(Ghorbani-Kalhor *et al.*, 2015)
ZIF-90-SH Thiol functionalized ZIF-90	--	22.4	(Bhattacharjee *et al.*, 2015)
UiO-66-NHC(S)NHMe	Langmuir isotherm	769	(Saleem *et al.*, 2016)
[Ni(3-bpd)$_2$(NCS)$_2$]$_n$ 3-bpd: 1,4-bis(3-pyridyl)-2,3-diaza-1,3-butadiene	--	713	(Halder *et al.*, 2017)
MOF-808-EDTA	Langmuir isotherm, Pseudo-second-order kinetic model	592	(Peng *et al.*, 2018)
JUC-62, MOF based on copper and 3,3', 5,5'-azo benzene tetra carboxylic acid	Langmuir isotherm, Pseudo-second-order kinetic model	836.7	(Wu *et al.*, 2016)

Table 6. Removal of Cu(II) Species from Aqueous Solution

MOFs	Adsorption isotherm and Kinetics	Adsorption capacity (mg/g)	Reference
		Cu(II)	
TMU-4; Zinc(II) metal-organic frameworks $[Zn_2(oba)_2(4\text{-bpdb})]\cdot(DMF)_x$	--	62	(Tahmasebi *et al.*, 2015)
TMU-6; $[Zn(oba)(4\text{-bpmb})_{0.5}]_n\cdot(DMF)_z$	--	60	(Tahmasebi *et al.*, 2015)
ZIF-8; Zeolitic imidazolate frameworks	Pseudo second-order kinetic model	454.72	(Huang *et al.*, 2018)
ZIF-67; Zeolitic imidazolate frameworks	Pseudo second-order kinetic model	617.51	(Huang *et al.*, 2018)
Chitosan-MOF Composite	Pseudo-second-order kinetic model	50.6	Wang *et al.*, 2016

Table 5. Removal of Cd(II) Species from Aqueous Solution

MOFs	Adsorption isotherm and Kinetics	Adsorption capacity (mg/g)	Reference
		Cd(II)	
$(Fe_3O_4$-ED)/MIL-101(Fe) $[(Fe_3O_4$-ethylenediamine)/MIL-101(Fe)]	Box-Behnken design (BBD) was applied for fitting a second-order response surface	155	(Babazadeh *et al.*, 2015)
Fe_3O_4@TAR	"	210	(Ghorbani-Kalhor, 2016)
MOF-DHz	"	188	(Taghizadeh *et al.*, 2013)
NH_2-Zr-MOFs	Pseudo second-order kinetic model	177.35	(Wang *et al.*, 2017)
TMU-16-NH_2	Langmuir and Freundlich isotherms models, Pseudo second-order kinetic model	126.6	(Roushani *et al.*, 2017)
HKUST-1-MW $[Cu_3$(1,3,5-benzene-tri-carboxilicacid$)_2]$ and HKUST-1-MW@ $H_3PW_{12}O_{40}$	Pseudo-second-order equation	32.45	Zou *et al.*, 2014
HS-mSi@MOF-5	Langmuir isotherm, Pseudo-second-order kinetic model	98	Zhang *et al.*, 2016

Table 7. Removal of Radionuclides Species from Aqueous Solution

MOFs	Adsorption isotherm and Kinetics	Adsorption capacity (mg/g)	Reference
		U(VI)	
CMPO-MIL-101(Cr); CMPO: N,N-Diisobutyl-2-(octylphenylphosphoryl)acetamide	Langmuir model, Pseudo second-order kinetic model	27.99	(De Decker *et al.*, 2017)
Co-SLUG-35; $[Co(H_2O)_4(4,4'\text{-bipy})_2](O_3S\text{-}C_2H_4\text{-}SO_3)\cdot 4H_2O]$	Pseudo second-order kinetic model	118	(Li *et al.*, 2017)
MIL-101-ED; ED Ethanediamine	Langmuir model, Pseudo second-order kinetic model	200	(Bai *et al.*, 2015)
MIL-101-DETA; (DETA ¼ Diethylenetriamine)	Langmuir model, Pseudo second-order kinetic model	350	(Bai *et al.*, 2015)
MOF-3; [MIL-101(Cr)-triazole-COOH]	Pseudo-second-order rate model	314	(Li *et al.*, 2016)
MOF-76; Synthesized using lanthanide ion and 1,3,5-benzenetricarboxylate (BTC) ligand	Langmuir isotherm	298	(Yang *et al.*, 2013)
Terbium(III)-based MOF	Langmuir model	179.8	(Liu *et al.*, 2017a)
UiO-66-NH_2	Langmuir isotherm, Pseudo-second-order model	114.9	(Luo *et al.*, 2016)
Zn(ADC)(4,4'-BPE)$_{0.5}$; synthesized using zinc ion, trans-1,2-bis-4-pyridylethene (4,4' -BPE) and azobenzene-4,4' - dicarboxylic acid (ADC) ligands	Langmuir and D-R (Dubinin-Radushkevich) isotherms, Pseudo-second-order kinetic model	312.32	(Liu *et al.*, 2016)
		Adsorption capacity (mg/g)	
		Se(IV)/Se(VI)	
NU-1000; Zirconium-based metal-organic frameworks	Langmuir isotherm model	Se(IV): 95	(Howarth *et al.*, 2015)
NU-1000; Zirconium-based MOFs	Langmuir isotherm model	Se(VI): 85	(Howarth *et al.*, 2015)
UiO-66-HCl; Hexanuclear zirconium MOFs	Pseudo-second-order kinetic model	Se(VI): 86.8	(Li *et al.*, 2017)
		Adsorption capacity (mg/g)	
		Re(VII)	
SBN; $[Ag(bipy)]NO_3$	Langmuir model	714	(Zhu *et al.*, 2017b)
SCU-100; $[Ag_2(tipm)]\cdot 2NO_3\cdot 1.5H_2O$ (SCU-100, SCU = Soochow University, tipm = tetrakis[4-(1-imidazol-yl)phenyl]methane	--	541	(Sheng *et al.*, 2017)
SLUG-21; Cationic material: $[Ag_2(4,40\text{-bipy})_2(O_3SCH_2CH_2SO_3)_3\ 4H_2O])$	--	602	(Fei *et al.*, 2011)
		Adsorption capacity (mg/g)	
		Th(IV)	
UiO-66-$(COOH)_2$	Langmuir isotherm model, Pseudo-second-order kinetic model	350	(Zhang *et al.*, 2017)

Removal of As(III) and As(V) Species From Aqueous Solution

MOF and MOF-composites have been used as potential adsorbent for the removal of As(III) and As(V) from aqueous solution. Table below describes briefly about types of MOF used and their properties.

CONCLUSION

Water-stable MOFs and MOF-based composites with superior adsorption capability have emerged as a novel method for the removal of heavy metal species (such as Cr(III)/Cr(VI), Cu(II), Hg(II), Cd(II), As(III)/As(V), Pb(II), radionuclides U(VI), Se(IV)/(VI), Cs(I) from aqueous solution. In most of the cases, the experimental adsorption data fits with Langmuir isotherm models and kinetics data follows the pseudo-second-order kinetic model. Exceptionally high selectivity, stability and reusability of MOFs makes it more advantageous compared to other conventional techniques used for the wastewater treatment. Thus, MOFs, multi-functional MOFs (hybrid MOFs) exhibits immense potential and can be explored more in near future for its use in various dimensions of water purifications.

REFERENCES

Abd El Salam, H. M., & Zaki, T. (2019). A Novel Microwave Synthesis of Manganese Based MOF for Adsorptive of Cd(II), Pb(II) and Hg(II) Ions from Aqua Medium. *Egyptian Journal of Chemistry*, *62*(5), 837–851. doi:10.21608/ejchem.2019.6524.1550

Aboutorabi, L., Morsali, A., Tahmasebi, E., & Büyükgüngor, O. (2016). Metal-Organic Framework based on isonicotinate N-oxide for fast and highly efficient aqueous phase Cr(VI) adsorption. *Inorganic Chemistry*, *55*(11), 5507–5513. doi:10.1021/acs.inorgchem.6b00522 PMID:27195982

Adriano, D. C. (2001). *Trace Elements in Terrestrial Environments*. Springer SBM., doi:10.1007/978-0-387-21510-5

Afshariazara, F., Morsalia, A., Wangb, J., & Junk, P. C. (2020). Highest and Fastest Removal Rate of Pb(II) Ions through Rational Functionalized decoration of a MOF cavity. *Chemistry (Weinheim an der Bergstrasse, Germany)*, *26*(6), 1355–1362. doi:10.1002/chem.201904436 PMID:31756257

Alloway, B. J. (1995). *Heavy Metals in Soils*. Blackie Academic and Professional. doi:10.1007/978-94-011-1344-1

Babazadeh, M., Hosseinzadeh-Khanmiri, R., Abolhasani, J., Ghorbani-Kalhor, E., & Hassanpour, A. (2015). Solid phase extraction of heavy metal ions from agricultural samples with the aid of a novel functionalized magnetic metal-organic framework. *RSC Advances*, *5*(26), 19884–19892. doi:10.1039/C4RA15532G

Babel, S., Kurniawan, T.A. (2004). Cr (VI) removal from synthetic wastewater using coconut shell charcoal and commercial activated carbon modified with oxidizing agents and/or chitosan, *Chemosphere*, 54, 951-967.

Bai, Z.-Q., Yuan, L.-Y., Zhu, L., Liu, Z.-R., Chu, S.-Q., Zheng, L.-R., Zhang, J., Chai, Z.-F., & Shi, W.-Q. (2015). Introduction of amino groups into acid-resistant MOFs for enhanced U(VI) sorption. *Journal of Materials Chemistry. A, Materials for Energy and Sustainability*, *3*(2), 525–534. doi:10.1039/C4TA04878D

Bakhtiari, N., & Azizian, S. (2015). Adsorption of copper ion from aqueous solution by nanoporous MOF-5: a kinetic and equilibrium study, *J. Mol. Liq.*, 206, 114-118.

Bakker, K. (2012). Water security: Research challenges and opportunities. *Science*, *337*, 914–915.

Barakat, M. A. (2011). New trends in removing heavy metals from industrial wastewater. *Arabian Journal of Chemistry*, *4*(4), 361–377. doi:10.1016/j.arabjc.2010.07.019

Bhattacharjee, S., Lee, Y.-R., & Ahn, W.-S. (2015). Post-synthesis functionalization of a zeolitic imidazolate structure ZIF-90: A study on removal of Hg(II) from water and epoxidation of alkenes. *CrystEngComm*, *17*(12), 2575–2582. doi:10.1039/C4CE02555E

Bhattacharyya, P., Chakrabarti, K., Chakraborty, A., Tripathy, S., & Powell, M. A. (2008). Fractionation and bioavailability of Pb in municipal solid waste compost and Pb uptake by rice straw and grain under submerged condition in amended soil. *Geosciences Journal*, *12*(1), 41–45. doi:10.100712303-008-0006-9

Boix, G., Troyano, J., Tovar, L. C. G., Camur, C., Bermejo, N., Yazdi, A., Piella, J., Bastús, N. G., Puntes, V., Imaz, I., & Maspoch, D. (2020). MOF-Beads containing inorganic nanoparticles for the simultaneous removal of multiple heavy metals from water. *ACS Applied Materials & Interfaces*, *12*(9), 10554–10562. doi:10.1021/acsami.9b23206 PMID:32026677

Bradl, H. B. (2005). *Heavy Metals in the Environment: Origin, Interaction and Remediation* (1st ed.). Elsevier Academic Press.

Chena, S., Fenga, F., Lia, S., Lia, X.-X., & Shu, L. (2018). Metal-organic framework DUT-67 (Zr) for adsorptive removal of trace Hg^{2+} and CH_3Hg^+ in water. *Chemical Speciation and Bioavailability*, *30*(1), 99–106. doi:10.1080/09542299.2018.1509020

De Decker, J., Folens, K., De Clercq, J., Meledina, M., Van Tendeloo, G., Du Laing, G., & Van Der Voort, P. (2017). Ship-in-a-bottle CMPO in MIL-101(Cr) for selective uranium recovery from aqueous streams through adsorption. *Journal of Hazardous Materials*, *335*, 1–9. doi:10.1016/j.jhazmat.2017.04.029 PMID:28414943

Drever, J. I. (1997). *The Geochemistry of Natural Waters: Surface and Groundwater Environments*. Prentice Hall.

Duruibe, J. O., Ogwuegbu, M. O. C., & Egwurugwu, J. N. (2007). Heavy metal pollution and human biotoxic effects. *International Journal of Physical Sciences*, *2*, 112–118.

Efome, J. E., Rana, D., Matsuura, T., & Lan, C. Q. (2018). Metal-Organic frameworks supported on nanofibers to remove heavy metals. *Journal of Materials Chemistry. A, Materials for Energy and Sustainability*, *6*(10), 4550–4555. doi:10.1039/C7TA10428F

Fei, H., Bresler, M. R., & Oliver, S. R. J. (2011). A new paradigm for anion trapping in high capacity and selectivity: Crystal-to-crystal transformation of cationic materials. *Journal of the American Chemical Society*, *133*(29), 11110–11113. doi:10.1021/ja204577p PMID:21707072

Feng, M., Zhang, P., Zhou, H.-C., & Sharma, V. K. (2018). Water-stable metal-organic frameworks for aqueous removal of heavy metals and radionuclides: A review. *Chemosphere*, *209*, 783–800. doi:10.1016/j.chemosphere.2018.06.114 PMID:29960946

Folens, K., Leus, K., Nicomel, N. R., Meledina, M., Turner, S., Tendeloo, G. V., Laing, G. D., & Voort, P. V. D. (2016). Fe_3O_4@MIL-101-A Selective and regenerable adsorbent for the removal of As species from water. *European Journal of Inorganic Chemistry*, *2016*(27), 4395–4401. doi:10.1002/ejic.201600160

Fu, F., & Wang, Q. (2011). Removal of heavy metal ions from wastewaters: a review. Journal of Environmental Management. *J. Environ. Manage.*, *92*, 407-418.

Ghorbani-Kalhor, E. (2016). A metal-organic framework nanocomposite made from functionalized magnetite nanoparticles and HKUST-1 (MOF-199) for preconcentration of Cd(II), Pb(II), and Ni(II). *Mikrochimica Acta*, *183*(9), 2639–2647. doi:10.100700604-016-1896-2

Ghorbani-Kalhor, E., Hosseinzadeh-Khanmiri, R., Abolhasani, J., Babazadeh, M., & Hassanpour, A. (2015). Determination of mercury(II) ions in seafood samples after extraction and pre-concentration by a novel functionalized magnetic metal-organic framework nanocomposite. *Journal of Separation Science*, *38*(7), 1179–1186. doi:10.1002/jssc.201401320 PMID:25631191

Gnanasekarana, G., Balagurua, S., Gangasalama, A., & Das, D. B. (2018). Removal of hazardous material from wastewater by using Metal-Organic Framework (MOFs) embedded polymeric membranes. *Separation Science and Technology*, *54*(3), 1–13.

Gu, Z., Song, W., Yanga, Z., & Zhou, R. (2018). Metal-organic framework as an efficient filter for the removal of heavy metal cations in water. *Physical Chemistry Chemical Physics*, *20*(48), 30384–30391. doi:10.1039/C8CP05129A PMID:30489583

Guala, S. D., Vega, F. A., & Covelo, E. F. (2010). The dynamics of heavy metals in plant-soil interactions. *Ecological Modelling*, *221*(8), 1148–1152. doi:10.1016/j.ecolmodel.2010.01.003

Halder, S., Mondal, J., Ortega-Castro, J., Frontera, A., & Roy, P. (2017). A Ni-based MOF for selective detection and removal of Hg^{2+} in aqueous medium: A facile strategy. *Dalton Transactions (Cambridge, England)*, *46*(6), 1943–1950. doi:10.1039/C6DT04722J PMID:28112319

Howarth, A. J., Katz, M. J., Wang, T. C., Platero-Prats, A. E., Chapman, K. W., Hupp, J. T., & Farha, O. K. (2015). High efficiency adsorption and removal of selenate and selenite from water using metal-organic frameworks. *Journal of the American Chemical Society*, *137*(23), 7488–7494. doi:10.1021/jacs.5b03904 PMID:26000611

Huang, Y., Zeng, X., Guo, L., Lan, J., Zhang, L., & Cao, D. (2018). Heavy metal ion removal of wastewater by zeolite-imidazolate frameworks. *Separation and Purification Technology*, *194*, 462–469. doi:10.1016/j.seppur.2017.11.068

Huo, J.-B., Xu, L., Yang, J.-C. E., Cui, H.-J., Yuan, B., & Fu, M. L. (2018). Magnetic responsive Fe_3O_4-ZIF-8 core-shell composites for efficient removal of As(III) from water. *Colloids and Surfaces. A, Physicochemical and Engineering Aspects*, *539*, 59–68. doi:10.1016/j.colsurfa.2017.12.010

Jian, M., Liu, B., Zhang, G., Liu, R., & Zhang, X. (2015). Adsorptive removal of arsenic from aqueous solution by zeolitic imidazolate framework-8 (ZIF-8) nanoparticles. *Colloids and Surfaces. A, Physicochemical and Engineering Aspects*, *465*, 67–76. doi:10.1016/j.colsurfa.2014.10.023

Jian, M., Wang, H., Liu, R., Qu, J., Wang, H., & Zhang, X. (2016). Self-assembled one-dimensional MnO_2@zeolitic imidazolate framework-8 nanostructures for highly efficient arsenite removal. *Environmental Science. Nano*, *3*(5), 1186–1194. doi:10.1039/C6EN00246C

Kazemipour, M., Ansari, M., Tajrobehkar, S., Majdzadeh, M., & Kermani, H. R. (2008). Removal of lead, cadmium, zinc, and copper from industrial wastewater by carbon developed from walnut, hazelnut, almond, pistachio shell, and apricot stone. *Journal of Hazardous Materials*, *150*(2), 322–327. doi:10.1016/j.jhazmat.2007.04.118 PMID:17548149

Lalor, G. C. (2008). Review of cadmium transfers from soil to humans and its health effects in the Jamaican environment. *The Science of the Total Environment*, *400*(1-3), 162–172. doi:10.1016/j.scitotenv.2008.07.011 PMID:18752835

Li, J., Liu, Y., Wang, X., Zhao, G., Ai, Y., Han, B., Wen, T., Hayat, T., Alsaedi, A., & Wang, X. (2017). Experimental and theoretical study on selenate uptake to zirconium metal-organic frameworks: Effect of defects and ligands. *Chemical Engineering Journal*, *330*, 1012–1021. doi:10.1016/j.cej.2017.08.038

Li, J. Q., Gong, L. L., Feng, X. F., Zhang, L., Wu, H. Q., Yan, C. S., Xiong, Y. Y., Gao, H. Y., & Luo, F. (2017). Direct extraction of U(VI) from alkaline solution and seawater via anion exchange by metal-organic framework. *Chemical Engineering Journal*, *316*, 154–159. doi:10.1016/j.cej.2017.01.046

Li, L., Feng, X., Han, R., Zang, S., & Yang, G. (2017). Cr(VI) removal via anion exchange on a silver-triazolate MOF. *Journal of Hazardous Materials*, *321*, 622–628. doi:10.1016/j.jhazmat.2016.09.029 PMID:27694026

Li, L., Ma, W., Shen, S., Huang, H., Bai, Y., & Liu, H. (2016). A combined experimental and theoretical study on the extraction of uranium by amino-derived metal-organic frameworks through post-synthetic strategy. *ACS Applied Materials & Interfaces*, *8*(45), 31032–31041. doi:10.1021/acsami.6b11332 PMID:27788576

Li, Z.-Q., Yang, J.-C., Sui, K.-W., & Yin, N. (2015). Facile synthesis of Metal-Organic Framework MOF-808 for arsenic removal. *Materials Letters*, *160*, 412–414. doi:10.1016/j.matlet.2015.08.004

Liu, F., Song, S., Cheng, G., Xiong, W., Shi, L., & Zhang, Y. (2018). MIL-101(Cr) Metal-Organic Framework functionalized with tetraethylenepentamine for potential removal of Uranium (VI) from waste water. *Adsorption Science and Technology*, *36*(7–8), 1550–1567. doi:10.1177/0263617418789516

Liu, S., Luo, M., Li, J., Luo, F., Ke, L., & Ma, J. (2016). Adsorption equilibrium and kinetics of uranium onto porous azo-metal–organic frameworks. *Journal of Radioanalytical and Nuclear Chemistry*, *310*(1), 353–362. doi:10.100710967-016-4852-z

Liu, W., Dai, X., Bai, Z., Wang, Y., Yang, Z., Zhang, L., Xu, L., Chen, L., Li, Y., Gui, D., Diwu, J., Wang, J., Zhou, R., Chai, Z., & Wang, S. (2017). Highly sensitive and selective uranium detection in natural water systems using a luminescent mesoporous metal-organic framework equipped with abundant Lewis basic sites: A combined batch, X-ray absorption spectroscopy, and first principles simulation investigation. *Environmental Science & Technology*, *51*(7), 3911–3921. doi:10.1021/acs.est.6b06305 PMID:28271891

Luo, B., Yuan, L., Chai, Z., Shi, W., & Tang, Q. (2016). U(VI) capture from aqueous solution by highly porous and stable MOFs: UiO-66 and its amine derivative. *Journal of Radioanalytical and Nuclear Chemistry*, *307*(1), 269–276. doi:10.100710967-015-4108-3

Maleki, A., Hayati, B., Naghizadeh, M., & Joo, S. W. (2015). Adsorption of hexavalent chromium by metal organic frameworks from aqueous solution. *Journal of Industrial and Engineering Chemistry*, *28*, 211–216. doi:10.1016/j.jiec.2015.02.016

Maranescu, B., Lupa, L., & Visa, A. (2017). Heavy metal removal from waste waters by phosphonate Metal Organic Frameworks. *Pure and Applied Chemistry*. Advance online publication. doi:10.1515/pac-2017-0307

Nasrollahpour, A., & Moradi, S. E. (2017). Hexavalent chromium removal from water by ionic liquid modified metal-organic frameworks adsorbent. *Microporous and Mesoporous Materials*, *243*, 47–55. doi:10.1016/j.micromeso.2017.02.006

O'Keeffe, M., & Yaghi, O. M. (2012). Review Deconstructing the Crystal Structures of Metal-Organic Frameworks and Related Materials into Their Underlying Nets. *Chemical Reviews*, *112*(2), 675–702. doi:10.1021/cr200205j PMID:21916513

Odum, H. T. (2000). *Heavy Metals in the Environment Using Wetlands for Their Removal*. Lewis Publishers.

Pan, B. J., Pan, B. C., Zhang, W. M., Lv, L., Zhang, Q. X., & Zheng, S. R. (2009). Development of polymeric and polymer-based hybrid adsorbents for pollutants removal from waters. *Chemical Engineering Journal*, *151*(1-3), 19–29. doi:10.1016/j.cej.2009.02.036

Peng, Y., Huang, H., Zhang, Y., Kang, C., Chen, S., Song, L., Liu, D., & Zhong, C. (2018). A versatile MOF-based trap for heavy metal ion capture and dispersion. *Nature Communications*, *9*(187), 1–9. doi:10.103841467-017-02600-2 PMID:29335517

Peng, Y., Huang, H., Zhang, Y., Kang, C., Chen, S., Song, L., Liu, D., & Zhong, C. (2018). A versatile MOF-based trap for heavy metal ion capture and dispersion. *Nature Communications*, *9*(1), 187. doi:10.103841467-017-02600-2 PMID:29335517

Wan Ngah, W. S., Hanafiah, M. A. K. M. (2008). Removal of heavy metal ions from wastewater by chemically modified plant wastes as adsorbents: A review. *Bioresource Technology*, *99*, 3935–3948. PMID:17681755

Pradeep, T., & Anshup, T. (2009). Noble metal nanoparticles for water purification: A critical review. *Thin Solid Films*, *517*(24), 6441–6478. doi:10.1016/j.tsf.2009.03.195

Rascio, N., & Izzo, F. N. (2011). Heavy metal hyperaccumulating plants: How and why do they do it? And what makes them so interesting? *Plant Science*, *180*(2), 169–181. doi:10.1016/j.plantsci.2010.08.016 PMID:21421358

Ricco, R., Konstas, K., Styles, M. J., Richardson, J. J., Babarao, R., Suzuki, K., Scopece, P., & Falcaro, P. (2015). Lead(II) uptake by aluminium based magnetic framework composites (MFCs) in water. *Journal of Materials Chemistry. A, Materials for Energy and Sustainability*, *3*(39), 19822–19831. doi:10.1039/C5TA04154F

Rivera, J. M., Rincón, S., Youssef, C. B., & Zepeda, A. (2016). Highly efficient adsorption of aqueous Pb(II) with mesoporous Metal-Organic Framework-5: An equilibrium and kinetic study. *Journal of Nanomaterials*, *2016*, 1–9. Advance online publication. doi:10.1155/2016/8095737

Roushani, M., Saedi, Z., & Baghelani, Y. M. (2017). Removal of cadmium ions from aqueous solutions using TMU-16-NH_2 metal organic framework. *Environmental Nanotechnology, Monitoring & Management*, *7*, 89–96. doi:10.1016/j.enmm.2017.01.003

Saleem, H., Rafique, U., & Davies, R. P. (2016). Investigations on post-synthetically modified UiO-66-NH2 for the adsorptive removal of heavy metal ions from aqueous solution. *Microporous and Mesoporous Materials*, *221*, 238–244. doi:10.1016/j.micromeso.2015.09.043

Saleem, H., Rafique, U., & Davies, R. P. (2016). Investigations on post-synthetically modified UiO-66-NH_2 for the adsorptive removal of heavy metal ions from aqueous solution. *Microporous and Mesoporous Materials*, *221*, 238–244. doi:10.1016/j.micromeso.2015.09.043

Sheng, D., Zhu, L., Xu, C., Xiao, C., Wang, Y., Wang, Y., Chen, L., Diwu, J., Chen, J., Chai, Z., Albrecht-Schmitt, T. E., & Wang, S. (2017). Efficient and selective uptake of TcO4- by a cationic metal-organic framework material with open Ag^+ sites. *Environmental Science & Technology*, *51*(6), 3471–3479. doi:10.1021/acs.est.7b00339 PMID:28211267

Shi, Z., Xu, C., Guan, H., Li, L., Fan, L., Wang, Y., Liu, L., Meng, Q., & Zhang, R. (2018). Magnetic metal organic frameworks (MOFs) composite for removal of lead and malachite green in wastewater. *Colloids and Surfaces. A, Physicochemical and Engineering Aspects*, *539*, 382–390. doi:10.1016/j.colsurfa.2017.12.043

Shooto, N. D., Dikio, E. D., Wankasi, D., & Sikhwivhilu, L. (2015). Synthesis, morphology and lead ion adsorption properties of metal organic frameworks of copper and cobalt. *Chemical Sciences Journal*, *6*(4), 113.

Siegel, F. R. (2002). *Environmental Geochemistry of Potentially Toxic Metals*. Springer., doi:10.1007/978-3-662-04739-2

Sobha, K., Poornima, A., Harini, P., & Veeraiah, K. A. (2007). study on biochemical changes in the fresh water fish, catla catla (hamilton) exposed to the heavy metal toxicant cadmium chloride, *Kathmandu Uni. J. Sci. Eng. Techno*, *1*(4), 1–11.

Stern, B. R., Solioz, M., Krewski, D., Aggett, P., Aw, T. C., Baker, S., Crump, K., Dourson, M., Haber, L., Hertzberg, R., Keen, C., Meek, B., Rudenko, L., Schoeny, R., Slob, W., & Starr, T. (2007). Copper and human health: biochemistry, genetics, and strategies for modelling doseresponse relationships. *J. Toxicol. and Env. Heal. B*, *10*, 157-222.

Sun, D. T., Peng, L., Reeder, W. S., Moosavi, S. M., Tiana, D., Britt, D. K., Oveisi, E., & Queen, W. L. (2018). Rapid, Selective heavy metal removal from water by a Metal-Organic Framework/Polydopamine Composite. *ACS Central Science*, *4*(3), 349–356. doi:10.1021/acscentsci.7b00605 PMID:29632880

Taghizadeh, M., Asgharinezhad, A. A., Pooladi, M., Barzin, M., Abbaszadeh, A., & Tadjarodi, A. (2013). A novel magnetic metal organic framework nanocomposite for extraction and pre concentration of heavy metal ions, and its optimization via experimental design methodology. *Mikrochimica Acta*, *180*(11-12), 1073–1084. doi:10.100700604-013-1010-y

Tahmasebi, E., Masoomi, M. Y., Yamini, Y., & Morsali, A. (2015, January 20). Application of mechanosynthesized azine-decorated zinc(II) metal-organic frameworks for highly efficient removal and extraction of some heavy-metal ions from aqueous samples: A comparative study. *Inorganic Chemistry*, *54*(2), 425–433. doi:10.1021/ic5015384 PMID:25548873

Vörösmarty, C. J., McIntyre, P. B., Gessner, M. O., Dudgeon, D., Prusevich, A., Green, P., Glidden, S., Bunn, S. E., Sullivan, C. A., Liermann, C. R., & Davies, P. M. (2010). Global threats to human water security and river biodiversity. *Nature*, *467*, 555–561.

Vu, T. A., Le, G. H., Dao, C. D., Dang, L. Q., Nguyen, K. T., Nguyen, Q. K., Dang, P. T., Tran, H. T. K., Duong, Q. T., Nguyen, T. V., & Lee, G. D. (2015). Arsenic removal from aqueous solutions by adsorption using novel MIL-53(Fe) as a highly efficient adsorbent. *RSC Advances*, *5*(7), 5261–5268. doi:10.1039/C4RA12326C

Wang, C., Liu, X., Chen, J. P., & Li, K. (2015). Superior removal of arsenic from water with zirconium metalorganic framework UiO-66. *Scientific Reports*, *5*(1), 16613. doi:10.1038rep16613 PMID:26559001

Wang, K., Gu, J., & Yin, N. (2017). Efficient removal of Pb(II) and Cd(II) using NH2-functionalized Zr-MOFs via rapid microwave-promoted synthesis. *Industrial & Engineering Chemistry Research*, *56*(7), 1880–1887. doi:10.1021/acs.iecr.6b04997

Wang, K., Tao, X. R., Xu, J. Z., & Yin, N. (2016). Novel Chitosan MOF Composite Adsorbent for the Removal of Heavy Metal Ions. *Chemistry Letters*, *45*(12), 1365–1368. doi:10.1246/cl.160718

Wu, S., Ge, Y., Wang, Y., Chen, X., Li, F., Xuan, H., & Li, X. (2017). Adsorption of Cr(VI) on nano UiO-66-NH_2 MOFs in water. *Environmental Technology*, 1–12. PMID:28625105

Wu, Y., Xu, G., Wei, F., Song, Q., Tang, T., Wang, X., & Hu, Q. (2016). Determination of Hg(II) in tea and mushroom samples based on metal-organic frameworks as solid phase extraction sorbents. *Microporous and Mesoporous Materials*, *235*, 204–210. doi:10.1016/j.micromeso.2016.08.010

Yang, J.-C., & Yin, X.-B. (2017). $CoFe_2O_4$@MIL-100(Fe) hybrid magnetic nanoparticles exhibit fast and selective adsorption of arsenic with high adsorption capacity. *Scientific Reports*, *7*(1), 40955. doi:10.1038rep40955 PMID:28102334

Yang, Q. X., Zhao, Q. Q., Ren, S. S., Lu, Q. Q., Guo, X. M., & Chen, Z. J. (2016). Fabrication of core-shell Fe_3O_4@ MIL-100(Fe) magnetic microspheres for the removal of Cr(VI) in aqueous solution. *Journal of Solid State Chemistry*, *244*, 25–30. doi:10.1016/j.jssc.2016.09.010

Yang, W., Bai, Z., Shi, W., Yuan, L., Tian, T., Chai, Z., Wang, H., & Sun, Z. (2013). MOF-76: From a luminescent probe to highly efficient UVI sorption material. *Chemical Communications (Cambridge)*, *49*(88), 10415–10417. doi:10.1039/c3cc44983a PMID:24079004

Yin, N., Wang, K., Xia, Y., & Li, Z. (2018). Novel melamine modified metal-organic frameworks for remarkably high removal of heavy metal Pb(II). *Desalination*, *430*, 120–127. doi:10.1016/j.desal.2017.12.057

Yusuf, A. S., Popoola, L. T., & Babatunde, E. O. (2019). Adsorption of cadmium ion from aqueous solutions by copper-based Metal Organic Framework: Equilibrium modeling and kinetic studies. *Applied Water Science*, *9*(4), 106. doi:10.100713201-019-0991-z

Zhang, H., Nai, J., Yu, L., & Lou, X. W. D. (2017). Metal-Organic-Framework based materials as platforms for renewable energy and environment applications. *Joule*, *1*(1), 77–107. doi:10.1016/j.joule.2017.08.008

Zhang, N., Yuan, L., Guo, W., Luo, S., Chai, Z., & Shi, W. (2017). Extending the use of highly porous and functionalized MOFs to Th(IV) capture. *ACS Applied Materials & Interfaces*, *9*(30), 25216–25224. doi:10.1021/acsami.7b04192 PMID:28699737

Zhang, Q., Yu, J., Cai, J., Zhang, L., Cui, Y., Yang, Y., Chen, B., & Qian, G. (2015). A porous Zrcluster-based cationic metal-organic framework for highly efficient $Cr_2O_7^{2-}$ removal from water. *Chemical Communications (Cambridge)*, *51*(79), 14732–14734. doi:10.1039/C5CC05927E PMID:26291500

Zhu, B.-J., Yu, X.-Y., Jia, Y., Peng, F.-M., Sun, B., Zhang, M.-Y., Luo, T., Liu, J.-H., & Huang, X.-J. (2012). Iron and 1,3,5-Benzenetricarboxylic Metal-Organic Coordination polymers prepared by Solvothermal method and their application in efficient As(V) removal from aqueous solutions. *The Journal of Physical Chemistry C*, *116*(15), 8601–8607. doi:10.1021/jp212514a

Zhu, K., Chen, C., Xu, H., Gao, Y., Tan, X., Alsaedi, A., & Hayat, T. (2017). Cr(VI) reduction and immobilization by core-double-shell structured magnetic polydopamine@zeolitic idazolate frameworks-8 microspheres. *ACS Sustainable Chemistry & Engineering*, *5*(8), 6795–6802. doi:10.1021/acssuschemeng.7b01036

Zhu, L., Xiao, C., Dai, X., Li, J., Gui, D., Sheng, D., Chen, L., Zhou, R., Chai, Z., Albrecht-Schmitt, T. E., & Wang, S. (2017). Exceptional perrhenate/pertechnetate uptake and subsequent immobilization by a low-dimensional cationic coordination polymer: Overcoming the hofmeister bias selectivity. *Environmental Science & Technology Letters*, *4*(7), 316–322. doi:10.1021/acs.estlett.7b00165

Zou, F., Yu, R., Li, R., & Li, W. (2013). Microwave-Assisted Synthesis of HKUST-1 and Functionalized HKUST-1-@$H_3PW_{12}O_{40}$: Selective Adsorption of Heavy Metal Ions in Water Analyzed with Synchrotron Radiation. *ChemPhysChem*, *14*(12), 2835–2832. doi:10.1002/cphc.201300215 PMID:23818161

Chapter 7
Remediation of Water and Wastewater Using Metal Organic Framework based Membranes

Anshul Yadav
Membrane Science and Separation Technology Division, CSIR-Central Salt & Marine Chemicals Research Institute, Bhavnagar, India

Pankaj Dnyandeo Indurkar
Membrane Science and Separation Technology Division, CSIR-Central Salt & Marine Chemicals Research Institute, Bhavnagar, India

ABSTRACT

Over the past few years, metal-organic frameworks have gained considerable attention. Metal-organic frameworks are widely regarded as promising materials for separation, catalysis, molecular recognition, and storage applications. The applications and potentials of engineered forms of metal-organic frameworks such as thin films and membranes supported on various substrates in the area of water purification are discussed in this chapter. Metal-organic frameworks applications in separation and purification techniques with metal-organic frameworks-based membrane matrix is underlined at their promising levels. Metal-organic frameworks polymer composite membranes, including polymer-supported metal-organic frameworks membranes, are identified and the recent development are discussed. This chapter will provide a comprehensive assessment of recent studies on the removal of various contaminants from water using metal-organic frameworks-based membranes.

INTRODUCTION

There is an increasingly global demand for safe drinking water. The world population and anthropogenic activities are growing, and the supplies of water are easily polluted. 2.1 billion people have no access to clean drinking water (water free of pollution) reported by the World Health Organization (WHO) (WHO and UNICEF 2017). This equivalent to 29% of the global population. In water and wastewater systems, membrane processes such as forward osmosis (FO), reverse osmosis (RO), nanofiltration

DOI: 10.4018/978-1-7998-4760-1.ch007

(NF), microfiltration (MF) and ultrafiltration (UF), membrane distillation (MD) are commonly used (Ezugbe and Rathilal 2020; Madhura et al. 2018). Membranes have taken on an ever more critical role in the separation and purification technology in recent decades. Polymer membrane production and operation technologies are relatively advanced due to the advantages such as cost-effectiveness, high processability and low power consumption. However, there is usually a trade-off between selectivity and permeability polymer membranes. To overcome this obstacle and due to their tunable and distinct pore size and porous nature, Metal-organic frameworks (MOFs) have been identified as perfect candidates for membrane purification applications. MOFs were first discovered in 1965 by Tomic (Tomic 1965). Earlier it was called supramolecules. The term "MOF" was introduced by by Yaghi et al. in 1995 (O M Yaghi, Li, and Li 1995). MOFs are organic and inorganic, hybrid crystalline materials that consist of collection of metal ions with a positive charge and are characterized by transparent, porous structures. In comparison to other porous materials, MOFs provide various advantages such as unusual structural complexity, including uniform pores, structural uniformity at atomic level, tuneable porosity, network topology, structure, proportions and chemical functionality versatility (Khan, Hasan, and Jhung 2013; Kitagawa, Kitaura, and Noro 2004; X. Li et al. 2017; Omar M Yaghi et al. 2003).

In the past, researchers have developed several hundred various MOFs have been studied for different applications (e.g. adsorption, membrane preparation, catalysis) including gas separation, gas storage, energy storage, and environmental applications (Rui et al. 2018; Seo et al. 2016; M. Zhang et al. 2018). MOFs can be synthesized using chemical or physical techniques by combining metallic and organic linker. MOFs, in the form of adsorbents, have also been used for the removal of water contaminants, such as lead (Pb) (Luo, Ding, and Luo 2015), arsenic (As) (C. Wang et al. 2015), *E coli* (X. ping Wang et al. 2020), nuclear waste (J. Cheng et al. 2018) etc. from aqueous phases. Different MOFs such as ZIFs, HKUST, UiO-66, MIL-101, MIL-100, MIL-125, etc. have been used for water purification in the past (Cook, Zheng, and Stang 2013; Elrasheedy et al. 2019; J. R. Li, Sculley, and Zhou 2012; Stock and Biswas 2012).

This chapter will focus on the purification of the aqueous stream through MOFs based membranes. The objectives of this chapter are: (i) to introduce types of contaminants and existing treatment technologies, (ii) familiarise with different types of MOFs based membranes and their synthesis techniques, (iii) outline applications of the aforementioned MOFs based membranes in water purification. The chapter has been organized into seven sections. At first, various contaminants and their treatment technologies are introduced. Different MOFs based membranes are discussed such as bare MOFs membranes, porous matrix membranes (PMM), mixed matrix membranes (MMM), and thin-film nanocomposite (TFN) membranes. Then synthesis methods of MOFs based membranes are outlined with various examples. Application of MOFs based membranes in FO, RO, NF, UF and MF with recent studies is provided. At last, an outlook from our point of view is given.

Contaminants in Water

A common problem worldwide is water contamination. These can be anthropogenic (human-made) or geological (geogenic). The contaminants are divided into two types associated with water pollution: inorganic contaminants, and organic contaminants (Kurwadkar 2019). Organic contaminants are mostly naturally occurring, while inorganic pollution is the result of human interaction. Biodegradable chemical pollutants are biodegradable waste. These sources of pollution are naturally found and caused by the environment. Inorganic contaminants include contaminant such as metals, and excess nutrients, present in water sources originate from either anthropogenic or geogenic. Table 1 shows the WHO guideline values of contaminants and its sources for drinking-water quality (Gorchev and Ozolins 2004; World Health Organization 201AD).

Table 1. WHO guideline values of contaminants and its sources for drinking-water quality (Gorchev and Ozolins 2004; World Health Organization 201AD)

Compound	Guideline value (mg/l)	Compound	Guideline value (mg/l)
Inorganic contaminants		**Organic contaminants**	
Arsenic	0.01	Benzene	10
Barium	0.7	Carbon tetrachloride	4
Chlorine	5	Di (2-ethylhexyl) phthalate	8
Boron	0.5	Dichlorobenzene, 1,2-	1000
Chromium	0.05	Dichlorobenzene, 1,4-	300
Fluoride	1.5	Dichloroethane, 1,2-	30
Manganese	0.4	Dichloroethene, 1,2-	50
Molybdenum	0.07	Dichloromethane	20
Selenium	0.01	Dioxane, 1,4-	50
Uranium	0.015	Edetic acid (EDTA)	600
Cadmium	0.003	Ethylbenzene	300
Cyanide	0.07	Hexachlorobutadiene	0.6
Mercury	0.006	Nitrilotriacetic acid (NTA)	200
Nitrate (NO_3^-)	50	Pentachlorophenol	9
Nitrite (NO_2^-)	3	Styrene	20
Antimony	20	Tetrachloroethene	40
Benzo[*a*]pyrene	0.7	Toluene	700
Copper	2000	Trichloroethene	20
Lead	10	Xylenes	500
Vinyl chloride	0.3	*Escherichia coli* (*E. coli*)	Must not be detectable in any 100 ml sample
Nickel	70		

TECHNOLOGIES FOR PURIFICATION OF WATER

Currently, many technologies are employed for water and wastewater treatment. These techniques should be known to get a better understanding of currently prevailing techniques used in water and wastewater treatment. Major technologies for water and wastewater treatment will be discussed in this section.

Coagulation-Flocculation

Inorganic salts (e.g. alum, $AlCl_3$, $FeCl_3$), polymers and organic polyelectrolytes are used for the coagulation-flocculation process (Matilainen, Vepsäläinen, and Sillanpää 2010). Because of its relatively low prices, it is growing in water treatment plants (WTPs). The efficiency of removal of this process is, however, relatively low.

Adsorption Processes

The term defines the adherence of a thin layer of molecules to fluid or solids surfaces with which they are in contact. Adsorption for organic material elimination, tertiary wastewater treatment, and groundwater remediation are commonly used in water treatment (Shen et al. 2019).

Advanced Oxidation Processes

Advanced oxidation methods (AOP) have gained more attention in the last decades in the research and development of wastewater treatment technologies and have been identified as technology using hydroxylic radical (^{o}OH) for oxidation (Bethi et al. 2016).

Electrodialysis Process

Electrodialysis (ED), under the influence of the relevant electrical potential separation, is used to transfer salt ions from one solution via the ion exchange membranes to another in an electrodialysis cell setup (Al-Amshawee et al. 2020).

Ion Exchange Resin

Ion exchange resins are small polymer beads with pores to capture and release ions. The reversible chemical reaction is an ion exchange where the dissolved ions have been withdrawn from the solution and replaced by other electric charge ions with the same or identical charges (Abdulgader, Kochkodan, and Hilal 2013).

Membrane Techniques

Membrane filtration is pressure-driven physical filtration processes where contaminated water is passed through membranes where the microorganisms and suspended particles are separated from the contaminated water. Various types of membrane filtration are MF (0.2 to 5 µm pore size), NF (0.001 to 0.01 µm pore size), UF (0.005 to 0.2 µm pore size) and RO (0.0001 to 0.001 µm pore size) (F. Xu et al. 2018). Other mentionable membrane techniques are FO and MD.

MOF BASED MEMBRANES

Membranes have taken on an ever more critical role in the separation and purification technology in recent decades. Polymer membrane production and operation technologies are relatively advanced due to the advantages such as cost-effectiveness, high processability and low power consumption. However, there is usually a trade-off between selectivity and permeability polymer membranes. To overcome this obstacle and due to their tunable and distinct pore size and porous nature, MOFs have been identified as perfect candidates for membrane purification applications. According to Li et al. c, MOFs membranes can be divided into; bare MOFs membranes, PMM, MMM, TFN membranes. These membranes have been successfully applied in various applications such as storage, chemical sensing, treatment, separation, and catalysis (Bétard and Fischer 2012; Canivet et al. 2014; Haldar, Duarah, and Purkait 2020; J. Li et al. 2020; Shekhah et al. 2011). Figure 1 shows the schematic illustration of the removal mechanism of contaminants in MOFs based membranes. In this section, we will only discuss MOFs membrane applications for water treatment and purification.

Pure MOFs Membranes

In this section, membranes made of pure MOFs are discussed. There has been little research on bare MOFs membranes used in water treatment because the pure, MOFs membranes have low mechanical strength and less stability (J. Li et al. 2020). These membranes are mostly prepared through *in situ* routes.

Porous Matrix Membranes (PMM)

In PMM, MOF MOF particles are filled in pores of the membranes which are removed when immersed in water. The pores thus formed increase the solvent diffusion upon the disintegration of the MOF particles. Lee et al. (J. Y. Lee et al. 2015; J. Y. Lee, Tang, and Huo 2014) have developed PMM in many of their studies for water treatment through UF and FO applications. In one of their studies, they reported MOFs for enhancing the porosity and interconnection of membranes for water purification. They achieved improved rejection due to increase in mass transfer (J. Y. Lee, Tang, and Huo 2014). They also fabricated MOFs filled polymer matrix through different processes viz. via phase inversions and layer-by-layer (LbL) deposition for FO applications. They were the first to control the internal concentration polarization (ICP) in the substrate layer in FO applications (J. Y. Lee et al. 2015). This study provides an additional dimension for ICP control in osmotically-driven membrane processes. Xu et al. (Y. Xu, Gao, Wang, Wang, Ji, Wang, et al. 2016) doped MIL-101(Cr) into dense selective polyamide (PA) layer on the polysulfone (PSf) for ultrafiltration (UF) application. The porous structures of MIL-101 (Cr) helped to create channels for water transport which led to an increase in permeance of water with high rejection. In another PMM study, Arjmadi et al. (Arjmandi et al. 2019) fabricated a MOFs based PMM using conventional casting method with magnetic MOFs in water as pore former for FO applications. Different polymers and MOFs loading with various size were studied. It was found that particle size, as well as loading % of MOFs, have an effect on membrane quality. The membranes prepared in this study could be used again, hence possess regeneration properties.

Figure 1. Schematic illustration of the removal mechanism of contaminants in MOFs based membranes

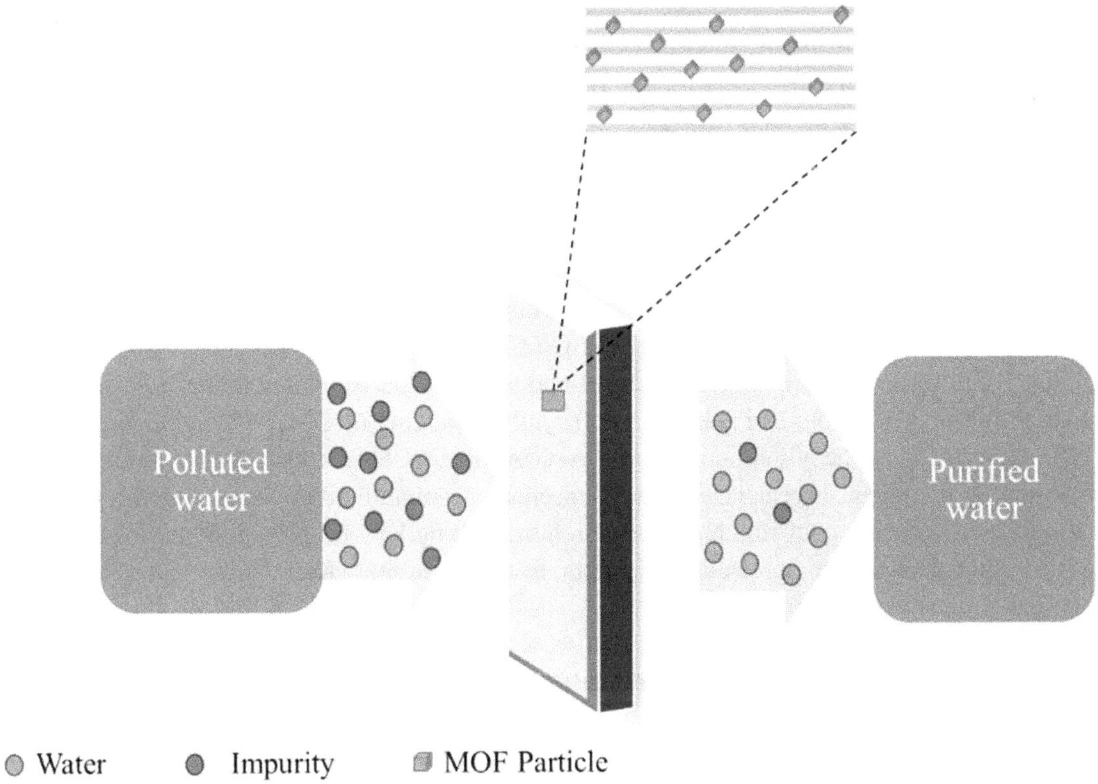

Mixed Matrix Membrane (MMM)

MOFs have been combined with polymers to give MMMs improved separation strength, as well as porous fillers with molecular sieving characteristics. A significant number of different membranes have been developed, and several MOFs/polymer hybrids tested and reported to demonstrate possible applications in water treatment and gas separation (Dechnik, Sumby, and Janiak 2017). In one of the first studies on MMM, Campbell et al. (Campbell et al. 2014) prepared HKUST-1 based MMM through *in situ* approach for separation of organic solvents. The support was submerged in HKUST-1 precursor solutions to grow HKUST-1 inside the pores of polymer membranes. They reported high rejections for styrene oligomers, high flux for these membranes. In another study, Ruan et al. (Ruan et al. 2016) precipitated MIL-53 (Al) in poly (m-phenylene isophthalamide) (PMIA) polymer matrix by phase inversion technique. The developed membranes had a high rejection for various dyes; nitroso-R salt (83.9%), xylenol orange (98.3%), and ponceau S (99.8%) mixed in Na_2SO_4 solution. Mohammadnezhad et al. (Mohammadnezhad, Feyzi, and Zinadini 2019) fabricated Ce(III) MOF/PES MMM via the phase inversion method. The MOFs based membrane showed high flux rates due to an increase in porosity and improved hydrophilicity of the membranes. The membranes were reported to have anti-fouling properties also.

Thin-Film Nanocomposite (TFN) Membranes

MOFs have been used to fabricate TFN membranes for improving membrane performance such as selectivity, flux and permeance. Duan et al. (Jintang Duan et al. 2015) fabricated TFN membranes by depositing ZIF-8 on PA film for desalination application. The membranes were reported to have better water permeance and high salt rejection compared to PA membranes. In another study, Goethem et al. (C. Van Goethem et al. 2016) (Cédric Van Goethem et al. 2018) used ZIF-8 as filler to fabricate a TFN membrane which showed a 220% increase in water permeance. Ma et al. (D. Ma et al. 2017) prepared UiO-66 NPs and embedded them into the PA layer to make a TFN membrane. The morphology was modified due to the introduction of UiO-66 NPs, which led to an improvement in rejection due to molecular sieving and superhydrophilic nature of UiO-66 NPs. He et al. (He et al. 2017) dispersed UiO-66 NPs into PA thin film to remove selenium (Se) and arsenic (As) from contaminated water. Recently, Golpur and Pakijeh (Golpour and Pakizeh 2018) synthesized UiO-66-NH_2 NPs and incorporated them onto PA layer over polyphenyl sulfone (PPSU)-graphene oxide (GO) support. The membranes were used for wastewater treatment. Thermal stability and mechanical strength improved considerably after incorporation of UiO-66-NH_2 confirming the formation of a strong UiO-66-NH_2-polymer matrix interface. The membrane showed a high rejection of kinetic hydrate inhibitor (KHI) (96%) and 57% increase in water flux compared to pristine membranes.

Synthesis of MOFs Based Membranes

This section discusses various fabrication methods for preparing MOF based membranes. It is to be noted that only those techniques will be discussed which are used to prepare MOFs membranes for water and wastewater treatment. Kadhom and Deng (Kadhom and Deng 2018) and Li (W. Li 2019) did an extensive review of different techniques of synthesis of MOFs membrane. The methods discussed in subsequent subsections are currently being used to fabricate MOFs based membranes for water and wastewater treatment. The advantages and disadvantages of these methods are outlined in Table 2.

MOFs Membranes Prepared by *in situ* Growth Route

The prepared route in situ is a membrane manufacturing method used to design continuous MOFs membranes for fluid separation. Zhang et al. (R. Zhang et al. 2014) developed ZIF-8/PSS membrane for NF application. The membrane was able to remove dye pollutants from the water with regeneration capacity. Recently, Wang et al. (X. ping Wang et al. 2020) fabricated ZIF-8 layer on PS membrane earlier modified by polydopamine (PDA). The introduction of ZIF-8 layer increase hydrophilicity of the membrane and increased water flux and antimicrobial performance (99% removal of *E. coli*).

MOFs Membranes Prepared by Blending Method

Blending is another key class of MOFs design strategy, in particular for MOFs-based MMMs. Ma et al. (J. Ma et al. 2017) developed UiO-66-GO/PES membranes through blending method. The developed membranes had high hydrophilic nature combine with high water flux and solute rejection. Similarly, Yang et al. (S. Yang et al. 2019) prepared HKUST-1-GO/CA membranes for UF application with 183.51 L m-2 h-1 water flux. The membranes also possessed anti-fouling properties. In another study, Ren et al. (Y. Ren et al. 2019) synthesized MIL-PVDF membranes for UF applications for removal of MB dye from real wastewater. Very recently, Li et al. (T. Li et al. 2020) designed PVA/Cu-iMOFs/PVDF UF membranes to remove perchlorate (ClO^{4-}) from the aqueous stream through the blending method. The membrane achieved removal rate of 99.6 5% over a wide pH range (3-10). The membrane was reported to have anti-fouling properties and highly stable.

MOFs Membranes Prepared by Interfacial Polymerization

The interfacial polymerization process used as a classic method for producing a thin, active film surface layer for NF and RO, by integrating the MOFs into the selective layer, technically improved to design advanced NF / RO composite membranes. Xu et al. (Y. Xu, Gao, Wang, Wang, Ji, and Gao 2016) fabricated MIL-101(Cr) TFN membranes through *in situ* interfacial polymerization on polyethersulfone (PES) support. The introduction of MIL-101(Cr) into PES support created water channels which led to an increase in permeation of membranes. In another study, Kadhom et al. (Kadhom, Hu, and Deng 2017) used UiO-66 and MIL-125 nanoparticles (NPs) to increase the performance of the TFC membrane for desalination applications. At optimum loading of UiO-66 and MIL-125 NPs, enhanced flux and high rejection were reported. Goethem et al. (C. Van Goethem et al. 2016) used ZIF-8 as filler to fabricate a TFN membrane through the interfacial polymerization, which showed a 220% increase in water permeance. The fillers were carefully positioned using the evaporation-controlled filler positioning (EFP) at the solvent/water interface.

MOFs Membranes Prepared by the Layer-By-Layer (LbL) Method

Khanjani and Morsali (Khanjani and Morsali 2014)synthesized MOF-5 on silk fibre through LbL method. The membrane was successfully used to remove Congo red dye with 50% rejection. Wang et al. (L. Wang et al. 2015)prepared nanocomposite membranes layers of ZIF-8 and PA on a porous substrate. The membrane achieved the flux of 11.2 kg/m2/h and rejection of 99.6% for dye removal with high selectivity. Basu and Balakrishnan (Basu and Balakrishnan 2017) synthesized ZIF-8/PSf membrane by layer by layer method for removal of paracetamol drug.

MOFs Membranes Prepared by Hydro/Solvothermal Method

The first suggested method of MOFs membrane synthesis was hydro/solvothermal production, and it is widely utilized. This approach is achieved by the immersion of substrates with both metal sources and linkers in precursor solutions. Liu et al. (X. Liu et al. 2015) fabricated polycrystalline membranes by depositing UiO-66 on alumina hollow fibres. The membrane was able to reject various ions (Ca^{2+}, Mg^{2+}, Al^{3+}). The membranes were found to be very stable even at high flow rates (170 l/h). MIL-68(Al)/PVDF membranes have been reported to have good adsorption capacity for p-nitrophenol (PNP) and methylene blue (MB) with high regeneration capacity (Tan et al. 2019). In another study, Gu et al. (Y. Gu et al. 2020) embedded NH^2-MIL-53 crystals onto wood-carbon (WC) substrate for removal of Pb(II) from water. The membrane was reported to show high uptake capacity and selectivity due to perfect integration of pore structures, a large number of binding sites, and affinity of amino functional groups towards Pb(II). Recently Yahaya et al. (Yahaya et al. 2020) embedded MIL-140B on a ceramic membrane surface by hydro/solvothermal synthesis. The membrane was used in FO applications, and excellent water flux (12.023 L/m2 h) were reported. Very recently, Xie et al. (Xie et al. 2020b) prepared PVDF/NH2-MIL-88B(Fe) membranes with anti-fouling properties for oil/water separation. The membrane was reported to have high mechanical strength, good hydrophilicity and recyclability which makes it a perfect candidate for potential wastewater treatment applications

MOFs Membranes Prepared by Electrodeposition Method

ZnO nanorods (NRs) have been electrodeposited onto the substrate surface to guide *in situ* growth of ZnZIF-8, CoZIF-67 and HKUST-1 (Zhou, Wei, Hou, et al. 2017) (Zhou, Wei, Zhuang, et al. 2017). The membranes prepared were reported to have reactivity comparable to organic functional groups while being cost-effective and environment friendly. Alizadeh and Nematollahi (Alizadeh and Nematollahi 2019) synthesized $Zn_3(BTC)_2$ and $Cu_3(BTC)_2$ thin films through the electrodeposition method. Wu et al. (Wu et al. 2017) prepared ZIF-8/ZnO/2-methylimidazole with large surface area and good thermal stability. The developed membrane showed high adsorption capacity for acidic drugs (e.g., ibuprofen, ketoprofen and acetylsalicylic acid) due to hydrophobic interactions and hydrogen bonding. In another study, Wu et al. (Wu et al. 2017) developed a ZIF based membrane with large surface area and good thermal stability through electrodeposition process. The developed membrane has high adsorption capacity for acidic drugs.

MOFs Based Membrane Prepared by Vacuum Filtration

Cheng et al. (J. Cheng et al. 2018) used a vacuum filtration technique to fabricate 2D Co-MOF/GO membranes for removal of Cs^+ from nuclear wastewater. The membrane proved to be efficient for elimination of radionuclide contamination. In another study, Cao et al. (Cao et al. 2018) synthesised hydrophilic UiO-66-NH_2/PAA *via the* vacuum-assisted process. The membrane exhibited high rejection (99%) for oil-water separation and anti-fouling properties. Zhang et al. (P. Zhang et al. 2018) prepared UiO-66-$(COOH)_2$/prGO membranes with nanochannels structure, high surface charge and stability for heavy metal ions removal. The membranes were able to remove congo red (98.2%), MB (92.55%), Cu^{2+} (96.5%) and for Cd^{2+} (92.6%). Xie et al. (Xie et al. 2020a) synthesised GO/MIL-88A(Fe) membrane by the vacuum filtration method. The membranes were used to treat real textile wastewater with high degradation efficiency for MB (98.81%) and BPA (97.27%). Recently, Gu et al. (Q. Gu et al. 2019) fabricated Zn/Co-ZIF-L based membranes for water and wastewater treatment. The membrane was reported to contain antibacterial and high-water flux.

Table 2. Advantages and disadvantages of different methods of fabricating MOFs

Method	Advantages	Disadvantages	Reference
in situ route	Good performance, compatibility, wide range of functionalization, uniform dispersion	High cost, weak intensity ofcharacteristic bands	(R. Zhang et al. 2014) (Denny and Cohen 2015) (Basu and Balakrishnan 2017)
Blending method	Non-destructive, cheap, high performance, easy	High cost, long reaction time	(S. Yang et al. 2019)(X. Li et al. 2017)
Interfacial polymerization	Tunable pore size and better interconnectivity of pores	High cost, long reaction time	(Seman, Khayet, and Hilal 2010)(Ameloot et al. 2011)
Layer by layer	compatibility, uniform dispersion	Poor stability, poor reproducibility	(Chen et al. 2015; Escobar-Ferrand et al. 2014; N. Wang et al. 2016)
Hydro/solvo-thermal	Promote nucleation growth, high crystallinity	Large amount solvents, long reaction time	(Jingui Duan et al. 2013, 2017, 2018)
Electrodeposition	Efficient, low-cost, and environmentally friendly	Problematic for large-scale membranes	(Zhao et al. 2015)
Vacuum filtration	Simple, facile	limited size and shape	(Zhao et al. 2015)

APPLICATIONS OF MOF BASED MEMBRANES

For membrane applications, MOFs membranes may constitute an important direction. Applications of MOFs-based membranes for liquid separation are growing due to their flexible compositions and controllable structures. Both MOFs membranes and MOFs MMM are multifunctional, have been developed for flexible and excellent performance in separation beyond the conventional applications for pervaporation and removal of colour and also contribute to advances in sectors ranging from catalytic deterioration, fouling resistance, oil-water separation. The successful production of water-stable MOFs has led to the radical separation of water treatment from MOFs-based membranes. The applications of MOFs based membranes for water and wastewater treatment may be divided into the following:

Forward Osmosis (FO)

In one of the first reports, Lee et al. (J. Y. Lee et al. 2015) fabricated FO PMM using MOFs particles as removable fillers in polyacrylonitrile (PAN) dope solution. The membrane showed improved mass transfer and ICP. In another study, Arjmadi et al. (Arjmandi et al. 2019) developed FO PMM membranes using MOFs as pore former with different polymers. The membrane showed great fluxes with Caspian seawater (117 L/m2.h), orange juice concentration (98 L/m2.h) and deionized (DI) water (141 L/m2.h) with high permeance. Ma et al. (D. Ma et al. 2017) prepared UiO-66 NPs and embedded them into the PA layer to make a TFN membrane. This membrane showed high rejection due to molecular sieving and superhydrophilic nature of UiO-66 NPs. The application of the membranes in FO resulted in an increase in water flux by 40% and 52% increase in permeability with high NaCl rejection (95%). Recent studies on the performance of the FO membrane in water treatment using MOFs are summarized in Table 3.

Table 3. Recent studies on the performance of FO membrane in water treatment using MOFs

MOF	Substrate	Target contaminant(s)	Rejection (%)	Reference
UiO-66	PSf	Phenol red	99	(T. Y. Liu et al. 2018)
UiO-66	PSf	Na_2SO_4 rejection	94–96	(T. Y. Liu et al. 2018)
UiO-66	PSf	NaCl	27	(T. Y. Liu et al. 2018)
ZIF-8	PES	Cu^{2+}, Ni^{2+}, Pb^{2+}	96	(Qiu and He 2019)

Reverse Osmosis (RO)

In MOFs based RO membranes, ZIF-8 has mostly been used due to advantages such as faster water transport within the framework and better compatibility with the polymer matrix. In one of the first studies in MOFs based RO membranes for water purification, Hu et al. (Hu, Chen, and Jiang 2011) performed a molecular simulation study in ZIF-8 membrane. Due to the sieving effect of small apertures in ZIF-8, Na+ and Cl- ions were not able to transport through the membrane. From this simulation study, it was suggested that ZIF-8 could be used in RO membranes for better water purification. Thereafter, Duan et al. (Jintang Duan et al. 2015) developed a hydrophobic, hydrothermally stable ZIF-8 RO membrane on to polyamide (PA) layer. They reported very high permeance (3.35 L m^{-2} h^{-1}) with high NaCl rejection from nano-sized (200 nm) ZIF-8 membranes. This reported value was much greater (162%) than the permeance obtained from pristine PA membranes. The reason attributed to the increase in permeance was better hydrophilicity and lesser cross-linking in ZIF-8 membranes as compared to the pristine PA membranes. ZIF-8 use in RO membranes for better water purification was experimentally verified from this study. In another study, Bonnett et al. (Bonnett et al. 2020) prepared membranes by binding myristic acid (MA) and NRs of PCN-222 through coordination modulation and further modified them via dispersion in the aqueous phase during interfacial polymerization. The PCN-222/MA membrane was characterized by increased water flux (up to 95%). The increased flux was due to improved water transport pathways at the NR-MA interface and through NR pore structures. Xu et al. (Y. Xu, Gao, Wang, Wang, Ji, Wang, et al. 2016) doped MIL-101(Cr) NPs in PA onto PSf support. The measured water permeance for the doped membrane was 2.2 L/m2·h·bar, which was 44% higher than the undoped PA membranes with 99% NaCl rejection. Recent studies on the performance of the RO membrane in water treatment using MOFs are summarized in Table 4.

Membrane Distillation (MD)

MD is a new technology that is a combination of membrane and distillation techniques. Several studies have been reported on MOFs based membranes in MD for improved purification of water. Among first such studies, Zuo and Chung (Zuo and Chung 2016) prepared a hydrophobic membrane by growing MOFs crystals on the alumina substrate and introduced perfluoro molecules on the surface of the membrane. The membrane possessed advantages of alumina substrate (high stability and high-water permeability) and perfluoro-MOFs (hydrophobicity). They obtained a very high flux of 32.3 L m^{-2} h^{-1}. Yang et al. (F. Yang et al. 2018) fabricated superhydrophobic (contact angle 138.06 ± 2.18°) poly(vinylidene fluoride)/ MOFs (iron 1,3,5- benzenetricarboxylate) membranes via electrospinning. They found the membrane

Table 4. Recent studies on the performance of RO membrane in water treatment using MOFs

MOFs	Substrate	Target contaminant(s)	Rejection (%)	Reference
MIL-101(Cr)	PSf	NaCl	99	(Y. Xu, Gao, Wang, Wang, Ji, Wang, et al. 2016)
ZIF-8	PSf	NaCl	99.2	(T. H. Lee et al. 2019)
ZIF-8	PSf	NaCl	-	(Jintang Duan et al. 2015)
ZIF-8	PSf	NaCl	99.4	(Aljundi 2017)
UiO-66 and MIL-125	PSf	NaCl	98.5	(Kadhom, Hu, and Deng 2017)
NH_2-MIL-88	PSf	Phenol	91 - 92	(L. Wang et al. 2016)

suitable for direct contact MD process with 2.87 kg/m2 h of water flux and 99.99% NaCl rejection. Donato et al. (Donato et al. 2020) prepared an MFI based membrane using a calcination process. The membrane was tested in vacuum MD for desalination application. They achieved 20.6 kgm-2h-1 flux combined with up to 99.9% NaCl rejection at 70 °C. Very recently, Kebria et al. (Kebria et al. 2019) prepared MD membranes by coating ZIF-8 on polyvinylidene fluoride (PVDF). They observed an increase in liquid entry pressure (LEP) of the membrane due to the development of ZIF-8 layer on the PVDF surface. Increase in water flux (350%) and NaCl rejection (99.5%) was observed as compared to unmodified membranes. The MOFs membrane was also found to possess anti-fouling property. From molecular simulations, it was found that there exists good compatibility of ZIF-8 with PVDF, hence resulting in good performance of the membrane. Recent studies on the performance of MD membrane in water treatment using MOFs are summarized in Table 5.

Table 5. Recent studies on the performance of MD membrane in water treatment using MOFs

MOFs	Substrate	Target contaminant(s)	Rejection (%)	Reference
ZIF-8	PVDF	NaCl	99.5	(Kebria et al. 2019)
Fe-BTC	PVDF	NaCl	99.99	(F. Yang et al. 2018)

Nanofiltration (NF)

Sorribas et al. (Sorribas et al. 2013) prepared TFN membranes by homogeneously dispersing ZIF-8, MIL-53, NH_2-MIL-53 and MIL-101 NPs in PA via *in situ* interfacial polymerizations. Permeance and rejection of the fabricated membranes were tested using methanol and tetrahydrofuran (THF). Increase in permeance was observed with an increase in pore size and porosity of the MOFs (fillers) used in preparing membranes. In a similar report, Ma et al. (X. H. Ma et al. 2017) fabricated an NF membrane by dispersing NH_2-MIL-101(Al) and NH_2-MIL-101(Cr) in the chitosan polymeric matrix. NH_2-MIL-101(Al) Membranes were able to provide twice the flux obtained from NH_2-MIL-101(Cr). The order of salt rejection from these membranes was the order of $MgCl_2$ > $CaCl_2$ > NaCl > Na_2SO_4. In another study, Campbell et al. (Campbell et al. 2015) fabricated TFN membranes using HKUST-1 on PA-P84 through interfacial synthesis. Trinh et al. (Trinh, Tran, and Taniike 2017) synthesized an NF membrane by depositing UiO-66 onto cellulose thin film. The developed membrane showed perfect rejection of methylene blue (MB) with high flux compared to commercial polymer membranes. Yang et al. (L. Yang, Wang, and Zhang 2017) developed an NF membrane by depositing ZIF-8 particles onto polyethyleneimine (PEI) film via interfacial reaction. The membranes showed 99.6% rejection for MB with up to 33.0 L m^{-2} h^{-1} bar^{-1} flux. Recent studies on the performance of NF membrane in water treatment using MOFs are summarized in Table 6.

Ultrafiltration (UF)

MOFs are intensively studied as polymer membrane filling materials primarily on account of their high polymer affinity, large pores volumes and increasing pore functionalities, but MOFs dependent ultrafiltration (UF) membranes have grown lagging behind for water treatment. In one of the first studies on MOFs based UF membranes, Yin et al. (Yin et al. 2016) prepared membranes from UiO-66-NH_2 for removal of Pb(II) from wastewater. They achieved 61.4% removal of Pb(II) with high flux. The mechanism of Pb(II) removal was due to the presence of coordination between Pd(II) and amino (-NH_2) group. The membrane also showed anti-fouling properties and 100% regeneration. In another study, Ma et a. (J. Ma et al. 2017) deposited UiO-66 on GO films in PES. The membrane showed high hydrophilicity and good water purification performance. The water flux increased by 351% and 78% compared to PES and GO/PES membranes. The fabricated membranes had anti-fouling properties also. Sun et al. (Sun, Tang, and Wu 2017) fabricated UiO-66 functionalized poly(sulfobetaine methacrylate) (PSBMA)/PSf membranes through the phase-inversion method. 602 L m^{-2} h^{-1} water flux was observed for this membrane which is 2.5 times that of the pristine PSf membrane and 2 times that of UiO-66-NH_2/PSf membrane. The increase in flux is due to superhydrophobic nature of UiO-66/(PSBMA. In one of the studies in UF membranes, Liu et al. (R. Liu, Sui, and Wang 2019) fabricated a MMM using NH2-MIL-101 in PES polymer matrix via *in situ* method. The membrane showed improvement in porosity with better hydrophilicity compared to PES membranes. They were able to reject phosphate with 100% rejection using this membrane. Recent studies on the performance of UF membrane in water treatment using MOFs are summarized in Table 7.

1.1 Microfiltration (MF)

MF technique is an important technique to remove micropollutants of the size range 0.2 to 5 μm. There have been very few studies on MOFs based membranes for making MF membranes for water treatment. Ragab et al. (Ragab et al. 2016) fabricated an MF membrane from ZIF-8 and polytetrafluoroethylene (PTFE). Due to the incorporation of ZIF-8, there was a 40% increase in the membrane adsorption capacity compared to unmodified PTFE membrane. In another study, Guo et al. (Guo et al. 2016) deposited ZIF-8 thin film PVDF hollow fibre to remove Rhodamine B dye from water. Few recent studies on the performance of MF membrane in water treatment using MOFs are summarized in Table 8.

Table 6. Recent studies on the performance of NF membrane in water treatment using MOFs

MOFs	Substrate	Target contaminant(s)	Rejection (%)	Reference
ZIF-8	PAN	Reactive black 5	99	(Zhu et al. 2017)
ZIF-8	PAN	Reactive blue 2	99	(Zhu et al. 2017)
ZIF-8	PAN	Methyl blue	98.6	(R. Zhang et al. 2014)
ZIF-8	PSf	Acetaminophen rejection	55	(Basu and Balakrishnan 2017)
UiO-66	PAN	Rose Bengal	100	(X. Cheng et al. 2017)
UiO-66	PAN	Azithromycin rejection	96.7	(X. Cheng et al. 2017)
MOF-5	PES, CA, and PVDF	Co^{2+}	75	(Gnanasekaran et al. 2019)
MOF-5	PES, CA, and PVDF	Cu^{2+}	55	(Gnanasekaran et al. 2019)
UiO-66	PES	SeO_3^{2-}	96.5	(He et al. 2017)
UiO-66	PES	SeO_4^{2-}	97.4	(He et al. 2017)
UiO-66	PES	$HAsO_4^{2-}$	98.6	(He et al. 2017)
ZIF-8	PES	Rose Bengal	85	(Y. Li et al. 2017)
ZIF-8	PSf	Na_2SO_4	95.2	(Liao et al. 2019)
Ce-MOF	PES	Dye removal	99	(Mohammadnezhad, Feyzi, and Zinadini 2019)
MIL-101(Cr)	m-phenylenediamine-PA P84	Rose Bengal	90	(Navarro et al. 2018)
HKUST-1 and ZIF-93	m-phenylenediamine-PA P84	diclofenac and naproxen	98	(Paseta et al. 2019)
IRMOF-3	PSf	Cu^{2+}	90	(Rao et al. 2017)
MIL-53(Al)	PMIA	Nitroso-R salt	83.9	(Ruan et al. 2016)
MIL-53(Al)	PMIA	Xylenol orange	98.3	(Ruan et al. 2016)
MIL-53(Al)	PMIA	ponceau S	99.8	(Ruan et al. 2016)
ZIF-8	Ceramic tube	Methyl blue rejection	98.6	(N. Wang et al. 2016)
BUT-8(A)	PAN	Methylene blue	98.3	(Meng et al. 2019)
BUT-8(A)	PAN	Congo red	99.8	(Meng et al. 2019)
BUT-8(A)	PAN	Acid fuchsin	89.3	(Meng et al. 2019)
BUT-8(A)	PAN	Methyl orange	82.1	(Meng et al. 2019)
UiO 66 NH_2	PES	KHI rejection	96	(Golpour and Pakizeh 2018)
ZIF-8	PSf	Congo red	99	(L. Wang et al. 2015)
MIL-88A(Fe)	PVDF	Methylene blue	98.81	(Xie et al. 2020a)
MIL-88A(Fe)	PVDF	Bisphenol A	97.27	(Xie et al. 2020a)
UiO-66-$(COOH)_2$	Nylon	Congo red	98.2	(P. Zhang et al. 2018)
UiO-66-$(COOH)_2$	Nylon	Methylene blue	92.55	(P. Zhang et al. 2018)
Zn-TCP(Fe)	Poly (ethylenimine) and Poly (di allyl dimethyl ammonium chloride)	Methyl red	-	(Ang and Hong 2017)
Zn-TCP(Fe)		Methyl orange	-	(Ang and Hong 2017)
Zn-TCP(Fe)		Brilliant blue G	-	(Ang and Hong 2017)
UiO-66	Regenerated cellulose	Methylene blue	100	(Trinh, Tran, and Taniike 2017)
Co-MOF	Organic membrane filter	Cs^{+}	88.4	(J. Cheng et al. 2018)
ZIF-8	PES	Malachite green	99.6	(Maroofi and Mahmoodi 2019)
ZIF-8	PES	Bovine serum albumin	100	(S. Zhang et al. 2020)
NH_2-MIL-101(Al) and NH_2-MIL-101(C	PSf	$MgCl_2$	93	(X. H. Ma et al. 2017)
	PSf	$CaCl_2$	86.5	(X. H. Ma et al. 2017)
ZIF-8	PEI	Methyl blue	99.6	(L. Yang, Wang, and Zhang 2017)

Table 7. Recent studies on the performance of UF membrane in water treatment using MOFs

MOFs	Substrate	Target contaminant(s)	Rejection (%)	Reference
ZIF-8	PVDF	Direct Yellow 12	89.65	(Karimi et al. 2019)
ZIF-8	PVDF	Reactive Blue 21	98.32	(Karimi et al. 2019)
UiO-66	PES	Humic acid	99	(Abdullah et al. 2018)
TMU-5	PES	Oil–water removal	98	(Gholami et al. 2018)
MIL-53, Cu-BTC, MIL-100	PAN	Humic acid Removal	90	(J. Y. Lee, Tang, and Huo 2014)
ZIF-8	PVDF	Ni(II) removal	-	(T. Li et al. 2018)
UiO-66	PES	Direct Red	80	(J. Ma et al. 2017)
Cu(TPA)	PES	Congo Red	80	(Makhetha and Moutloali 2018)
Cu(TPA)	PES	Methyl Orange	50	(Makhetha and Moutloali 2018)
Cu(TPA)	PES	Methylene Blue	20	(Makhetha and Moutloali 2018)
MIL-101(Cr)	PAN	Bisphenol A	92	(Z. Ren, Luo, and Wan 2018)
MIL-53(Fe)	PVDF	Methylene blue	95	(Y. Ren et al. 2019)
UiO-66	PSf	Bovine serum albumin	98.5	(Sun, Tang, and Wu 2017)
ZIF-8	PSf	Bovine serum albumin	98	(Sun, Tang, and Wu 2018)
HKUST-1	CA	Bovine serum albumin	98	(S. Yang et al. 2019)
UiO-66-NH_2	Zirconia ceramic	Pb^{2+}	100	(Yin et al. 2016)
Cu-MOFs	PVDF	ClO^{4-}	99.6	(T. Li et al. 2020)
NH2-MIL-101	PES	Phosphate	100	(R. Liu, Sui, and Wang 2019)

Table 8. Recent studies on the performance of MF membrane in water treatment using MOFs

MOFs	Substrate	Target contaminant(s)	Rejection (%)	Reference
ZIF-8	PTFE	Micropollutants	95%	(Ragab et al. 2016)
MOF-808(Zr)	PAN	Cd^{2+} and Zn^{2+}	-	(Efome et al. 2018)
ZIF-8	PVDF	Rhodamine B	97.5	(Guo et al. 2016)

CONCLUSION

The MOFs have shown some impressive results to improve the performance of membranes because of their structure and properties. There have been increasing attempts to use MOFs for water purification. Only a limited portion of these works display an increased permeability and selectivity. While the MOFs based membranes have all the advantages, limitations such as they have their own complexities and difficulties, limiting their large-scale use and manufacture. Some of which are the high cost, difficult synthesis techniques, stability and regeneration. A deeper understanding of interface chemistry is needed to develop MOFs for membrane applications, so that membrane properties can be correlated more closely with the functionality of MOFs. In various water treatment procedures, we expect the rapid growth of MOF applications and could lead to further improvement of membrane efficiency in new type of MOFs.

ACKNOWLEDGEMENT

The authors are thankful to CSIR-CSMCRI for the support in preparation of this work. CSIR-CSMCRI PRIS no. 27/2021.

REFERENCES

Abdulgader, H. A., Kochkodan, V., & Hilal, N. (2013). Hybrid Ion Exchange - Pressure Driven Membrane Processes in Water Treatment: A Review. *Separation and Purification Technology*, *116*, 253–264. doi:10.1016/j.seppur.2013.05.052

Abdullah, N., Rahman, M. A., Dzarfan Othman, M. H., Jaafar, J., & Aziz, A. A. (2018). Preparation, Characterizations and Performance Evaluations of Alumina Hollow Fiber Membrane Incorporated with UiO-66 Particles for Humic Acid Removal. *Journal of Membrane Science*, *563*(May), 162–174. doi:10.1016/j.memsci.2018.05.059

Abu Seman, M. N., Khayet, M., & Hilal, N. (2010). Nanofiltration Thin-Film Composite Polyester Polyethersulfone-Based Membranes Prepared by Interfacial Polymerization. *Journal of Membrane Science*, *348*(1–2), 109–116. doi:10.1016/j.memsci.2009.10.047

Al-Amshawee, S., Yunus, M. Y. B. M., Azoddein, A. A. M., Hassell, D. G., Dakhil, I. H., & Hasan, H. A. (2020). Electrodialysis Desalination for Water and Wastewater: A Review. *Chemical Engineering Journal*, *380*, 122231. doi:10.1016/j.cej.2019.122231

Alizadeh, S., & Nematollahi, D. (2019). Convergent and Divergent Paired Electrodeposition of Metal-Organic Framework Thin Films. *Scientific Reports*, *9*(1), 1–13. doi:10.103841598-019-50390-y PMID:31586078

Aljundi, I. H. (2016, December). Desalination Characteristics of TFN-RO Membrane Incorporated with ZIF-8 Nanoparticles. *Desalination*, *420*, 12–20. doi:10.1016/j.desal.2017.06.020

Ameloot, R., Vermoortele, F., Vanhove, W., Roeffaers, M. B. J., Sels, B. F., & De Vos, D. E. (2011). Interfacial Synthesis of Hollow Metalâ€"organic Framework Capsules Demonstrating Selective Permeability. *Nature Chemistry*, *3*(5), 382–387. doi:10.1038/nchem.1026 PMID:21505497

Ang, H., & Hong, L. (2017). Polycationic Polymer-Regulated Assembling of 2D MOF Nanosheets for High-Performance Nanofiltration. *ACS Applied Materials & Interfaces*, *9*(33), 28079–28088. doi:10.1021/acsami.7b08383 PMID:28752999

Arjmandi, M., Peyravi, M., Chenar, M. P., & Jahanshahi, M. (2019). A New Concept of MOF-Based PMM by Modification of Conventional Dense Film Casting Method: Significant Impact on the Performance of FO Process. *Journal of Membrane Science*, *579*(February), 253–265. doi:10.1016/j.memsci.2019.02.020

Basu, S., & Balakrishnan, M. (2017). Polyamide Thin Film Composite Membranes Containing ZIF-8 for the Separation of Pharmaceutical Compounds from Aqueous Streams. *Separation and Purification Technology*, *179*, 118–125. doi:10.1016/j.seppur.2017.01.061

Bétard, A., & Fischer, R. A. (2012). Metal-Organic Framework Thin Films: From Fundamentals to Applications. *Chemical Reviews*, *112*(2), 1055–1083. doi:10.1021/cr200167v PMID:21928861

Bethi, Sonawane, Bhanvase, & Gumfekar. (2016). Nanomaterials-Based Advanced Oxidation Processes for Wastewater Treatment: A Review. *Chemical Engineering and Processing - Process Intensification, 109*, 178–89. https://linkinghub.elsevier.com/retrieve/pii/S0255270116303312

Bonnett, B. L., Smith, E. D., De La Garza, M., Cai, M., Haag, J. V. IV, Serrano, J. M., Cornell, H. D., Gibbons, B., Martin, S. M., & Morris, A. J. (2020). PCN-222 Metal-Organic Framework Nanoparticles with Tunable Pore Size for Nanocomposite Reverse Osmosis Membranes. *ACS Applied Materials & Interfaces*, *12*(13), 15765–15773. doi:10.1021/acsami.0c04349 PMID:32163702

Campbell, J., Davies, R. P., Braddock, D. C., & Livingston, A. G. (2015). Improving the Permeance of Hybrid Polymer/Metal-Organic Framework (MOF) Membranes for Organic Solvent Nanofiltration (OSN)-Development of MOF Thin Films via Interfacial Synthesis. *Journal of Materials Chemistry. A, Materials for Energy and Sustainability*, *3*(18), 9668–9674. doi:10.1039/C5TA01315A

Campbell, J., Székely, G., Davies, R. P., Braddock, D. C., & Livingston, A. G. (2014). Fabrication of Hybrid Polymer/Metal Organic Framework Membranes: Mixed Matrix Membranes versus in Situ Growth. *Journal of Materials Chemistry. A, Materials for Energy and Sustainability*, *2*(24), 9260–9271. doi:10.1039/C4TA00628C

Canivet, J., Fateeva, A., Guo, Y., Coasne, B., & Farrusseng, D. (2014). Water Adsorption in MOFs: Fundamentals and Applications. *Chemical Society Reviews*, *43*(16), 5594–5617. doi:10.1039/C4CS00078A PMID:24875439

Cao, J., Su, Y., Liu, Y., Guan, J., He, M., Zhang, R., & Jiang, Z. (2018). Self-Assembled MOF Membranes with Underwater Superoleophobicity for Oil/Water Separation. *Journal of Membrane Science*, *566*, 268–277. doi:10.1016/j.memsci.2018.08.068

Chen, Q., Yu, P., Huang, W., Yu, S., Liu, M., & Gao, C. (2015). High-Flux Composite Hollow Fiber Nanofiltration Membranes Fabricated through Layer-by-Layer Deposition of Oppositely Charged Cross-linked Polyelectrolytes for Dye Removal. *Journal of Membrane Science*, *492*, 312–321. doi:10.1016/j.memsci.2015.05.068

Cheng, J., Liang, J., Dong, L., Chai, J., Zhao, N., Ullah, S., Wang, H., Zhang, D., Imtiaz, S., Shan, G., & Zheng, G. (2018). Self-Assembly of 2D-Metal-Organic Framework/Graphene Oxide Membranes as Highly Efficient Adsorbents for the Removal of Cs+ from Aqueous Solutions. *RSC Advances*, *8*(71), 40813–40822. doi:10.1039/C8RA08410F

Cheng, X., Jiang, X., Zhang, Y., Lau, C. H., Xie, Z., Ng, D., Smith, S. J. D., Hill, M. R., & Shao, L. (2017). Building Additional Passageways in Polyamide Membranes with Hydrostable Metal Organic Frameworks to Recycle and Remove Organic Solutes from Various Solvents. *ACS Applied Materials & Interfaces*, *9*(44), 38877–38886. doi:10.1021/acsami.7b07373 PMID:29022696

Cook, T. R., Yao, R. Z., & Stang, P. J. (2013). Metal-Organic Frameworks and Self-Assembled Supramolecular Coordination Complexes: Comparing and Contrasting the Design, Synthesis, and Functionality of Metal-Organic Materials. *Chemical Reviews*, *113*(1), 734–777. doi:10.1021/cr3002824 PMID:23121121

Dechnik, J., Sumby, C. J., & Janiak, C. (2017). Enhancing Mixed-Matrix Membrane Performance with Metal-Organic Framework Additives. *Crystal Growth & Design*, *17*(8), 4467–4488. doi:10.1021/acs.cgd.7b00595

Denny, M. S. Jr, & Cohen, S. M. (2015). In Situ Modification of Metal-Organic Frameworks in Mixed-Matrix Membranes. *Angewandte Chemie International Edition*, *54*(31), 9029–9032. doi:10.1002/anie.201504077 PMID:26073065

Donato, L., Garofalo, A., Drioli, E., Alharbi, O., Aljlil, S. A., Criscuoli, A., & Algieri, C. (2020). Improved Performance of Vacuum Membrane Distillation in Desalination with Zeolite Membranes. *Separation and Purification Technology*, *237*, 116376. doi:10.1016/j.seppur.2019.116376

Duan, J., Higuchi, M., Horike, S., Foo, M. L., Rao, K. P., Inubushi, Y., Fukushima, T., & Kitagawa, S. (2013). Methane Separation: High CO2 /CH4 and C2 Hydrocarbons/CH4 Selectivity in a Chemically Robust Porous Coordination Polymer (Adv. Funct. Mater. 28/2013). *Advanced Functional Materials*, *23*(28), 3481. doi:10.1002/adfm.201370138

Duan, J., Higuchi, M., Zheng, J., Noro, S., Chang, I.-Y., Hyeon-Deuk, K., Mathew, S., Kusaka, S., Sivaniah, E., Matsuda, R., Sakaki, S., & Kitagawa, S. (2017). Density Gradation of Open Metal Sites in the Mesospace of Porous Coordination Polymers. *Journal of the American Chemical Society*, *139*(33), 11576–11583. doi:10.1021/jacs.7b05702 PMID:28747050

Duan, J., Pan, Y., Liu, G., & Jin, W. (2018). Metal-Organic Framework Adsorbents and Membranes for Separation Applications. *Current Opinion in Chemical Engineering*, *20*, 122–131. doi:10.1016/j.coche.2018.04.005

Duan, J., Pan, Y., Pacheco, F., Litwiller, E., Lai, Z., & Pinnau, I. (2015). High-Performance Polyamide Thin-Film-Nanocomposite Reverse Osmosis Membranes Containing Hydrophobic Zeolitic Imidazolate Framework-8. *Journal of Membrane Science*, *476*, 303–310. doi:10.1016/j.memsci.2014.11.038

Efome, J. E., Rana, D., Matsuura, T., & Lan, C. Q. (2018). Insight Studies on Metal-Organic Framework Nanofibrous Membrane Adsorption and Activation for Heavy Metal Ions Removal from Aqueous Solution. *ACS Applied Materials & Interfaces*, *10*(22), 18619–18629. doi:10.1021/acsami.8b01454 PMID:29763287

Elrasheedy, A., Nady, N., Bassyouni, M., & El-Shazly, A. (2019). Metal Organic Framework Based Polymer Mixed Matrix Membranes: Review on Applications in Water Purification. *Membranes*, *9*(7), 88. doi:10.3390/membranes9070088 PMID:31330993

Escobar-Ferrand, L., Li, D., Lee, D., & Durning, C. J. (2014). All-Nanoparticle Layer-by-Layer Surface Modification of Micro- and Ultrafiltration Membranes. *Langmuir*, *30*(19), 5545–5556. doi:10.1021/la5002489 PMID:24568094

Ezugbe, E. O., & Rathilal, S. (2020). Membrane Technologies in Wastewater Treatment: A Review. *Membranes*, *10*(5). PMID:32365810

Gholami, F., Zinadini, S., Zinatizadeh, A. A., & Abbasi, A. R. (2018). TMU-5 Metal-Organic Frameworks (MOFs) as a Novel Nanofiller for Flux Increment and Fouling Mitigation in PES Ultrafiltration Membrane. *Separation and Purification Technology*, *194*, 272–280. doi:10.1016/j.seppur.2017.11.054

Gnanasekaran, G., Balaguru, S., Arthanareeswaran, G., & Das, D. B. (2019). Removal of Hazardous Material from Wastewater by Using Metal Organic Framework (MOF) Embedded Polymeric Membranes. *Separation Science and Technology (Philadelphia)*, *54*(3), 434–446. doi:10.1080/01496395.2018.1508232

Golpour, M., & Pakizeh, M. (2018). Preparation and Characterization of New PA-MOF/PPSU-GO Membrane for the Separation of KHI from Water. *Chemical Engineering Journal*, *345*, 221–232. doi:10.1016/j.cej.2018.03.154

Gorchev, H. G., & Ozolins, G. (2004). *Guidelines for Drinking-Water Quality (3rd ed.).* WHO.

Gu, Q., Albert Ng, T. C., Sun, Q., Kotb Elshahawy, A. M., Lyu, Z., He, Z., Zhang, L., Ng, H. Y., Zeng, K., & Wang, J. (2019). Heterogeneous ZIF-L Membranes with Improved Hydrophilicity and Anti-Bacterial Adhesion for Potential Application in Water Treatment. *RSC Advances*, *9*(3), 1591–1601. doi:10.1039/C8RA08758J

Gu, Y., Wang, Y., Li, H., Qin, W., Zhang, H., Wang, G., Zhang, Y., & Zhao, H. (2020). Fabrication of Hierarchically Porous NH2-MIL-53/Wood-Carbon Hybrid Membrane for Highly Effective and Selective Sequestration of Pb2+. *Chemical Engineering Journal*, *387*, 124141. doi:10.1016/j.cej.2020.124141

Guo, Y., Wang, X., Hu, P., & Peng, X. (2016). ZIF-8 Coated Polyvinylidenefluoride (PVDF) Hollow Fiber for Highly Efficient Separation of Small Dye Molecules. *Applied Materials Today*, *5*, 103–110. doi:10.1016/j.apmt.2016.07.007

Haldar, D., Duarah, P., & Purkait, M. K. (2020). MOFs for the Treatment of Arsenic, Fluoride and Iron Contaminated Drinking Water: A Review. *Chemosphere*, *251*, 126388. doi:10.1016/j.chemosphere.2020.126388 PMID:32443223

He, Y., Yu, P. T., Ma, D., & Chung, T. S. (2017). UiO-66 Incorporated Thin-Film Nanocomposite Membranes for Efficient Selenium and Arsenic Removal. *Journal of Membrane Science*, *541*(March), 262–270. doi:10.1016/j.memsci.2017.06.061

Hu, Z., Chen, Y., & Jiang, J. (2011). Zeolitic Imidazolate Framework-8 as a Reverse Osmosis Membrane for Water Desalination: Insight from Molecular Simulation. *The Journal of Chemical Physics*, *134*(13), 134705. doi:10.1063/1.3573902 PMID:21476767

Kadhom, M., & Deng, B. (2018). Metal-Organic Frameworks (MOFs) in Water Filtration Membranes for Desalination and Other Applications. *Applied Materials Today*, *11*, 219–230. doi:10.1016/j.apmt.2018.02.008

Kadhom, M., Hu, W., & Deng, B. (2017). Thin Film Nanocomposite Membrane Filled with Metal-Organic Frameworks UiO-66 and MIL-125 Nanoparticles for Water Desalination. *Membranes*, *7*(2), 31. doi:10.3390/membranes7020031 PMID:28613247

Karimi, A., Vatanpour, V., Khataee, A., & Safarpour, M. (2019). Contra-Diffusion Synthesis of ZIF-8 Layer on Polyvinylidene Fluoride Ultrafiltration Membranes for Improved Water Purification. *Journal of Industrial and Engineering Chemistry*, *73*, 95–105. doi:10.1016/j.jiec.2019.01.010

Kebria, M. R. S., Rahimpour, A., Bakeri, G., & Abedini, R. (2018, October). 2019. "Experimental and Theoretical Investigation of Thin ZIF-8/Chitosan Coated Layer on Air Gap Membrane Distillation Performance of PVDF Membrane. *Desalination*, *450*, 21–32. doi:10.1016/j.desal.2018.10.023

Khan, N. A., Hasan, Z., & Jhung, S. H. (2013). Adsorptive Removal of Hazardous Materials Using Metal-Organic Frameworks (MOFs): A Review. *Journal of Hazardous Materials*, *244–245*, 444–456. doi:10.1016/j.jhazmat.2012.11.011 PMID:23195596

Khanjani, S., & Morsali, A. (2014). Ultrasound-Promoted Coating of MOF-5 on Silk Fiber and Study of Adsorptive Removal and Recovery of Hazardous Anionic Dye 'Congo Red.'. *Ultrasonics Sonochemistry*, *21*(4), 1424–1429. doi:10.1016/j.ultsonch.2013.12.012 PMID:24412182

Kitagawa, S., Kitaura, R., & Shin-ichiro, N. (2004). Functional Porous Coordination Polymers. *Angewandte Chemie International Edition*, *43*(18), 2334–2375. doi:10.1002/anie.200300610 PMID:15114565

Kurwadkar, S. (2019). Occurrence and Distribution of Organic and Inorganic Pollutants in Groundwater. *Water Environment Research*, *91*(10), 1001–1008. doi:10.1002/wer.1166 PMID:31230394

Lee, Tang, & Huo. (2014). Fabrication of Porous Matrix Membrane (PMM) Using Metal-Organic Framework as Green Template for Water Treatment. *Scientific Reports*, *4*, 1–5.

Lee, J. Y., She, Q., Huo, F., & Tang, C. Y. (2015). Metal-Organic Framework-Based Porous Matrix Membranes for Improving Mass Transfer in Forward Osmosis Membranes. *Journal of Membrane Science*, *492*, 392–399. doi:10.1016/j.memsci.2015.06.003

Lee, T. H., Oh, J. Y., Hong, S. P., Lee, J. M., Roh, S. M., Kim, S. H., & Park, H. B. (2019). ZIF-8 Particle Size Effects on Reverse Osmosis Performance of Polyamide Thin-Film Nanocomposite Membranes: Importance of Particle Deposition. *Journal of Membrane Science*, *570–571*, 23–33. doi:10.1016/j.memsci.2018.10.015

Li. (2018). Efficient Removal of Nickel(II) from High Salinity Wastewater by a Novel PAA/ZIF-8/PVDF Hybrid Ultrafiltration Membrane. *Water Research, 143*(2), 87–98.

Li, J., Wang, H., Yuan, X., Zhang, J., & Chew, J. W. (2020). Metal-Organic Framework Membranes for Wastewater Treatment and Water Regeneration. *Coordination Chemistry Reviews, 404*, 213116. doi:10.1016/j.ccr.2019.213116

Li, J. R., Sculley, J., & Hong, C. Z. (2012). Metal-Organic Frameworks for Separations. *Chemical Reviews, 112*(2), 869–932. doi:10.1021/cr200190s PMID:21978134

Li, T. (2019, July). Integrating Cationic Metal-Organic Frameworks with Ultrafiltration Membrane for Selective Removal of Perchlorate from Water. *Journal of Hazardous Materials*, 381.

Li, W. (2018, September). Metal–Organic Framework Membranes: Production, Modification, and Applications. *Progress in Materials Science, 100*, 21–63. doi:10.1016/j.pmatsci.2018.09.003

Li, X., Liu, Y., Wang, J., Gascon, J., Li, J., & Van der Bruggen, B. (2017). Metal-Organic Frameworks Based Membranes for Liquid Separation. *Chemical Society Reviews, 46*(23), 7124–7144. doi:10.1039/C7CS00575J PMID:29110013

Li, Y., Wee, L. H., Martens, J. A., & Vankelecom, I. F. J. (2017). Interfacial Synthesis of ZIF-8 Membranes with Improved Nanofiltration Performance. *Journal of Membrane Science, 523*, 561–566. doi:10.1016/j.memsci.2016.09.065

Liao, Z., Fang, X., Xie, J., Li, Q., Wang, D., Sun, X., Wang, L., & Li, J. (2019). Hydrophilic Hollow Nanocube-Functionalized Thin Film Nanocomposite Membrane with Enhanced Nanofiltration Performance. *ACS Applied Materials & Interfaces, 11*(5), 5344–5352. doi:10.1021/acsami.8b19121 PMID:30638012

Liu, R., Sui, Y., & Wang, X. (2019). Metal–Organic Framework-Based Ultrafiltration Membrane Separation with Capacitive-Type for Enhanced Phosphate Removal. *Chemical Engineering Journal, 371*(February), 903–913. doi:10.1016/j.cej.2019.04.136

Liu, T. Y., Yuan, H.-G., Liu, Y.-Y., Ren, D., Su, Y.-C., & Wang, X. (2018). Metal-Organic Framework Nanocomposite Thin Films with Interfacial Bindings and Self-Standing Robustness for High Water Flux and Enhanced Ion Selectivity. *ACS Nano, 12*(9), 9253–9265. doi:10.1021/acsnano.8b03994 PMID:30153418

Liu, X., Demir, N. K., Wu, Z., & Li, K. (2015). Highly Water-Stable Zirconium Metal-Organic Framework UiO-66 Membranes Supported on Alumina Hollow Fibers for Desalination. *Journal of the American Chemical Society, 137*(22), 6999–7002. doi:10.1021/jacs.5b02276 PMID:26023819

Luo, X., Ding, L., & Luo, J. (2015). Adsorptive Removal of Pb(II) Ions from Aqueous Samples with Amino-Functionalization of Metal–Organic Frameworks MIL-101(Cr). *Journal of Chemical & Engineering Data, 60*(6), 1732–1743. doi:10.1021/je501115m

Ma, D., Peh, S. B., Han, G., & Chen, S. B. (2017). Thin-Film Nanocomposite (TFN) Membranes Incorporated with Super-Hydrophilic Metal-Organic Framework (MOF) UiO-66: Toward Enhancement of Water Flux and Salt Rejection. *ACS Applied Materials & Interfaces, 9*(8), 7523–7534. doi:10.1021/acsami.6b14223 PMID:28186405

Ma, J., Guo, X., Ying, Y., Liu, D., & Zhong, C. (2017). Composite Ultrafiltration Membrane Tailored by MOF@GO with Highly Improved Water Purification Performance. *Chemical Engineering Journal*, *313*, 890–898. doi:10.1016/j.cej.2016.10.127

Ma, X. H., Yang, Z., Yao, Z.-K., Xu, Z.-L., & Tang, C. Y. (2016, July). A Facile Preparation of Novel Positively Charged MOF/Chitosan Nanofiltration Membranes. *Journal of Membrane Science*, *525*, 269–276. doi:10.1016/j.memsci.2016.11.015

Madhura, L., Kanchi, S., Sabela, M. I., Singh, S., Bisetty, K., & Inamuddin. (2018). Membrane Technology for Water Purification. *Environmental Chemistry Letters*, *16*(2), 343–365. doi:10.100710311-017-0699-y

Makhetha, T. A., & Moutloali, R. M. (2018). Antifouling Properties of Cu(Tpa)@GO/PES Composite Membranes and Selective Dye Rejection. *Journal of Membrane Science*, *554*(January), 195–210. doi:10.1016/j.memsci.2018.03.003

Maroofi, S. M., & Mahmoodi, N. M. (2019). Zeolitic Imidazolate Framework-Polyvinylpyrrolidone-Polyethersulfone Composites Membranes: From Synthesis to the Detailed Pollutant Removal from Wastewater Using Cross Flow System. *Colloids and Surfaces. A, Physicochemical and Engineering Aspects*, *572*(March), 211–220. doi:10.1016/j.colsurfa.2019.03.093

Matilainen, A., Vepsäläinen, M., & Sillanpää, M. (2010). Natural Organic Matter Removal by Coagulation during Drinking Water Treatment: A Review. *Advances in Colloid and Interface Science*, *159*(2), 189–197. doi:10.1016/j.cis.2010.06.007 PMID:20633865

Meng, Y., Shu, L., Liu, L., Wu, Y., Xie, L.-H., Zhao, M.-J., & Li, J.-R. (2019). A High-Flux Mixed Matrix Nanofiltration Membrane with Highly Water-Dispersible MOF Crystallites as Filler. *Journal of Membrane Science*, *591*(August), 117360. doi:10.1016/j.memsci.2019.117360

Mohammadnezhad, F., Feyzi, M., & Zinadini, S. (2019). A Novel Ce-MOF/PES Mixed Matrix Membrane; Synthesis, Characterization and Antifouling Evaluation. *Journal of Industrial and Engineering Chemistry*, *71*, 99–111. doi:10.1016/j.jiec.2018.09.032

Navarro, M., Benito, J., Paseta, L., Gascón, I., Coronas, J., & Téllez, C. (2018). Thin-Film Nanocomposite Membrane with the Minimum Amount of MOF by the Langmuir-Schaefer Technique for Nanofiltration. *ACS Applied Materials & Interfaces*, *10*(1), 1278–1287. doi:10.1021/acsami.7b17477 PMID:29243908

Paseta, L., Antorán, D., Coronas, J., & Téllez, C. (2019). 110th Anniversary: Polyamide/Metal-Organic Framework Bilayered Thin Film Composite Membranes for the Removal of Pharmaceutical Compounds from Water. *Industrial & Engineering Chemistry Research*, *58*(10), 4222–4230. doi:10.1021/acs.iecr.8b06017

Qiu, M., & He, C. (2018, December). Efficient Removal of Heavy Metal Ions by Forward Osmosis Membrane with a Polydopamine Modified Zeolitic Imidazolate Framework Incorporated Selective Layer. *Journal of Hazardous Materials*, *367*, 339–347. doi:10.1016/j.jhazmat.2018.12.096 PMID:30599406

Ragab, D., Gomaa, H. G., Sabouni, R., Salem, M., Ren, M., & Zhu, J. (2016). Micropollutants Removal from Water Using Microfiltration Membrane Modified with ZIF-8 Metal Organic Frameworks (MOFs). *Chemical Engineering Journal*, *300*, 273–279. doi:10.1016/j.cej.2016.04.033

Rao, Z., Feng, K., Tang, B., & Wu, P. (2017). Surface Decoration of Amino-Functionalized Metal-Organic Framework/Graphene Oxide Composite onto Polydopamine-Coated Membrane Substrate for Highly Efficient Heavy Metal Removal. *ACS Applied Materials & Interfaces*, *9*(3), 2594–2605. doi:10.1021/acsami.6b15873 PMID:28035795

Ren, Y., Li, T., Zhang, W., Wang, S., Shi, M., Shan, C., Zhang, W., Guan, X., Lv, L., Hua, M., & Pan, B. (2019). MIL-PVDF Blend Ultrafiltration Membranes with Ultrahigh MOF Loading for Simultaneous Adsorption and Catalytic Oxidation of Methylene Blue. *Journal of Hazardous Materials*, *365*, 312–321. doi:10.1016/j.jhazmat.2018.11.013 PMID:30447639

Ren, Z., Luo, J., & Wan, Y. (2018). Highly Permeable Biocatalytic Membrane Prepared by 3D Modification: Metal-Organic Frameworks Ameliorate Its Stability for Micropollutants Removal. *Chemical Engineering Journal*, *348*, 389–398. doi:10.1016/j.cej.2018.04.203

Ruan, H., Guo, C., Yu, H., Shen, J., Gao, C., Sotto, A., & Van der Bruggen, B. (2016). Fabrication of a Mil-53(Al) Nanocomposite Membrane and Potential Application in Desalination of Dye Solutions. *Industrial & Engineering Chemistry Research*, *55*(46), 12099–12110. doi:10.1021/acs.iecr.6b03201

Rui, K., Wang, X., Du, M., Zhang, Y., Wang, Q., Ma, Z., Zhang, Q., Li, D., Huang, X., Sun, G., Zhu, J., & Huang, W. (2018). Dual-Function Metal–Organic Framework-Based Wearable Fibers for Gas Probing and Energy Storage. *ACS Applied Materials & Interfaces*, *10*(3), 2837–2842. doi:10.1021/acsami.7b16761 PMID:29286235

Seo, P. W., Khan, N. A., Hasan, Z., & Jhung, S. H. (2016). Adsorptive Removal of Artificial Sweeteners from Water Using Metal–Organic Frameworks Functionalized with Urea or Melamine. *ACS Applied Materials & Interfaces*, *8*(43), 29799–29807. doi:10.1021/acsami.6b11115 PMID:27723294

Shekhah, O., Liu, J., Fischer, R. A., & Wöll, C. (2011). MOF Thin Films: Existing and Future Applications. *Chemical Society Reviews*, *40*(2), 1081–1106. doi:10.1039/c0cs00147c PMID:21225034

Shen, C., Zhao, Y., Li, W., Yang, Y., Liu, R., & Morgen, D. (2019). Global Profile of Heavy Metals and Semimetals Adsorption Using Drinking Water Treatment Residual. *Chemical Engineering Journal*, *372*, 1019–1027. doi:10.1016/j.cej.2019.04.219

Sorribas, S., Gorgojo, P., Téllez, C., Coronas, J., & Livingston, A. G. (2013). High Flux Thin Film Nanocomposite Membranes Based on Metal-Organic Frameworks for Organic Solvent Nanofiltration. *Journal of the American Chemical Society*, *135*(40), 15201–15208. doi:10.1021/ja407665w PMID:24044635

Stock, N., & Biswas, S. (2012). Synthesis of Metal-Organic Frameworks (MOFs): Routes to Various MOF Topologies, Morphologies, and Composites. *Chemical Reviews*, *112*(2), 933–969. doi:10.1021/cr200304e PMID:22098087

Sun, H., Tang, B., & Wu, P. (2017). Development of Hybrid Ultrafiltration Membranes with Improved Water Separation Properties Using Modified Superhydrophilic Metal-Organic Framework Nanoparticles. *ACS Applied Materials & Interfaces*, *9*(25), 21473–21484. doi:10.1021/acsami.7b05504 PMID:28594542

Sun, H., Tang, B., & Wu, P. (2017, October). Hydrophilic Hollow Zeolitic Imidazolate Framework-8 Modified Ultrafiltration Membranes with Significantly Enhanced Water Separation Properties. *Journal of Membrane Science*, *551*, 283–293. doi:10.1016/j.memsci.2018.01.053

Tan, Y., Sun, Z., Meng, H., Han, Y., Wu, J., Xu, J., Xu, Y., & Zhang, X. (2019). A New MOFs/Polymer Hybrid Membrane: MIL-68(Al)/PVDF, Fabrication and Application in High-Efficient Removal of p-Nitrophenol and Methylene Blue. *Separation and Purification Technology*, *215*(January), 217–226. doi:10.1016/j.seppur.2019.01.008

Tomic, E. A. (1965). Thermal Stability of Coordination Polymers. *Journal of Applied Polymer Science*, *9*(11), 3745–3752. doi:10.1002/app.1965.070091121

Trinh, D. X., Thuy, P. N. T., & Taniike, T. (2017). Fabrication of New Composite Membrane Filled with UiO-66 Nanoparticles and Its Application to Nanofiltration. *Separation and Purification Technology*, *177*, 249–256. doi:10.1016/j.seppur.2017.01.004

Van Goethem, C., Verbeke, R., Hermans, S., Bernstein, R., & Vankelecom, I. F. J. (2016). Controlled Positioning of MOFs in Interfacially Polymerized Thin-Film Nanocomposites. *Journal of Materials Chemistry. A, Materials for Energy and Sustainability*, *4*(42), 16368–16376. doi:10.1039/C6TA05175H

Van Goethem, C., Verbeke, R., Pfanmöller, M., Koschine, T., Dickmann, M., Timpel-Lindner, T., Egger, W., Bals, S., & Vankelecom, I. F. J. (2018, March). The Role of MOFs in Thin-Film Nanocomposite (TFN) Membranes. *Journal of Membrane Science*, *563*, 938–948. doi:10.1016/j.memsci.2018.06.040

Wang, C., & Liu, X. (2015). Superior Removal of Arsenic from Water with Zirconium Metal-Organic Framework UiO-66. *Scientific Reports, 5*, 16613. https://pubmed.ncbi.nlm.nih.gov/26559001

Wang, L., Duan, S., Fang, M., Liu, J., He, J., Li, J., & Lei, J. (2016). Surface Modification Route to Prepare Novel Polyamide@NH2-MIL-88B Nanocomposite Membranes for Water Treatment. *RSC Advances*, *6*(75), 71250–71261. doi:10.1039/C6RA09080J

Wang, L., Fang, M., Liu, J., He, J., Li, J., & Lei, J. (2015). Layer-by-Layer Fabrication of High-Performance Polyamide/ZIF-8 Nanocomposite Membrane for Nanofiltration Applications. *ACS Applied Materials & Interfaces*, *7*(43), 24082–24093. doi:10.1021/acsami.5b07128 PMID:26485228

Wang, N., Liu, T., Shen, H., Ji, S., Li, J.-R., & Zhang, R. (2016). Ceramic Tubular MOF Hybrid Membrane Fabricated through in Situ Layer-by-Layer Self-Assembly for Nanofiltration. *AIChE Journal. American Institute of Chemical Engineers*, *62*(2), 538–546. doi:10.1002/aic.15115

Wang, Hou, Chen, & Meng. (2020). In-Situ Growth of Metal-Organic Framework Film on a Polydopamine-Modified Flexible Substrate for Antibacterial and Forward Osmosis Membranes. *Separation and Purification Technology, 236*, 116239. . doi:10.1016/j.seppur.2019.116239

WHO & UNICEF. (2017). *Progress on Drinking Water, Sanitation and Hygiene: 2017 Update and SDG Baseline.* World Health Organization. https://www.unicef.org/publications/files/Progress_on_Drinking_Water_Sanitation_and_Hygiene_2017.pdf

World Health Organization. (2010). *WHO Housing and Health Guidelines.* https://www.who.int/phe%0Ahttp://apps.who.int/bookorders

Wu, M., Ye, H., Zhao, F., & Zeng, B. (2017). High-Quality Metal-Organic Framework ZIF-8 Membrane Supported on Electrodeposited ZnO/2-Methylimidazole Nanocomposite: Efficient Adsorbent for the Enrichment of Acidic Drugs. *Scientific Reports*, *7*(January), 1–9. https://dx.doi.org/10.1038/srep39778

Xie, A. (2019, July). Photo-Fenton Self-Cleaning PVDF/NH2-MIL-88B(Fe) Membranes towards Highly-Efficient Oil/Water Emulsion Separation. *Journal of Membrane Science*, *595*, 117499. https://doi.org/10.1016/j.memsci.2019.117499

Xie, A. (2020). Graphene Oxide/Fe(III)-Based Metal-Organic Framework Membrane for Enhanced Water Purification Based on Synergistic Separation and Photo-Fenton Processes. *Applied Catalysis B: Environmental*, *264*(July), 118548. https://doi.org/10.1016/j.apcatb.2019.118548

Xu, F. (2018). *Phosphorus Removal and Recovery From Anaerobic Digestion Residues*. Elsevier. https://linkinghub.elsevier.com/retrieve/pii/S2468012518300038

Xu, Y., Gao, X., Wang, Q., Wang, X., Ji, Z., & Gao, C. (2016). Highly Stable MIL-101(Cr) Doped Water Permeable Thin Film Nanocomposite Membranes for Water Treatment. *RSC Advances*, *6*(86), 82669–82675.

Xu, Y., Gao, X., Wang, X., Wang, Q., Ji, Z., & Wang, X. (2016). Highly and Stably Water Permeable Thin Film Nanocomposite Membranes Doped with MIL-101 (Cr) Nanoparticles for Reverse Osmosis Application. *Materials (Basel)*, *9*(11).

Yaghi, O. M., Li, G., & Li, H. (1995). Selective Binding and Removal of Guests in a Microporous Metal–Organic Framework. *Nature*, *378*(6558), 703–706. https://dx.doi.org/10.1038/378703a0

Yaghi, O. M. (2003). Reticular Synthesis and the Design of New Materials. *Nature, 423*(6941), 705–14. doi:10.1038/nature01650

Yahaya, N. Z. S. (2020). Synthesis and Characterizations of MIL-140B-Al2O3/YSZ Ceramic Membrane Using Solvothermal Method for Seawater Desalination. *Journal of the Australian Ceramic Society*, *56*(1), 291–300.

Yang, F. (2018). Metal-Organic Frameworks Supported on Nanofiber for Desalination by Direct Contact Membrane Distillation. *ACS Applied Materials & Interfaces*, *10*(13), 11251–11260.

Yang, L., Wang, Z., & Zhang, J. (2017). Zeolite Imidazolate Framework Hybrid Nanofiltration (NF) Membranes with Enhanced Permselectivity for Dye Removal. *Journal of Membrane Science*, *532*(March), 76–86. https://dx.doi.org/10.1016/j.memsci.2017.03.014

Yang, S., Zou, Q., Wang, T., & Zhang, L. (2019). Effects of GO and MOF@GO on the Permeation and Antifouling Properties of Cellulose Acetate Ultrafiltration Membrane. *Journal of Membrane Science*, *569*, 48–59. https://doi.org/10.1016/j.memsci.2018.09.068

Yin, N., Wang, K., Wang, L., & Li, Z. (2016). Amino-Functionalized MOFs Combining Ceramic Membrane Ultrafiltration for Pb (II) Removal. *Chemical Engineering Journal*, *306*, 619–628. https://dx.doi.org/10.1016/j.cej.2016.07.064

Zhang, M. (2018). Insights into the Use of Metal–Organic Framework As High-Performance Anticorrosion Coatings. *ACS Applied Materials & Interfaces*, *10*(3), 2259–2263. https://dx.doi.org/10.1021/acsami.7b18713

Zhang, P. (2018). Ultrathin Reduced Graphene Oxide/MOF Nanofiltration Membrane with Improved Purification Performance at Low Pressure. *Chemosphere*, *204*, 378–389. https://doi.org/10.1016/j.chemosphere.2018.04.064

Zhang, R. (2014). Coordination-Driven in Situ Self-Assembly Strategy for the Preparation of Metal-Organic Framework Hybrid Membranes. *Angewandte Chemie International Edition*, *53*(37), 9775–9779.

Zhang, S. (2020). Water-Soluble MOF Nanoparticles Modified Polyethersulfone Membrane for Improving Flux and Molecular Retention. *Applied Surface Science*, *505*, 144553. https://doi.org/10.1016/j.apsusc.2019.144553

Zhao, X. (2015). Recent Advances in the Fabrication and Structure-Specific Applications of Graphene-Based Inorganic Hybrid Membranes. *Nanoscale*, *7*(12), 5080–5093.

Zhou, S., Wei, Y., & Hou, J. (2017). Self-Sacrificial Template Strategy Coupled with Smart in Situ Seeding for Highly Oriented Metal-Organic Framework Layers: From Films to Membranes. *Chemistry of Materials*, *29*(17), 7103–7107.

Zhou, S., Wei, Y., & Zhuang, L. (2017). Introduction of Metal Precursors by Electrodeposition for the in Situ Growth of Metal-Organic Framework Membranes on Porous Metal Substrates. *Journal of Materials Chemistry. A, Materials for Energy and Sustainability*, *5*(5), 1948–1951.

Zhu, J. (2017). Elevated Performance of Thin Film Nanocomposite Membranes Enabled by Modified Hydrophilic MOFs for Nanofiltration. *ACS Applied Materials & Interfaces*, *9*(2), 1975–1986.

Zuo, J., & Chung, T. S. (2016). Metal-Organic Framework-Functionalized Alumina Membranes for Vacuum Membrane Distillation. *Water (Switzerland)*, *8*(12), 1–15.

APPENDIX

Abbreviations

AOP	Advanced oxidation methods
CA	Cellulose acetate
DI	Deionized
ED	Electrodialysis
EFP	Evaporation-controlled filler positioning
FO	Forward osmosis
GO	Graphene oxide
ICP	Internal concentration polarization
KHI	Kinetic hydrate inhibitor
LbL	layer-by-layer
MOFs	Metal-organic frameworks
MB	Methylene blue
MA	Myristic acid
MMM	Mixed matrix membranes
NF	Nanofiltration
NRs	Nanorods
PES	Polyethersulfone
PDA	Polydopamine
PAN	Polyacrylonitrile
PTFE	Polytetrafluoroethylene
PMIA	Poly(m-phenylene isophthalamide)
PPSU	Polyphenylsulfone
PMIA	poly (m-phenylene isophthalamide)
PNP	p-nitrophenol
PA	Polyamide
PMM	Porous matrix membranes
RO	Reverse osmosis
THF	Tetrahydrofuran
TFN	Thin-film nanocomposite
UF	Ultrafiltration
WTPs	Water treatment plants
WC	Wood-carbon

Chapter 8
Metal-Organic Frameworks for Drug Delivery Applications

Doaa Ahmed Ghareeb
Independent Researcher, Egypt

Nessma Magdy Nasr
University Student's Hospital, Egypt

ABSTRACT

Metal-organic frameworks (MOFs) are based on metals and organic linkers; they possess large surface areas, suitable pore size and shape, wide range of chemical composition, and functionalized pore surface, which enable them for possible applications as delivery vehicles for therapeutic agents. The challenges include not only the development of new solids but also continuous improvements in the formulation and processing of the materials, including modifying the morphology and shape of the frameworks to fit the proposed applications of drug delivery. This chapter discussed enormous MOF-based stimuli responsive drug delivery systems, and considerable achievements have been made as a new avenue for drug delivery, their structural aspects, their applications in the controlled release of the drugs, and future view for development of drug controlled release researches using MOFs. Among the properties that must be developed and approved are the materials' toxicology, stability, their reproducibility of manufacture of MOFs in body's liquid, and pharmacokinetics of drug-loaded MOFs.

INTRODUCTION

Drug delivery systems are one of the best applications for the treatment of human diseases and biomedical materials (Gupta et al., 2012; Torchilin, 2005; Vallet-Regí et al., 2007; Yoo & Lee, 2006). In the field of drug delivery, looking for non-toxic carriers for drugs delivery to the body is a challenge (Davis et al., 2008; Dyson & Sava, 2006; Fathabadi et al., 2012). To solve these problems, metal organic frameworks (MOFs) are used as carriers for biomedical applications. The features of MOFs offer great chance for their uses as drug carriers, as high porosity, adjustable size and high loading capacity, which make it possible to use these frameworks in therapeutic projects such as drug delivery(Haydar et al., 2017), but

DOI: 10.4018/978-1-7998-4760-1.ch008

also some properties need to be improves: toxicology of materials for use in healthcare applications. MOFs are required some conditions to have easier engineering the surface of materials to control in vivo fate such as controllable degradation of the solids, efficient encapsulation of therapeutic drugs with high loading amounts, controllable release of encapsulated cargos (Horcajada et al., 2010) and non-toxic building blocks, so chromium, cadmium, nickel, cobalt, and other metals are not suitable due to their high toxicity, some metals exist in appreciable amounts in the body, for example, iron is an main component of hemoglobin and is approximately 22 mM in blood plasma (Bertini, 2007) and also, Ca, Cu, Mn, Mg, Zn, Fe, Ti, their toxicity are determined by oral lethal dose LD50, Ca, 1 g/kg; Cu, 25 μg/kg; Mn, 1.5 g/kg; Mg, 8.1 g/kg; Zn, 350 μg/kg; Fe, 30 g/kg; Ti, 25 g/kg and Zr. 4.1 g/kg (Horcajada et al., 2012). The toxicity data (LD50) for some organic linkers are 5 g/kg (terephthalic acid), 8.4 g/kg (trimesic acid), 5 g/kg (2,6 napthalenedicarboxylic acid), 1.13 g/kg (1-methylimidazole), 1.4 g/kg (2-methylimidazole), 5 g/kg (isonicotinic acid), and 1.6 g/kg (5-aminoisophthalic acid), which are accepted for bio application. In vivo toxicity tests of iron trimesic MIL-100 (MIL = Material Institut Lavoisier) and trimesic acid were performed through intravenous administrating to rats, a symmetrical cyclic oligosaccharide, g-cyclodextrin (g-CD), was considered as a ligand to synthesize two MOFs, [(g-CD) $(KOH)_2$] and [(g-CD)$(RbOH)_2$ (Smaldone et al., 2010).

MOFs are insoluble in water, (Burrows et al., 2013). The extent of solubility of the MOFs in water can be increased by modifying surface or introducing additional functionalities to MOFs such as, attaching water attracting ligands or polymers, water-soluble or water-adsorbing MOFs are being useful for the CR of water-soluble drug release ()

The structure of MOFs is more open and less dense than a zeolite, but their nature and the extent of their flexibility are not clearly understood (Fletcher et al., 2005). MOFs are anionic, and can be designed to obtain hydrophilic pores to carry either the positive or negative charges, and can be used to encapsulate the drugs that contain the opposite charges to MOFs (Huxford et al., 2010). MOFs have high drug loading capacity, and versatile functionality (An et al., 2009), and are efficient of releasing drug to a physiological medium related to their pore characteristics (McKinlay et al., 2010) as well as based on the extent of host-guest interactions that take place. A series of biological evaluations on MOFs were examined such as, biocompatibility, bio-toxicity, cellular uptakes, tissue responses, and intracellular drug delivery. If MOFs are effectively able to display cell targeting, diagnostic and therapeutic abilities along with biological or toxicological properties, then these MOFs could be useful for oncological or cancer treatments. Some MOFs were conjugated with targeting peptides to facilitate cell targeting and treatments through releasing the drugs in the targeted area, and the drugs were encapsulated. The size, shape and morphology of MOFs impact on their capacity of loading and releasing drugs, and their toxicity in a biological environment. The loading of drugs into MOFs takes long time, and large amount of drugs used to feed for loading, so the drug loading efficiencies of MOFs are low, and it varies from MOFs to MOFs (Cai et al., 2019). This chapter presents an overview of MOFs used in drug controlled release and also, illustrates the perspectives, insights, and critical reflections on the controlled release of drugs from MOFs, and the biological evaluations of MOFs being used in CR of drugs.

METAL–ORGANIC FRAMEWORK-BASED STIMULI-RESPONSIVE SYSTEMS FOR DRUG DELIVERY

Stimuli-responsive MOFs have been used for controllable drug release and can be divided into two categories: single and multiple-stimuli-responsive MOFs, which are able to achieve regulated delivery of loaded drugs upon activation by diverse stimuli, such as pH, magnetic field, ions, temperature, light, and pressure. The mechanisms of stimuli-responsive drug delivery systems can be described as follows: after injection, the stimuli responsive drug delivery system (nano carriers) accumulates at a tumor site via passive or active (receptor–ligand affinity) targeting. When activated by specific triggers, the chemical composition or physical structure of these systems undergoes a given transformation, which induces drug release **Figure 1**. By comparing with healthy tissues, tumor microenvironments at specific conditions

Figure 1. Schematic representation of metal–organic frameworks (MOFs)-based stimuli-responsive system for drug delivery

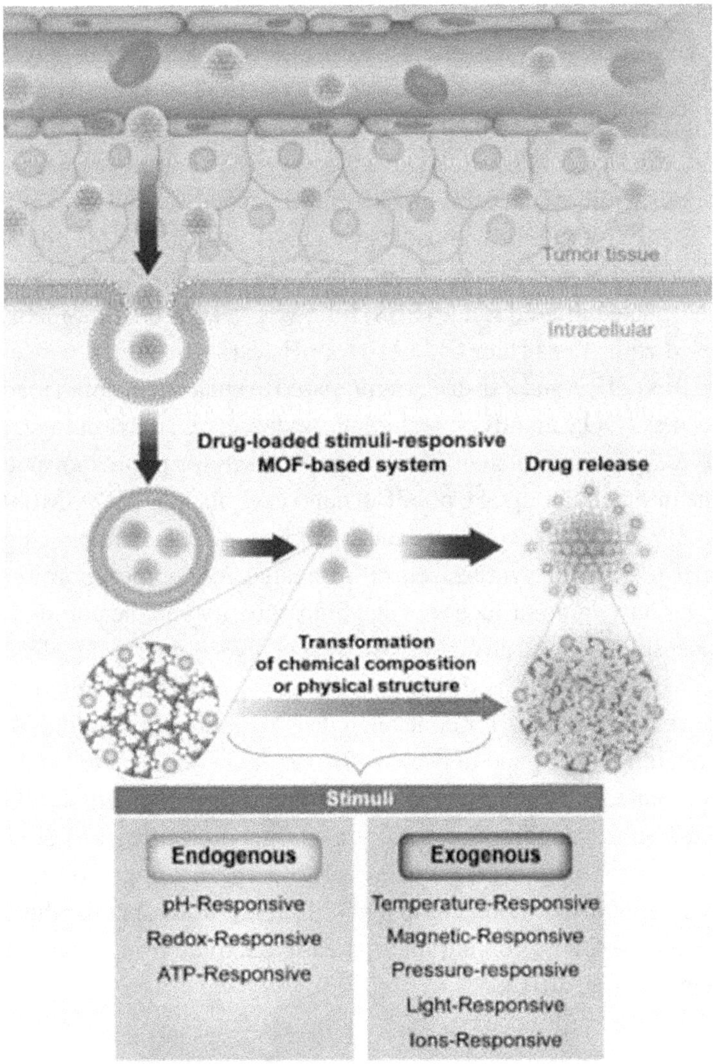

at lower pH and higher temperature that can be utilized for improvement of internal stimuli-responsive drug delivery systems (e.g., pH, redox, ATP), while external stimuli-responsive systems (e.g., magnetic field, temperature, ions, pressure, light) determine the nanoparticles at the tumor site, which represent the best drug release used for clinical application (Cai et al., 2019)

SINGLE STIMULUS-RESPONSIVE MOFS

pH-Responsive MOFs

Zinc-Based MOFs

Zeolitic imidazolate framework-8 (ZIF-8) is the most used MOF for pH-responsive drug release; shows improved drug loading capacity and pH-responsive drug release. The polyacrylic acid@ZIF-8 (PAA@ ZIF-8) nano carrier was synthesized via ion exchange between Zn^{2+} ions and polyacrylic acid sodium salt with 2-ethylimidazole in a methanol solution.(Ren et al., 2014) .A confocal laser scanning microscopy analysis showed that the DOX (DOX; an anticancer drug)-loaded PAA@ZIF-8 released the drug faster in a buffer solution at pH 5.5, which showed a low toxicity in the live cells. Another pH-sensitive nano carrier, fluorescein camptothecin (CPT)-loaded ZIF-8 NPs (70 nm) (Zhuang et al., 2014). The negatively charged CPT molecules were successfully loaded in the ZIF-8 NPs, the drug released by the dissociation of the ZIF-8 framework in acidic media exhibited increasing in cell death compared with those with free CPT. Consequently CC-ZIFs NPs (ZIF-8 coloading protein (Cas9) and single guide RNA (sgRNA)) are used for pH-responsive drug delivery. The CC-ZIFs NPs were synthesized by mixing Cas9 and sgRNA at 1:1 molar ratio in phosphate buffered saline (PBS) followed by addition of 2-methylimidazole and zinc nitrate. Cas9 was labeled with Alexa fluor 647 (AF) for pH-responsive release investigation (Alsaiari et al., 2018) . Similarity, ZIFs (ZIF-8 and ZIF-67) was prepared to encapsulate high loading amounts (14–20 wt%) of organic molecules (DOX and dyes, including rhodamine B, ethyl orange, and methylene blue. The DOX-loaded ZIF-8 crystals were an efficient nanoplatform for pH-responsive drug release. TEM imaging confirmed the intracellular uptake of ZIF-8 nano crystals into MDA-MB-468 cancer cells, the effect of DOX-loaded ZIF-8 crystals on breast cancer cell lines was shown to be higher than that of free DOX (Zheng et al., 2016). Zou et al, synthesized ZIF-8-coated hollow mesoporous silica NPs (HMSNs; HMSN@ZIF-8), ZIF-8's have large surface area and high porosity and the complexation of Zn^{2+}–DOX, the DOX-loaded nanocapsules showed an obvious pH-responsive release under physiological conditions; over 37 h, less than 23% of the DOX was slowly released at pH 7.4. This result provided a new route for preparing pH-responsive nanocapsules for controlled drug delivery (Zou et al., 2017). ZIF-8 modified with folic acid (FA) as a targeting ligand to provide for simultaneous targeted cellular

Chowdhuri et al, presented a carboxymethyl chitosan-modified magnetic NMOF (IRMOF-3) composed of Zn^{2+} ions and 2-amino terephthalic acid with a target molecule (folic acid) on its surface. The carboxymethyl chitosan increased the drug loading efficiency and improved the performance of the pH-responsive drug release. Furthermore, compared with L929 normal cells, this folate-targeted magnetic NMOF produced more specific cellular internalization in folateover expressing HeLa cancer cells (Chowdhuri, Singh, Ghosh et al, 2016).

Iron-Based MOFs:

Iron-based MOF composed of Fe^{2+} ions and 1,1′-(1,4-butanediyl) bis(imidazole) (bbi) for encapsulating DOX, the sensitivity of the MOF coordination bonds to external pH allowed this DOX-encapsulating MOF to release its load in acidic media. Also, FA was conjugated to its surface to allow the nano composite to be selectively taken up by cancer cells (Gao et al., 2013). Another pH-sensitive nanocarrier, mussel adhesive proteins (MAPs)-based polymeric NPs (80–130 nm) containing Fe(III)-3,4-dihydroxyphenylalanine (DOPA) complex with MOF-like structure, these MAP-based Fe(III)-DOPA NPs showed significant increase of DOX release in a buffer at pH 6, these NPs gave effective cytotoxicity on cancer cells through efficient cellular uptake and cytosolic release (Kim et al., 2015).

Wang et al, designed the novel Fe_3O_4@C@MIL-100(Fe) nano composite for the synchronous delivery of hydrophobic dihydroartemisinin (DHA) and Fe(III) for cancer therapy, this delivery system demonstrated a pH-responsive biodegradation (MIL-100(Fe) dissociates in acidic solutions) with the synchronous release of DHA and Fe(III). This system may be a potential alternative antitumor modality in the clinical setting (Wang, Zhou, Chen, Shi, Xia et al, 2016). Also, MOF nano composite was synthesized by modifying Fe_3O_4 nanoparticles with a layer of polyvinyl pyrrolidone (PVP) and polyetherimide (PEI) to obtain Fe3O4@PVP-PEI nano spheres, the nano composite selectively and pH-responsively captured and released glycoproteins under moderate pH conditions, this nanocomposite exhibits great potential for glycoprotein separation (Yang et al., 2017).

Moreover, DOX was loaded into the MIL-100(Fe) shell of the core–shell structure polypyrrole (PPy)@MIL-100(Fe) with a 12.8% loading ratio. The PPy@MIL-100(Fe) composite illustrated a pH-responsive drug release, which was examined in PBS buffers with different pH values (7.4 and 5.0) at 37 °C (Zhu et al., 2016).

Zirconium and Hafnium -Based MOFs

A pH-responsive zirconium-based MOF (UiO-66) was applied to the delivery of alendronate (AL), the release of AL from loaded UiO-66 was tested in PBS buffer at 37°C. Over 60 h, exactly, 59% and 42.7% of the DOX was released at pHs 5.5 and 7.4, respectively because AL is protonated in acidic environments, which weakens the interaction between AL and the Zr–O clusters in UiO-66. However, The AL-loaded UiO-66 inhibited the growth of cancer cells more efficiently than free AL, indicating it is a potential drug carrier for therapeutic applications (Cavka et al., 2008; Zhu et al., 2014). Consequently, a cationic NMOF (ZJU-101) built from zirconium and 2,2′-bipyridine-5,5′- dicarboxylate was used for the delivery of the anionic drug diclofenac sodium (DS) in PBS at pH 5.4 (inflamed tissues) than at pH 7.4 (normal tissues), because ion exchange between the drug and anions occurs in acidic media, This pH-responsive DS-loaded ZJU-101 NMOF is a promising candidate for anti-inflammatory drug delivery (Yang et al., 2016). Also, porphyrin-based MOF (PCN-221) was synthesized using $ZrCl_4$ and 5,10,15,20-tetrakis(4-carboxyphenyl) porphyrin for using for oral drug carrier (Lin, Hu, Jiang et al, 2016). Moreover, 200 nm zirconium terephthalate UiO-66 NPs that were coated with functionalized modulators, loaded with the drug calcein, and surface modified with polyethylene glycol (PEG) chains. At pH 7.4, the PEG chains enhanced the stability of the MOF toward phosphates and overcame the burst release by blocking interaction with the NP exteriors, but at pH 5.5, a stimuli-responsive drug release (Lzaro et al., 2017).

Interestingly, nano scale coordination polymers (NCPs) with NMOF-like structures are preferring carriers for drug delivery. NCPs based on high Z element hafnium (Hf) ions and an acidic-responsive benzoic-imine linker. These NCPs were modified with PEG for physiological stability and loaded with the chemotherapeutic drug chloro (triphenyl phosphine) gold(I) (TPPGC). The TPPGC-loaded NCP-PEG NPs dissociated at the lower pH found in tumors due to the acid-triggered cleavage of the organic linker, which resulted in the release of the TPPGC. The Hf ions could also act as radio sensitizers by absorbing X-rays for radiotherapy. After the efficient accumulation of TPPGC loaded NCP–PEG NPs in a tumor after intravenous injection, a remarkable synergistic therapeutic efficacy was achieved in chemo radiotherapy. These TPPGC-loaded NCP–PEG NPs are favorable agents for the chemo radiotherapy of cancer (Liu et al., 2017).

Europium and Gadolinium -Based MOF

It was reported (Peller et al., 2018) an antigen-loaded MOF for cancer therapy that was assembled by Eu^{+3}ions and guanine monophosphate. The maximal encapsulating efficiency of the antigens into this MOF was 55 wt%, and the antigen-loaded MOF showed pH-dependent drug-release behavior. Over 48 h,minimum amount of antigens were released at pH 7.4, while the most antigen about 60% were released at pH 5.0 due to the MOF's dissociation in the acidic media(Duan et al., 2017).

Interestingly, grinding was used to transform the gadolinium(III)- based MOF Gd-pDBI [pDBI, 1,4-bis(5-carboxy-1H-benzimidazole- 2-yl)benzene] to the nanosized MG-Gd-pDBI (≈140 nm), MG-Gd-pDBI showed low blood toxicity and a higher DOX loading capacity (12 wt%) than Gd-pDBI. The DOX-loaded MG-Gd-pDBI exhibited 22% of the loaded DOX released after 5 d at pH 7.4, while 44% was released at pH 5, illustrating MG-Gd-pDBI's potential as a pH-responsive drug carrier (Kundu et al., 2014).

Magnetically Responsive MOFs

Magnetic-responsive drug delivery is used to drive the drug-loaded magnetic nano particles to the intended site to increase therapeutic efficiency and also, act as an external stimulus for a controllable release of the drug and. Moreover, MOF-based magnetic structure could be utilized in imaging-guided therapy (Freeman et al., 1960; Kumar & Mohammad, 2011; Peller et al., 2018). MOF-based magnetic nanocomposite $Fe_3O_4@Cu_3(BTC)_2$ (BTC,benzene-1,3,5-tricarboxylate) was designed by the incorporation of Fe_3O_4 in $Cu_3(BTC)_2$ nano crystals of a favorable targeted drug delivery system. The anticancer drug nimesulide (NIM) was encapsulated into this magnetic nano composite by inserting $Fe_3O_4@Cu_3(BTC)_2$ in a NIM trichloromethane solution. The loaded NIM was completely released after 11 d in physiological saline at 37°C (Ke et al., 2011).However, the synthesis of γ-Fe_2O_3@MIL-53(Al) by integrating superparamagnetic γ-Fe_2O_3 nano particles into the MOF MIL-53(Al) this magnetic nano composite illustrated a controlled drug release in saline at 37 °C (Wu et al., 2014). Moreover, the superpar amagnetic property of γ-Fe_2O_3 endowed this nano composite with excellent magnetic performance for targeted drug delivery and separation. Also, iron carboxylate NMOF (F-NMOF) is used with high saturation magnetization for magnetically aided drug delivery. The DOX-loaded F-NMOF exhibited sustained DOX release in PBS, releasing 88% in 15 d (Sethi et al., 2016). Similarly, Sharma et al, reported magnetic NMOFs (M-NMOFs) with iron carboxylate for the magnetically aided co delivery of DOX and the photosensitizer methylene blue (MB; a photodynamic therapy agent) to cancer cells. The drug-loaded M-NMOFs showed a sustained release of both DOX and MB. In a human pancreatic carcinoma cell (Panc-1) viability assay, the cytotoxicity of the co loaded M-NMOFs was much larger than that of the free drugs or with mono loading under light irradiation. The application of magnetic-field guidance and light-triggered therapy would improve the efficiency of cancer therapy (Sharma et al., 2017).

Ion-Responsive MOF

In the ion responsive MOF-based drug delivery system, the diffusion and release of drugs is depended on the strong electrostatic interactions between the ionic drugs and MOFs. A zinc-based anionic MOF (bioMOF-1) was synthesized by adding biphenyldicarboxylic acid into a mixture of adenine and zinc acetate dihydrate. The resulting anionic bioMOF-1 was employed to store procainamide HCl (a cationic anti-arrhythmia drug), with the electrostatic interaction between the anions and cations. The release behavior was examined by putting procainamide HCl-loaded bio-MOF-1 in PBS buffer (pH 7.4), which showed that the ionic interactions between bio-MOF-1 and procainamide allowed drug release to be stimulated by the buffer cations (An et al., 2009).

However, MOF MOF-74-Fe(III), which was synthesized via the oxidation of neutral MOF-74-Fe(II). An MTT (3-(Vallet-Regí et al., 2007)-2,5 diphenyl tetrazolium bromide) assay illustrated that this cationic MOF had very low cytotoxicity on PC12 cells. IBU anions were loaded into the cationic MOF with an efficient encapsulating capacity (≈15.9 wt%) via ion exchange and salt penetration procedures. The release process occur in two pathways, the first was mainly sodium IBU and other coordinated free anions and the rate of release governed by the rate of diffusion or ion exchange between the drug-loaded MOF and PBS solution, while the second pathway, the coordinated drug was triggered by phosphate anions (PO_4^{3-}) due to their coordination with iron cations. These two pathways made g the drug release more flexible (Hu et al., 2014).

Interestingly, the anticancer drug 5-fluorouracil (5-FU) was captured by MOFIn1, which consists of In^{3+} ions and tris (para-carboxylphenyl) phosphine oxide. The release behavior was examined, when the 5-FUloaded MOF-In1 was immersed in PBS with various Zn^{2+} ion concentrations (500×10^{-9} to 10×10^{-3} m), 65% of the 5-FU was released when the Zn^{2+} concentration was 500×10^{-9} m, with more 5-FU released as the Zn^{2+} concentration increased. Moreover, because the host–guest interactions between 5-FU and MOF-In1 are weakened with increasing temperature, external heating also increased the release of 5-FU(Du et al., 2017).

Temperature-Responsive MOFs

Poly(N-isopropyl acrylamide) (PNIPAM) is a favorable building block for thermo sensitive drug nano carriers that have a lower critical solution temperature. When the temperature is decreased under its cloud point (Tc, around 32 °C), PNIPAM demonstrate hydrophilicity and tends to dissolve in water; consequently, resorufin, caffeine, and procainamide were loaded in the nano carrier by adding the UiO-66-PNIPAM (Figure2a) in guest solutions, then the release behaviors were assessed at 25 °C or 40 °C (Figure 5b), the amount of released drug increased at 25 °C and arrested at 40 °C (Figure 2c), which indicated that controlled release was driven by temperature variation (Nagata et al., 2015). Moreover, two zinc-based MOFs, $Zn_{16}(ad)_8(TP)_8O_2(H_2O)_8{\cdot}4HTP{\cdot}36DMF{\cdot}16H_2O$ (ad = adenine, H2TP = [1,1′:4′,1′′-terphenyl]-4,4′′- dicarboxylic acid for ZJU-64, and 2′,5′-dimethyl-[1,1′:4′,1′′-terphenyl]-4,4′′-dicarboxylic acid for ZJU-64-CH_3), anticancer drug MTX was encapsulated into the two MOFs with loading a capacity of 13.45 and 10.63 wt% for ZJU-64 and ZJU-64-CH_3, respectively. The same amount of MTX was released from MTX-loaded ZJU-64 and ZJU-64-CH_3 at 37 °C for 72 h, but only

Figure 2. a) Schematic showing the controlled release profiles of UiO-66-PNIPAM. b) Release behavior of drug-loaded UiO-66-PNIPAM in water at 25 °C and 40 °C for seven days. c) Temperature-responsive release behavior of resorufin from UiO-66-PNIPAM in water at 572 nm (Nagata et al., 2015)

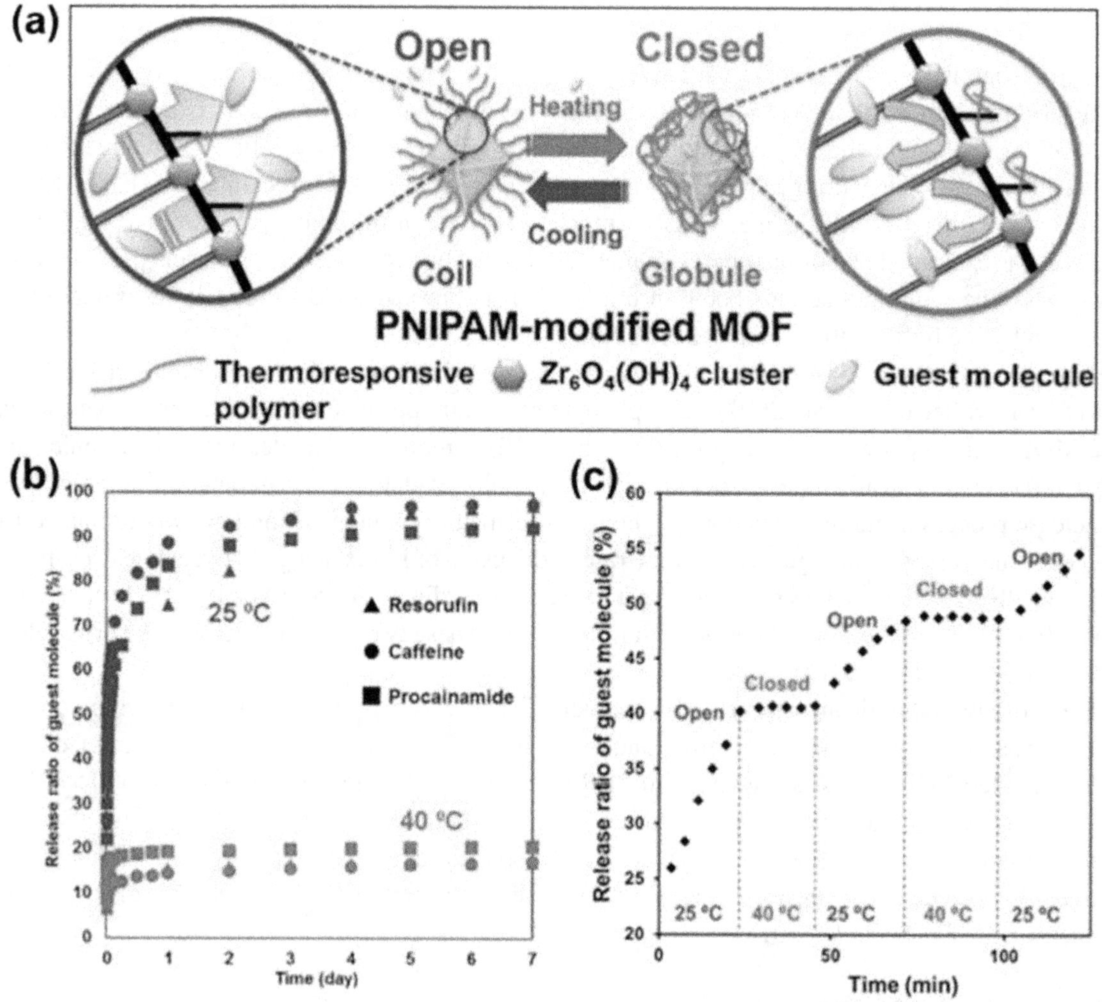

took approximately 1.5 and 6 h at 60 °C, promising a potential application of ZJU-64 and ZJU-64-CH_3 as temperature-responsive drug carriers (Lin, Hu, Yu et al, 2016).

Pressure-Responsive MOF

Zirconium-based MOF (ZJU-800) consist of zirconium and (2E,2E′)-3,3′-(2-fluoro-1,4-phenylene) diacrylic acid (F-H2PDA), which act as pressure-responsive drug release. The ZJU-800 MOF has a homogeneous shape with an average size of 300 nm both before and after drug loading illustrating its stability and lower cytotoxicity of this MOF was also seen by confocal microscopy. An adjustable release time of the drug from 2 to 8 d was checked by controlling the degree of compaction between the MOF and drug (Jiang et al., 2016).

Multiple Stimulus-Responsive MOFs

The complexity of the human body environment limits abilities to deliver drugs in the body using single stimulus-responsive MOF drug carriers, so multiple stimuli-responsive MOFs can be used for improving the load capacities and chemotherapeutic efficiencies. Pillararenes, are rapidly developed supra molecular hosts, have special structure, facile functionalization and desirable host–guest performance (Ogoshi et al., 2008; Si et al., 2015; Strutt et al., 2014; Xue et al., 2012; Zhang & Zhao, 2013) . The combination of MOFs with pillararenes will reveal a new procedure for the bio applications of MOFs (Aznar et al., 2016). UMCM-1-NH_2 was used as the host material to store cargoes of guest molecules with modified carboxylatopillar 5 arene (CP5) as a dual stimuli-responsive gatekeeper linked by positively-charged pyridinium (Py), (CP5-capped UMCM1-NH–Py) (Tan et al., 2015). In acidic tumor tissues, the gatekeeper of CP5 tended to be neutral, which decrease the force of non-covalent binding interactions between CP5 and the stalk, resulting the release of cargo. Also, addition of a competitive binding agent (methylviologen salts) caused the release of cargo (Figure 3), where the CP5-capped drug nano carrier showed high encapsulation efficiency. Another CP5-capped UiO-66-NH_2 5-Fu nano carrier, UiO-66-NH_2 was modified with positively charged quaternary ammonium salt as stalks (Q) to thread the negatively charged CP5 rings, to form 2 pseudorotaxanes as stimuli-responsive supramolecular gates that regulate drug release. Zn^{2+} can be a competitive binding agent to release of 5-Fu. So, the dual stimuli-responsive drug nano carrier based on CP5-gated MOF can be used in the therapy of central nervous system diseases.

Figure 3. Dual-stimuli responsive DDS based on UMCM-1-NH_2 NMOF by pillararenes (Tan et al., 2015)

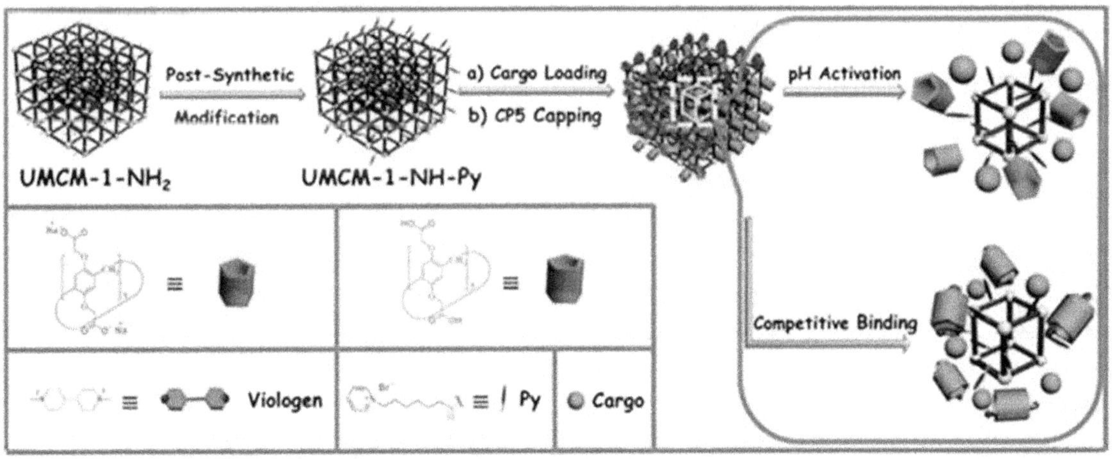

A triply-responsive theranostic hybrid platform based on CP5-gated Zr-MOFs Positively charged quaternary ammonium salt stalks were tethered on UiO-66-NH_2 via post-synthetic modification .Therefore, herein, a triple modal therapeutic trigger was integrated in one system (Figure 4). Controlled release experiments illustrated that the prepared therapeutic platform was favorable for potential bone regeneration and bone cancer therapy due to lower pH and higher concentration of Ca^{2+} in osteoclasts and tumor cells could increase the release of the therapeutic agents via pH- and Ca^{2+}competitive responsiveness.

Moreover, high temperatures could damage and kill cancer cells as well as weaken host–guest interactions to prompt gradually release of the therapeutic molecule (5-Fu) (Tan et al., 2016).

Figure 4. Triple-stimuli-responsive DDS based on CP5-capped UiO-66 (Tan et al., 2016)

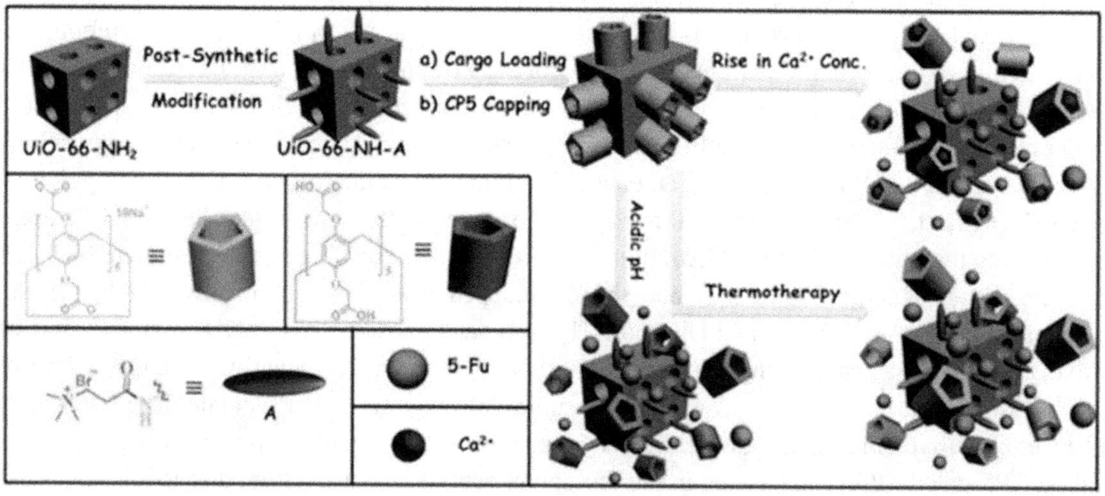

Consequently,β-cyclodextrin (β-CD) is favorable supra molecular macro cycles used to regulate targeted drug release, a multifunctional MOF platform for drug delivery of MIL-101 (Figure 5a). After

Figure 5. a) The multifunctional DDS based on MIL-101 and b) its cancer therapy procedure [62]

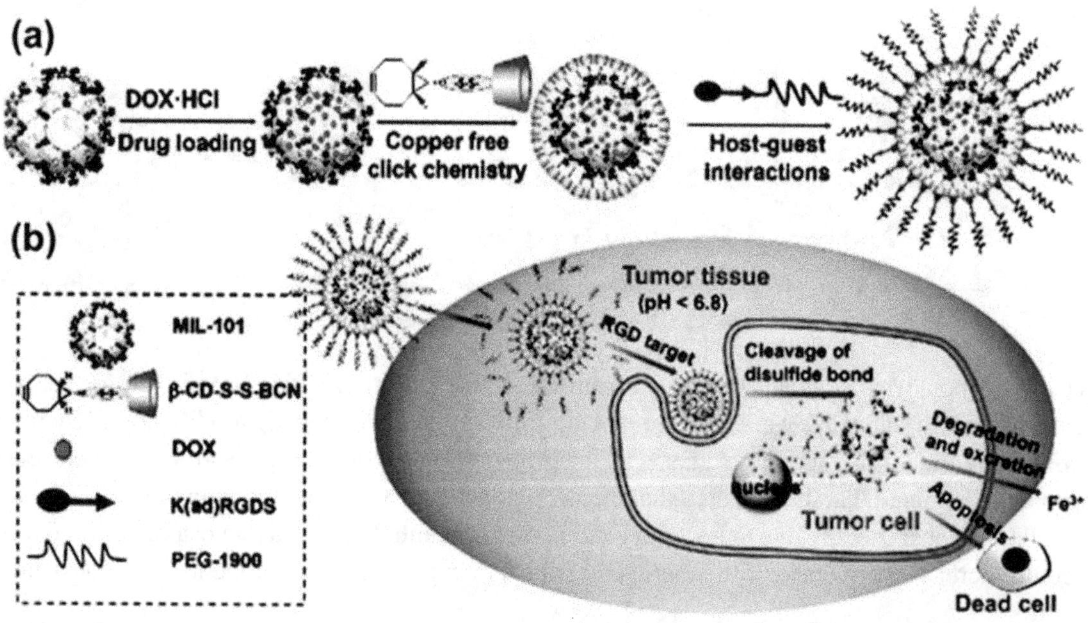

encapsulating DOX, MIL-101-N3 (Fe) was coated with β-CD derivative (β-CD-SS-BCN), the DDS had enhanced tumor cell uptake and that reducing agents triggered drug release, resulting in efficient drug delivery and cancer therapy (Figure 5b). Also, surface modification lowered the cytotoxicity of the loaded DOX on normal cells, however, in vivo experiments showed high chemotherapy efficiency of the DDS (Wang et al., 2015).

Moreover, Light-responsive release and competitive binding agent-responsive release were carried out in the system by binding β-CD on the surface of UiO-68-azobenzene (β-CD–capped UiO-68-azo) through supra molecular complexation (Figure 6a). Azobenzene groups gave photo-controllable cis/trans isomerizations and the isomerization of azobenzene from trans to cis to arrange cargo release (Figure 6b), a competitive binding agent such as amantadine could trigger the dissociation of the gatekeeper from the azobenzene stalks due to its higher binding affinity toward β-CD (Figure 6c) (Meng et al., 2016).

Figure 6. a) Schematic illustration of the synthesis of dual-stimuli-responsive DDS based on UiO-68-azo MOF capped by β-CD. b) Schematic representation of light triggered release. c) Schematic representation of step-by-step release (Meng et al., 2016)

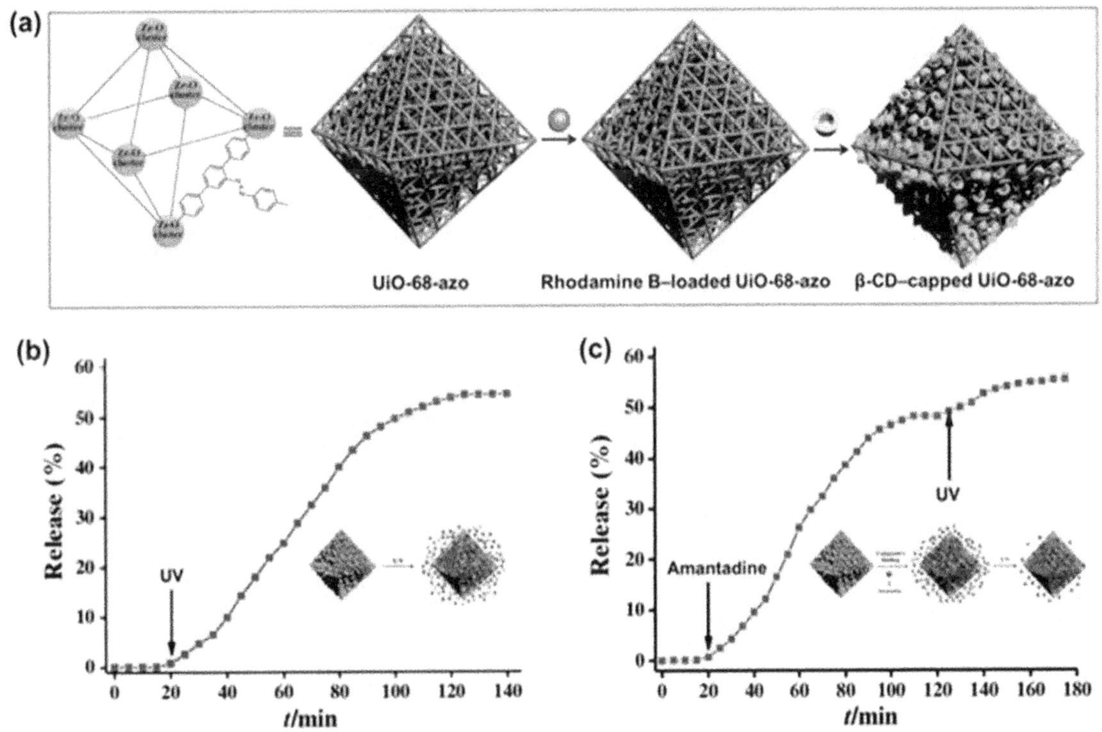

A list of the controlled release of drugs with various release mechanisms, and biological evaluations performed on the unmodified and modified MOFs used as drug carrying substrates are illustrated in Table 1.

Table 1. The controlled release of drugs with various release mechanisms, and biological evaluations performed on the unmodified and modified mofs used as drug carrying substrates

Drug	Type of drug	Type of MOFs used	Drug release
Procainamide HCl [20]	Antiarrythmia drug	Zinc-based bio-MOFs	Cation-triggered release
5-fluorouracil [64]	Anticancer drug	Zinc-based bio-MOFs	3-stages of drug release
Deferiprone [13]	Iron–chelator drug	Zinc-based bio-MOFs	Acid-induced drug release (pH-dependent)
Caffeine [65]	Amphiphlic drug	Zeolite imidazolate ZIF-8	Slow release in water
Nitric Oxide (NO) [66]	Bio-active gas	Ca- tetracarboxylate MOFs (BioMIL-3)	Slow release at biologically active concentration
Iodine (I_2) [67]	Bio-active gas	Zn-lactate and pybz containing porous MOFs	Controlled release to ethanol
5-fluorouracil [68]	Anticancer drug	Cu-BTC MOFs	Slow release in physiological environment
Ibuprofen [69]	Analgesic and antiinflammatory drug	MIL-53 (Cr, Fe)	Zero-order and slow release as a consequence of the framework flexibility
Ethoxysuccinatocisplatin (ESCP) [70]	Pro-drug of cisplatin	Modified MIL-101 (Fe) MOFs	Degradation of the modified MOFs
5-fluorouracil [71]	Anticancer drug	Zn (II) carboxylate MOFs	Diffusion controlled and pH-induced degradation
Ibuprofen [72]	Analgesic and anti-inflammatory drug	Cr-based MIL-100 and MIL-101 MOFs	Combined effects of diffusion, cation-anion
Rhodamine 6G and Doxorubicin hydrochloride [60]	Luminescent dye and Anticancer drug	Drug loaded carboxylatopillar 5 arenecapped UMCM-1-NH–Py	pH-responsive drug release
c, c, t-(diamminedichlorodisuccinato)Pt(IV) [73]	Anticancer drug	Tb^{3+} containing nanoscale coordination polymer coated with silica	Reduction induced release
Busulfan, Azidothymidine triphosphate, Cidofovir, Doxorubicin [9]	Anticancer or antiviral drugs	MIL 88A, MIL-53, MIL-89 and MIL-100	Governed by diffusion and/or drug-matrix interactions
Methotrexate [74]	Antineoplastic drug	RAFT copolymer modified Gd-MOFs	Dose-dependent drug release
Ovalbumin, ribonu-Clease A, human serum albumin, glucose dehydrogenase, lipase, haemoglobin, lysozyme, insulin, horseradish peroxidase, trypsin, urease, and oligonucleotide [75]	Proteins, enzyme, and DNA	ZIF-8, HKUST-1, Eu/Tb-BDC, and MIL-88A MOFs	Biomimetic mineralisation
Paclitaxel [76]	Hydrophobic anticancer drug	Iso-reticular magnetic nanoscale MOFs conjugated with folic acid (Fe3O4@IRMOF-3/FA)	Controlled release at physiological conditions (pH 7.4)
5-fluorouracil [77]	Anticancer drug	ZIF-90 nanoparticles conjugated with rat serum albumin	Stimuli-responsive drug delivery under an ELFAMF (20 Hz)
Artemisinin [78]	Traditional Chinese anticancer medicine	Core-shell containing PB@MIL-100 (Fe) dual MOFs nanoparticles	pH-responsive drug delivery

APPLICATIONS OF MOFS IN THE CONTROLLED RELEASE OF THE DRUGS:

MOFs and NMOFs are commonly used for drug delivery as drug carriers for clinical applications. A series of biocompatibility and cytotoxicity researchs were examined on the modified and unmodified MOFs in various cells. The cell killing efficacy by apoptosis, target-specific cellular uptake and internalization of the MOFs by cells via endocytosis were evident. The conjugation of targeted moieties to MOFs elevated the delivery of drugs, and helped detect particular cells or tissues e.g., cancer. These conjugations were achieved using affinity ligands, peptides, oligonucleotides, antibodies, drugs, and bioactive compounds by binding them to the targeted cell surface markers, the nano-sized MOFs are able to carry multiple drugs due to their high surface-area-to-volume ratio (Kamaly et al., 2010).

1.1 MOFs as Anti-cancer Drug Delivery Systems

N-MOFs have come out as favorable candidates for biomedical applications, especially in cancer-targeted therapy, because of their low toxicity and high efficiency (Horcajada et al., 2010). Cisplatin and its derivatives are used as anti-cancer therapies based on platinum metal and used effectively to treat testicular and ovarian cancer, but their level of toxicity can cause severe side-effects including nausea, nephrotoxicity and bone marrow suppression (Graf & Lippard, 2012). To reduce side effects, one of the cisplatin derivatives, ethoxy succinato-cisplatin (ESCP), which treats carcinomas including cancers of soft tissue, bones, muscles, sarcoma cancers and blood vessels (Florea & Büsselberg, 2011), was loaded in a functionalized Fe-MIL-101 as shown in (Figure 7) (Taylor-Pashow et al., 2009).

Moreover, busulfan is a chemotherapeutic agent used in high doses to cure chronic leukemia (Andersson, 2008) and osteosarcoma. The mechanism of encapsulating this drug in porous MOFs to prevent its degradation in aqueous solutions and prohibit its accumulation in the liver (Chalat et al., 2011). Four MOFs (MIL-88A, MIL-89, MIL-53 and MIL-100) were synthesized, then the drug was loaded in the MOFs through mixing and the release and loading were measured.

Another application of MOFs as anti-cancer drug carriers designed to load doxorubicin (DOX) in MIL-100 (Fe) (Reschiglian et al., 2014). DOX is a cytotoxic anti-cancer agent used for treatment of childhood solid tumors, breast cancer, soft tissue sarcomas, leukemias and lymphomas (Takemura & Fujiwara, 2007). But, it has much side effects including vomiting, nausea, hair loss, necrosis and oral mucositis, all of which limit its use (Liu et al., 2001) as a result of the cardiotoxicity during the first two or three days from its use (Chatterjee et al., 2010) which causes congestive heart failure and heart muscle disease (cardiomyopathy) (Chen et al., 2011). To overcome the above mentioned disadvantages DOX, it was loaded in MIL-100(Fe), (Figure 8) (Krishna & Mayer, 2000). Also, DOX was loaded in ZIF-8 particles as shown in (Figure 9) (Adhikar et al., 2015). Moreover DOX was encapsulated in another MOF that was prepared from gadolinium (III) salt and 1,4-bis(5-carboxy-1H-benzimidazole-2-yl)benzene)) linker (pDBI) shows the linker rotamers and space-filling model of Gd-pDBI as shown in (Figure10) [41].

Consequently, topotecan (TPT) is a chemotherapeutic agent loaded in nano MOFs, It is a hydrophobic drug is used to treat lung, cervical and ovarian tumors. The TPT was loaded in porous Fe-MIL-100 to control the release as shown in (Figure 11) (Nunzio & Agostoni, 2014).

Interestingly, 5-Fluorouracil (5-FU) is anti-cancer drug loaded in ZIF-8 (Sun et al., 2012). The anti-neoplastic agent is used to treat various cancers including liver, breast, brain, gastrointestinal tract, pancreas and other solid tumors. The hematological, gastrointestinal and dermatological side effects of

Figure 7. Attachment of ethoxysuccinato-cisplatin through amino groups on NH_2-BDC particles (Taylor-Pashow et al., 2009)

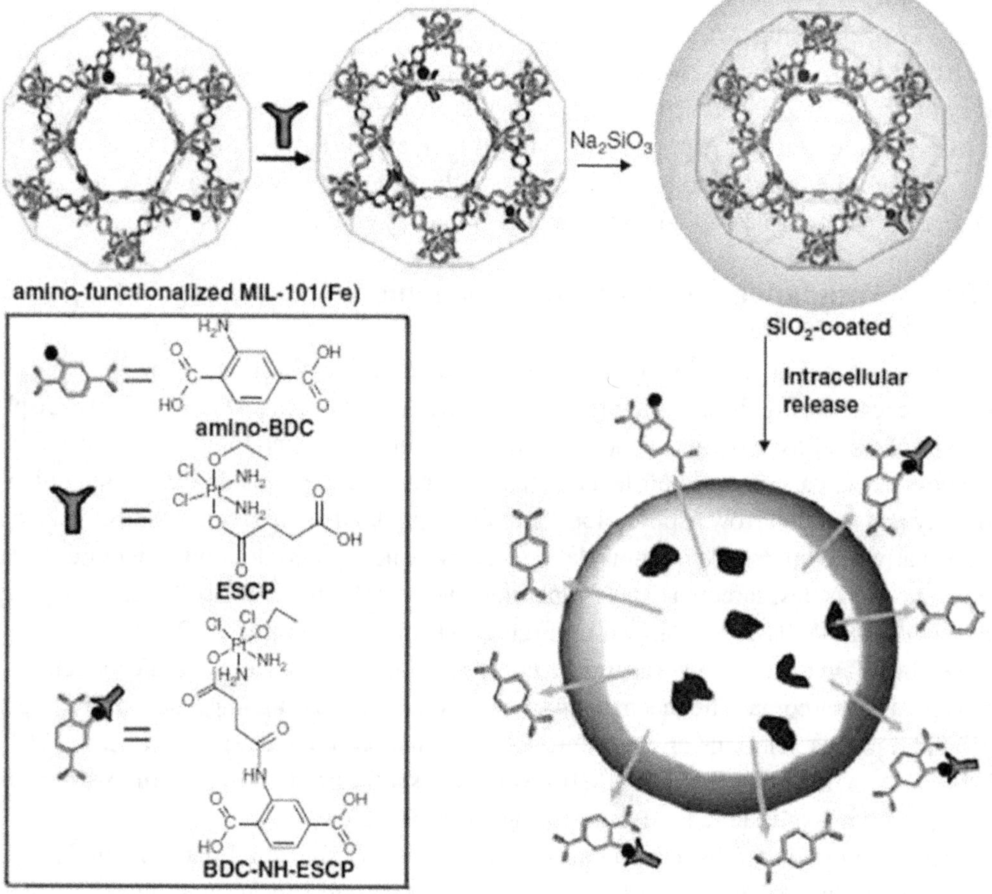

Figure 8. A schematic of MIL-100(Fe) with DOX (Reschiglian et al., 2014)

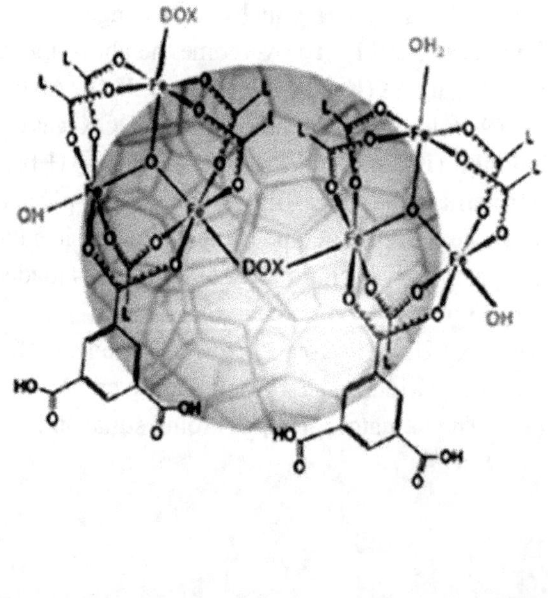

Figure 9. A schematic of doxorubicin encapsulation inside ZIF-8 framework (Adhikar et al., 2015)

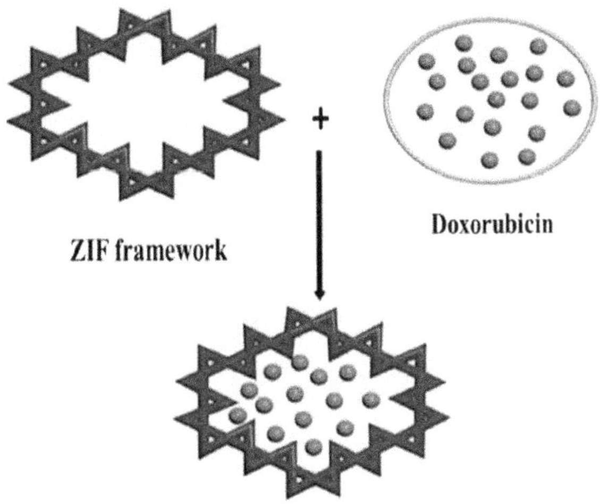

Figure 10. A representation of the linker rotamers and spacefilling model of gd-pdbi along the a-axis [41]

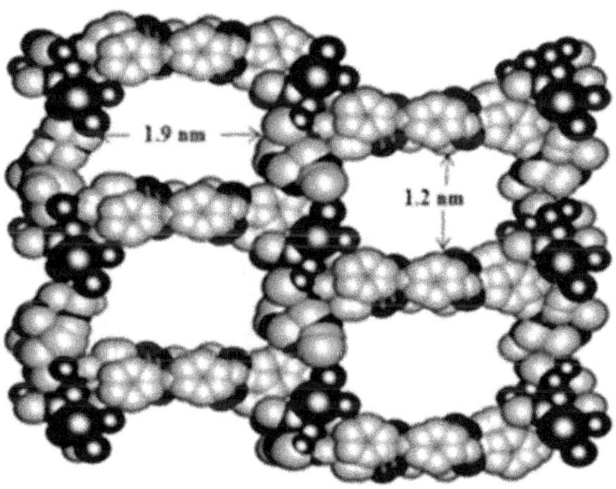

Figure 11. The scheme of entrapping TPT within MIL-100 Nano MOFs (Nunzio & Agostoni, 2014)

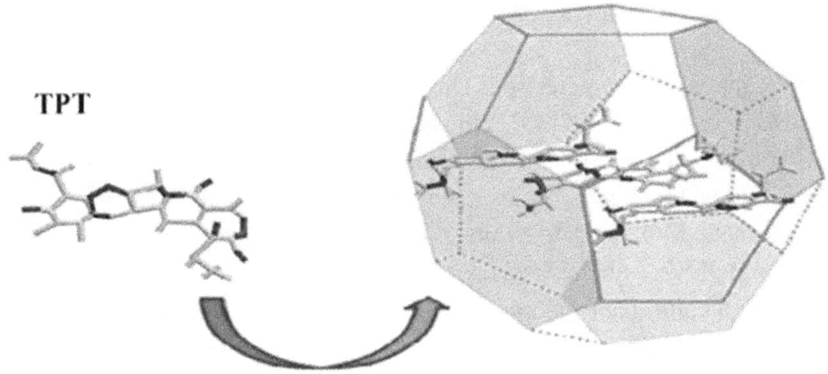

Figure 12. A schematic representation of fluorescein encapsulated inside ZIF-8 frameworks [24]

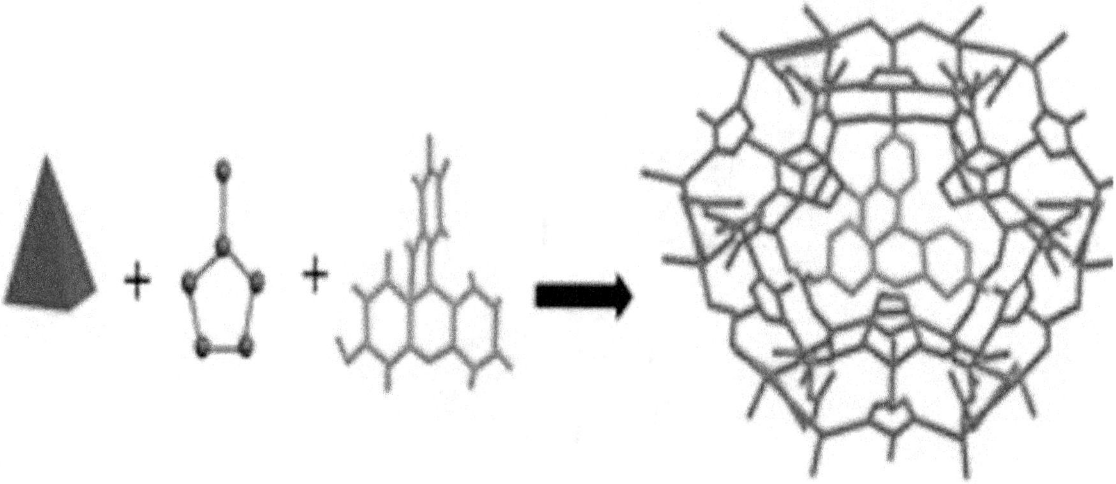

Figure 13. A schematic of Camptothecin encapsulated inside ZIF-8 frameworks [24]

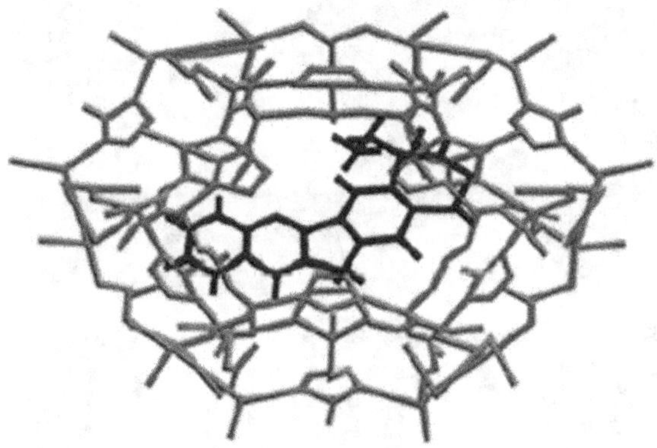

this cytotoxic drug limit its clinical applications, thus it was encapsulated in ZIF-8 to overcome these drawbacks, (Figure 12) (Arias, 2008).

Another chemotherapeutic drug, camptothecin (CPT), was loaded in ZIF-8, this drug was encapsulated in the MOFs to decreased the side effects. The loading of CPT molecules in ZIF-8 is shown in (Figure **13**) [24].

MOFs that exhibit magnetic characteristics can be utilized to target the drug to the desired location by applying an external magnetic field [45]. This method is applicable in the treatment of tumors with a known location (Tran & Wilson, 2011). Nimesulide (NIM), an anti-cancer agent used in the treatment of pancreatic [45], colorectal, breast, prostate and other tumors (Rainsford, 2005) was encapsulated in a three dimensional magnetic MOF [45], to reduce its adverse side effects (mainly gastrointestinal intolerance and ulcers). The MOF was synthesized using iron oxide (Fe_3O_4) and $Cu_3(BTC)_2$ [45].

Curcumin (CCM) is a natural anti-neoplastic agent was encapsulated in ZIF-8 (Zheng et al., 2015). It was found that the anti-cancer activity of CCM would prevent the progression of many tumors including breast, pancreatic, colon, head and neck, prostate and other cancers (Shanmugam et al., 2015). The disadvantage of CCM is the difficulty encountered upon its dministration in the systemic circulation due to its poor absorption and rapid metabolism (Anand et al., 2007; Zheng et al., 2015). To overcome these shortcomings, several nano carriers including silica nanoparticles (Zheng et al., 2015), liposomes (Mourtas et al., 2014), micelles (Sahu et al., 2008) and other particles were used to encapsulate CCM.

MOF in the Control Release of Anti-Bacterial Drug

Ciprofloxacin (CIP) is a fluoroquinolone antibiotic used to treat skin, bone, joint and respiratory infections . CIP, after penetrating into the cell's membrane, stops DNA synthesis by preventing the effect of DNA gyrase in the bacteria, when the DNA opens for replication, it will lack the enzyme and it will disappear (Zhanet et al., 2006). metal-organic framework Fe_3O_4@PAA@ZIF-8 as carrier for the drug delivery of CIP. Antibacterial properties of drug-loaded nano structures were compared with pure CIP against Escherichia coli and Staphylococcus aureus bacteria .It was found that ciprofloxacin drug loaded on the framework further inhibited the growth of the bacterium. So, Fe_3O_4@PAA@ZIF-8@CIP has a stronger antibacterial property than pure CIP. It can be concluded that the use of Fe_3O_4@PAA@ZIF-8 as a carrier, in addition to slow release in physiological conditions of the body, can also improve the antibacterial properties of the drug, (Figure 14) (Esfahanian et al., 2019).

Figure 14. Simulation of the preparation of the Fe_3O_4@PAA@ZIF-8 framework, drug loading, release and antibacterial testing (Esfahanian et al., 2019)

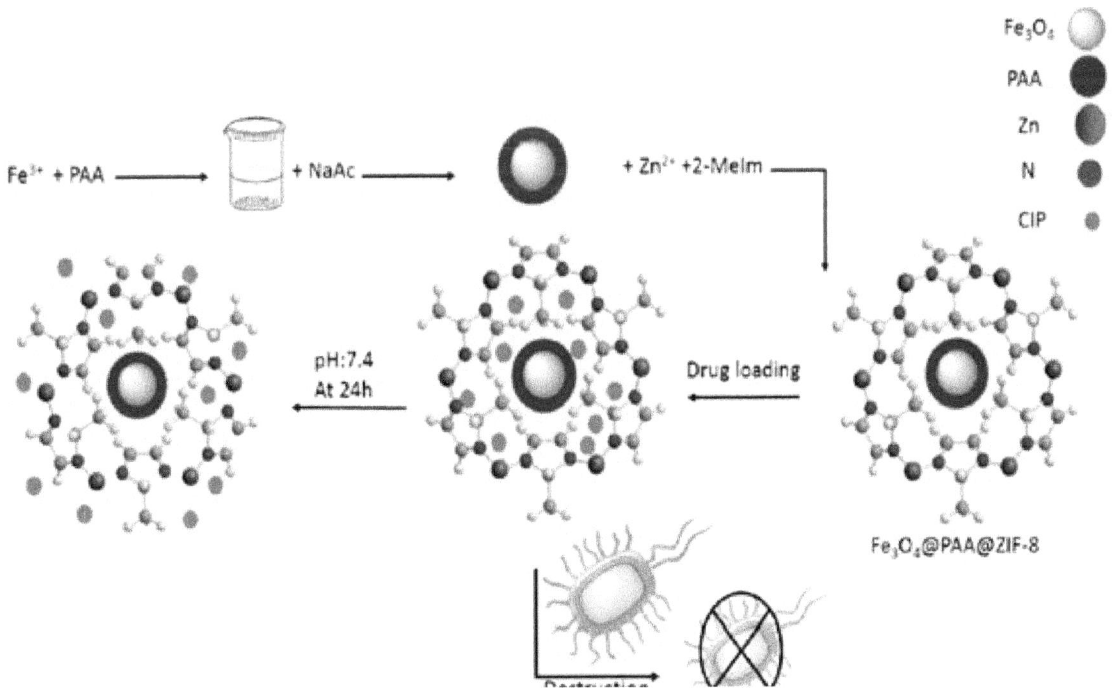

Sr /PTA MOF as a Drug Delivery System for Osteoarthritis Treatment

Osteoarthritis (OA) is main cause of disability and functional incapacity in adults with episodic pain, physical disability, inflammation and unbalanced subchondral bone (Gul et al., 2018).

Inflammatory factors inhibit collagen formation to degrade extracellular matrix, while pain decreased mobility in OA patients, which causes the loss of bone in the subchondral bone. Ketoprofen, 2-(3-benzoylphenyl)- propionic acid is one of the non-steroidal anti-inflammatory drugs (NSAIDs), has strong analgesic and antipyretic properties. Also, strontium may exert a positive effect on OA pathophysiology by inhibiting the synthesis of key factors leading to bone resorption associated with the (Tat et al., 2011). A Sr-based metal-organic framework (MOF) is introduced as ketoprofen carrier to form a comprehensive system for treating osteoarthritis (OA), and the drug loading amount and release rate is investigated. Structural characterization of the samples showed that Sr/PT A-MOF had good crystal morphology and structure, and chemical and thermal stability. Therefore, Sr/PTA-MOF-Ketoprofen (Figure 15) could be used for the treatment of osteoarthritis, which was a favorable compound for anti-inflammation, analgesia and maintaining bone balance (Li et al., 2019).

Figure 15. Sr/PTA-MOF-ketoprofen was considered as comprehensive treatment system of osteoarthritis (Li et al., 2019)

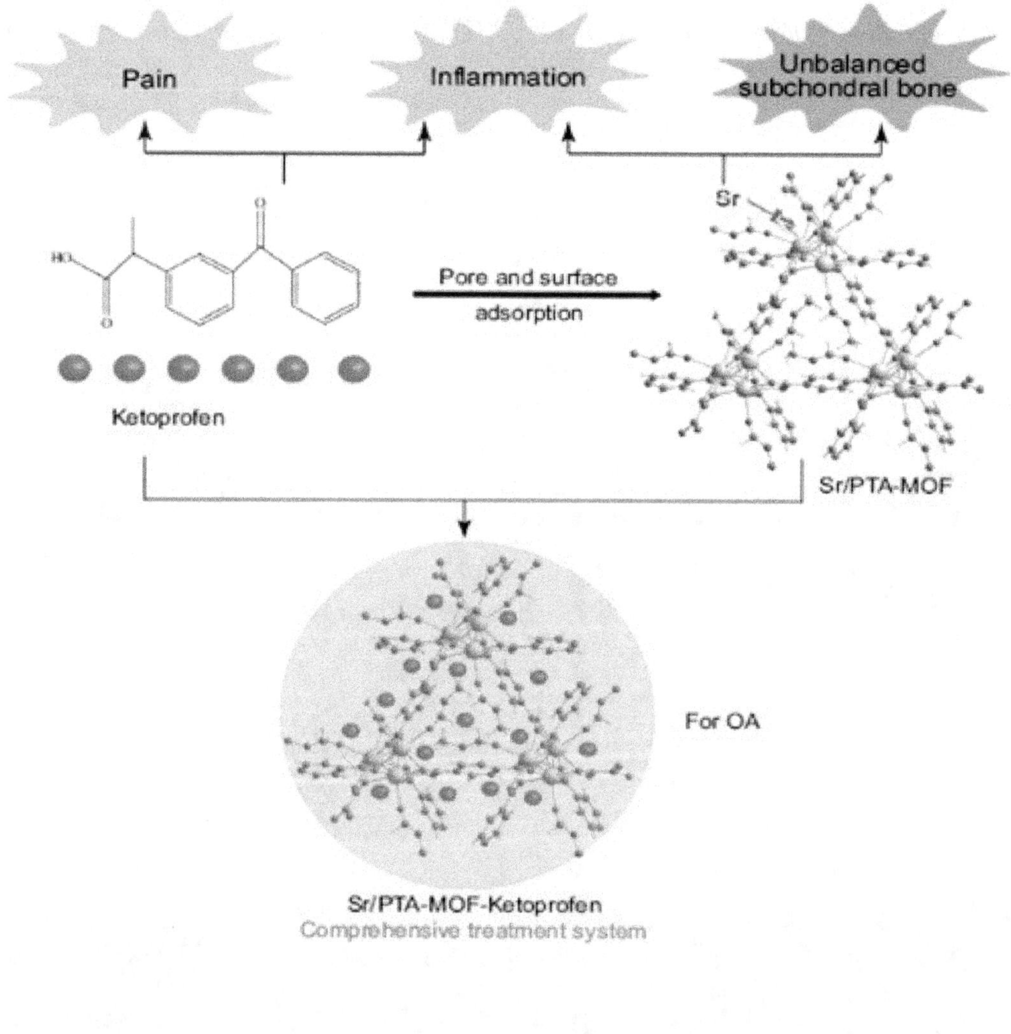

Iron-Based MOF as a Theranostic Carrier for Tuberculosis Therapy

Tuberculosis is an infectious disease caused by Mycobacterium tuberculosis (MTB) Isoniazid (Isonicotinic Acid Hydrazide, INH) is compatible antibiotic for use in theranostic drug delivery .It is effective against mycobacteria, especially Mycobacterium tuberculosis it prevent biosynthesis of the mycolic acid but by long-term therapy causes various side effects such as hepatitis (Ormerod & Horsfield, 1996), hepatic injury and neuropsychiatric disturbances (Gulbay et al., 2006), seizures (Bray, 1984) or polyneuropathy (Stettner et al., 2015). MOF Fe-MIL-101-NH_2 is used as a theranostic carrier of antituberculous drug. It has been noted that Fe-MIL-101-NH_2 particles were concentrated in the cell cytoplasmic area and able to release drug inside cells, which makes them favorable drug delivery system for TB therapy and inhibit clinically side effects of traditional drug administration. Designingof other anti-TB drugs into MOF structure seems to be favorable to prevent the development of MTB bacilli resistance.

Application of nanoMOF to Lung Drug Delivery

It has been detected that Iron-carboxilated nano MOF had a unique pH- and reversible aggregation behavior, they form aggregates at the neutral pH of the blood to be retained within the lung capillaries. Such royalty has been an advantage for the delivery of the anticancer drug gemcitabine monophosphate to this tissue, where aggregates were present in the lung as fast as 5 minutes post injection, the release of the drug payload occurred following particle degradation and surface charge changes. Antitumor efficacy was noted in an experimental model of metastatic lung carcinoma compared to the free drug (Figure16) (Simon-Yarza et al., 2017).

Figure 16. Schematic representation ofnano MOF reversible aggregation in vivo (Simon-Yarza et al., 2017)

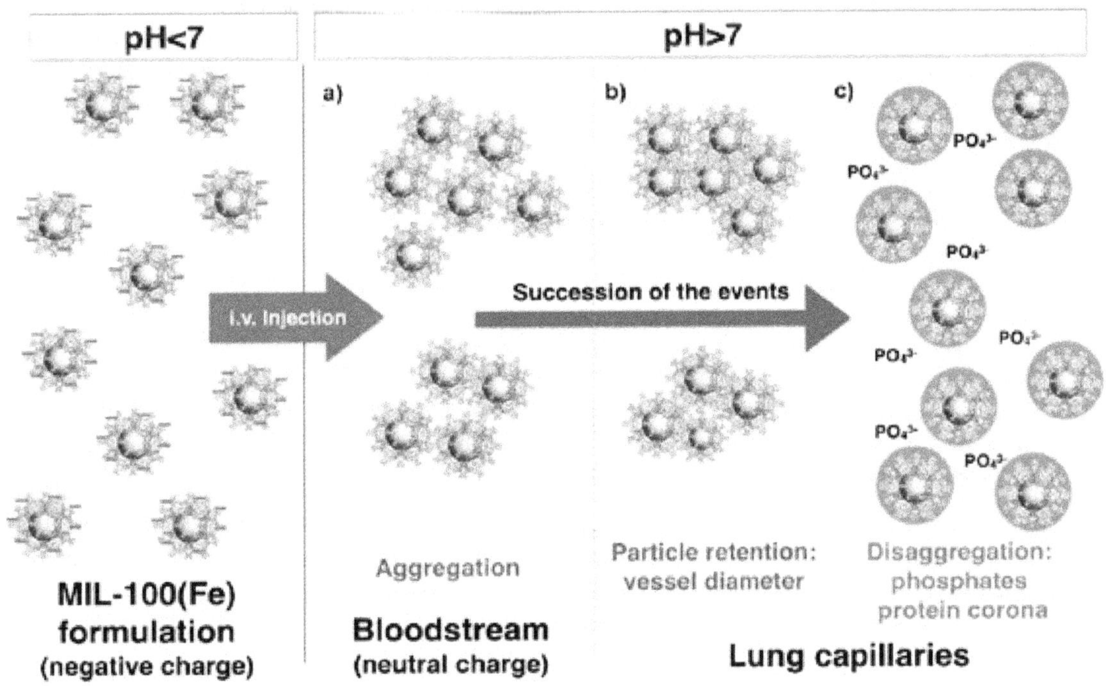

MOF as a Drug Delivery System for Anti-Inflammatory Drug

Flurbiprofen (FBP) is a nonsteroidal anti-inflammatory drug to decrease the side effects of FBP, numerous delivery systems have been synthesized by employing Ca-MOF and different Fe-MILs (MIL-53, MIL-100, and MIL-101), it was found that the highest FBP loading capacity with Fe-MIL-100 and lowest loading capacity with Ca-MOF (Thommes & Cychosz, 2014). The MOFs have a variety of loading capacities. Fe-MIL-100 and 101 were the best MOFs as drug delivery systems for FBP. The overall release profile was slow for FBP to be formulated as a controlled-release drug. So, the MOFs are a favorable material for FBP to be a carrier (Haydar et al., 2017).

Bone infections are accompanied by inflammation, soft tissue destruction and bone loss. The infection and inflammation are commonly controlled by antibiotics and analgesic drugs, while bone fillers are used to exchange the lost bone. As oral delivery drugs suffers from low bioavailability and low concentration at the infection site, so combinatorial drug delivery system (DDS) composed of monophasic calcium deficient hydroxyapatite (CDHA) nano carriers gave best results as antibacterial, anti-inflammatory and bone regenerative aspects of periodontal infection. The CDDS showed nearly 65-75% antibiotic release and an ibuprofen release of about25%. at the end of 72 h and 24 h, respectively. Compared to individual drug release profile from CDHA carriers, CDDS showed a more controlled release of antibiotics and higher ibuprofen release. The CDDS also demonstrated significant in vitro antibacterial activity, anti-inflammatory activity and biocompatibility suggesting possible application of the CDHA based combinatorial DDS for controlling of periodontitis as well as bone infections (Kalidoss et al., 2017).

MOFs in the Control Release of Antiarrythmia Drug

The zinc-based MOFs were designed for the controlled release of drugs. A purine nucleobase containing bio-MOFs were prepared by the reaction of bi-phenyldicarboxylic acid (BPDC), zinc acetate dehydrate and adenine, and then a bio-molecular ligand was attached to MOFs. The procainamide HCl drug was loaded into bio-MOFs through a cation-exchange process and the drug release was examined in water and PBS buffer. The PBS buffer permit the cation-exchange process, and thus procainamide was delivered with a significant burst release, where 80% of the loaded drugs were released within 20 h and the remaining drugs were completely released after 72 h (An et al., 2009).

MOFs in the Control Release of Analgesic Drug

Two chromium-based cubic MIL-100 and MIL-101 MOFs synthesized from di- or tri-carboxylic acid, ibuprofen drug was loaded into both MIL-100 and MIL-101, it showed an increase in drug loading. Three stages of drug release in SBF were observed from both MIL-100 and MIL-101 which were related to the position of drug within the cages of MOFs. When the drugs were close to walls, host guest interactions were occurred as was governed by anion-cation electrostatic and π-π interactions between the aromatic rings of ibuprofen and the organic part of MOFs. The release of ibuprofen from MIL-100 was completed after 3 days *in vitro*. The ibuprofen delivery from MIL-101 in the first 8 hours was controlled by diffusion and a complete drug release was achieved in 6 days (Horcajada et al., 2006). Also, chromium or iron containing flexible nano porous MIL-53 (Cr, Fe) MOFs showed *in vitro* delivery of ibuprofen in SBF solution. The ibuprofen was immersed in MIL-53 (Cr,Fe) in which 20 wt % of ibuprofen was adsorbed in 3 days and illustrated slow delivery of ibuprofen that up to 3 weeks (Horcajada et al., 2008).

MOFs in the Control Release of Amphiphilic Drug

Zeolite based MOFs was able to load and release caffeine, so caffeine was encapsulated with zeolite imidazolate ZIF-8 using two approaches. First by encapsulation during the synthesis of ZIF-8, which achieved 28.1% drug loading in 2h, and second, encapsulation, where the pre-synthesised ZIF-8 was used to load the drug, and consequently achieved 25.4 wt.% drug loading in 3 days. The caffeine-loaded ZIF-8 released 94% of loaded caffeine in 15 days. The release pattern and the amount of caffeine released from ZIF-8 were the same to ZIF-8. The interactions between the drug and the matrix of the substrate via van der Waals interactions between the caffeine and the imidazole methyl groups of ZIF-8 and or hydrogen bonding between the hydrogen atoms of methyl groups present in ZIF-8 with the carbonyl groups present in caffeine, and these two interactions caused to decrease the release of caffeine. It was found that zeolite-based MOFs may be considered as a drug substrate for control release of other drugs (Liedana et al., 2012).

MOFs in the Control Release of Bio-Active Gas

The MIL MOFs have been also used to load and release biologically active gas such as. NO and I_2 at a biological environment. It was synthesized a 1D-microporous 3D calcium tetracarboxylate MOFs (BioMIL-3) with Ca^{2+} Lewis acid sites (Figure17). The NO gas was adsorbed over the dried BioMIL-3 and the adsorbed amount of NO was low compared to the quantity from a theoretical calculation, thus not all NO were loading into the MOFs. The delivery of NO from MOFs was studied at room temperature over wet nitrogen and a minimum release of NO was achieved over 10 hours at a biological active concentration, which indicated a strong binding between the NO and the Ca^{2+} through chemisorption

Figure 17. Crystal structure of BioMIL-3. Top: view of the inorganic chains connected through the tetracarboxylate ligand;bottom (Miller et al., 2013)

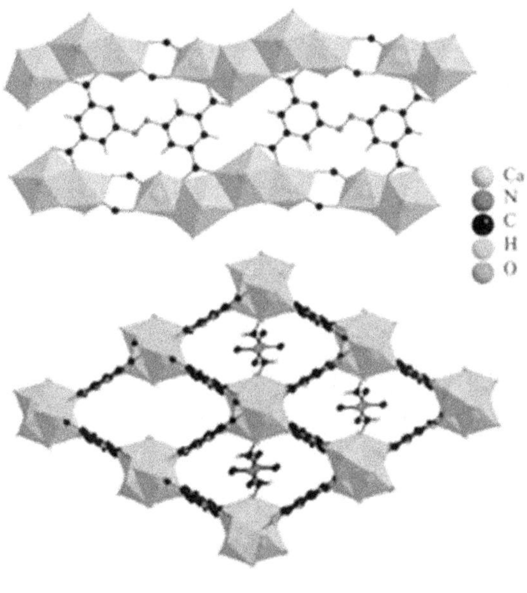

process caused difficulties to exchange the NO by water molecules in unsaturated sites of BioMIL-3 (Miller et al., 2013).

Moreover, Zn-based MOFs released the loaded I_2. These MOFs were constructed with Zn-lactate metal organic pillars binded by 4-pyridylbenzoate (pybz), when the MOFs were activated, they displayed best affinity to I_2, and the I_2 was slowly delivered. The delivery of I_2 was increased linearly with time, and the release was controlled by host-guest interactions. The release of I_2 showed a zero-order release behavior which is quite favorable for biological applications (Zeng et al., 2010).

FUTURE VIEW ON THE CONTROLLED RELEASE OF DRUGS FROM MOFS

The surface modification of MOFs is important in controlled release of drugs. The surface modification can help to decrease the interaction of MOFs with the biological medium or unwanted healthy cells and facilitates the drugs crossing the physical barriers, thus helps to achieve targeted drug delivery (Chowdhury, 2016; Fang et al., 2016). the chemical modifications applied on surfaces such as surface modification during MOFs synthesis through coordination modulation; post-synthetic surface modification and MOFs that are grown directly on the surfaces of others in the form of epitaxial and surface growth. These surface modification techniques are necessary for improvement the drug delivery systems using MOFs(McGuire & Forgan, 2015). Consequently, one of the barriers of using MOFs in the controlled drug release is the metal ions that are coordinated to the framework may cause toxicity, if the MOFs can be modified by attachment of appropriate polymers to the MOFs framework to develop the stability of MOFs and reduce the leak of metal ions, by introducing both cell targeting ligand and drug to the MOFs realize simultaneous cell targeting, and disease treatment. In addition, After the syntheses of MOFs some portion of the solvents remain inside the structure of MOFs and thus may affect the loading and releasing of drugs from MOFs. Because it may hinder the coordination on the unsaturated sites, and make them less attainable. In addition to steric hindrance and diffusion constraints can make the loading and releasing of drugs from MOFs more difficult and complicated. One ionic exchange drug was reported, which helped to eliminate the coordinated DMF without making any structural variations in the MOFs structure (Huxford et al., 2010).

Also, by depending on the structure, property, nature, and functional groups present in the MOFs and drugs, a series of interactions between drugs and MOFs are observed such as, electrostatic, hydrogen bonding, π-π interactions, chemisorption, and vander waals interactions slowed down the drug release from MOFs. Consequently, chirality of MOFs and delivery of chiral drug, zinc based chiral MOFs showed low amount of drug loading (33 wt.%) and total 85% of the loaded-drugs were released due to the structural differences in size, if any chiral matrices MOFs are properly selected, so a lot of chiral drugs are loaded and released easily, Also, there are various factors that controlled the drug release from MOFs the such as decomposition of MOFs, diffusion-controlled, pH-controlled and MOFs flexibility. However, stability of MOFs under humid condition, long-term stability, long-term toxicological, effects on body or tissues and level of biocompatibility of MOFs must take into account (Keskin & Kizilel, 2011).

Moreover, The size of MOFs is related to toxicity in the control release of the drug (Huxford et al., 2010; Keskin & Kizilel, 2011), If the sizes of MOFs can be minimized to the size of nano materials, it may help to develop pharmacokinetic properties of the encapsulated drugs, control toxicity and systemic circulation in blood (Huxford et al., 2010). In the case of silica nanoparticles (Slowing et al., 2008), if the particle size of MSNs in the range from 50–300 nm, it may allow a facile endocytosis by cells without causing any significant cytotoxicity. So, the particle sizes of MOFs may effect the load and release of drugs, and their cytotoxic manner. It is expected that in the near future, series of modeling on the MOFs-based drug control release, explain the structure-property relationships as a basis of giving us information of synthesizing new MOFs and nano MOFs and their required modifications, and how to choose the right MOFs for effective control release of drugs.

Each stimuli-responsive modality has its own conditions. For instance, the magnetic-responsive modality takes a large amount of sample to work, while the light-responsive modality suffers from shallow penetration. Although the multiple stimuli-responsive modality is used to overcome these limitations, more reseaches should be done on the synthesis of MOF-based stimuli-responsive drug delivery systems with low-toxicity and good colloidal stability. In addition of the investigations for the synthesis of biocompatible MOF-based stimuli responsive nano particles with good stability is still deficient. To decrease the toxicity of nanoparticles, employing "green" synthesis routes and choosing endogenous components as organic linkers to prepare metal biomolecule frameworks (bioMOFs) and by surface modification with functional molecules to increase stability of nano particles .(Li et al., 2018; Wang, 2018; Zhang et al., 2018)

Furthermore, scientists must test the efficiency of MOF-based stimuli-responsive drug delivery systems corresponding to clinical application include the examining of their stability, degradation mechanics and side effects on normal organs .Most of the studies on MOF-based stimulus-responsive nano particles for drug delivery were still in the in vitro step, more systematic in vivo examinations should be achieved for preclinical evaluation (Baati et al., 2013; Duan et al., 2017; Kundu et al., 2014; Liu et al., 2017).

REFERENCES

Adhikar, C., Das, A., & Chakrabort, A. (2015). Article. *Molecular Pharmaceutics*, *12*(9), 3158–3166. doi:10.1021/acs.molpharmaceut.5b00043 PMID:26196058

Alsaiari, S. K., Patil, S., Alyami, M., Alamoudi, K. O., Aleisa, F. A., Merzaban, J. S., Li, M., & Khashab, N. M. (2018). Endosomal Escape and Delivery of CRISPR/Cas9 Genome Editing Machinery Enabled by Nanoscale Zeolitic Imidazolate Framework. *Journal of the American Chemical Society*, *140*(1), 143–146. doi:10.1021/jacs.7b11754 PMID:29272114

An, J., Geib, S. J., & Rosi, N. L. (2009). Cation-Triggered Drug Release from a Porous Zinc−Adeninate Metal−Organic Framework. *Journal of the American Chemical Society*, *131*(24), 8376–8377. doi:10.1021/ja902972w PMID:19489551

Anand, P., Kunnumakkara, A. B., Newman, R. A., & Aggarwal, B. B. (2007). Bioavailability of Curcumin: Problems and Promises. *Molecular Pharmaceutics*, *4*(6), 807–818. doi:10.1021/mp700113r PMID:17999464

Andersson, B. S. (2008). *Biology of Blood and Marrow Transplantation*. doi:10.1016/j.bbmt.2008.03.009 PMID:18489993

Arias, J. L. (2008). Novel Strategies to Improve the Anticancer Action of 5-Fluorouracil by Using Drug Delivery Systems. *Molecules (Basel, Switzerland)*, *13*(10), 2340–2369. doi:10.3390/molecules13102340 PMID:18830159

Aznar, E., Oroval, M., Pascual, L., Murguia, J. R., Manez, R. M., & Sancenon, F. (2016). Gated Materials for On-Command Release of Guest Molecules. *Chemical Reviews*, *116*(2), 561–718. doi:10.1021/acs.chemrev.5b00456 PMID:26730615

Baati, T., Njim, L., Neffati, F., Kerkeni, A., Bouttemi, M., Gref, R., Najjar, M. F., Zakhama, A., Couvreur, P., Serre, C., & Horcajada, P. (2013). In depth analysis of the in vivo toxicity of nanoparticles of porous iron(iii) metal–organic frameworks. *Chemical Science (Cambridge)*, *4*(4), 1597. doi:10.1039/c3sc22116d

Banerjee, R., Sahoo, S. C., & Kundu, T. (2013). *World Patent WO2013051035*. Academic Press.

Bertini, I. (2007). *Biological inorganic chemistry:structure and reactivity*. University Science Books.

Bray, P. F. (1984). Isoniazid-induced acute toxic encephalopathy. *Neurology*, *34*(5), 703. doi:10.1212/WNL.34.5.703-a PMID:6538670

Burrows, A. D., Jurcic, M., Keenan, L. L., Lane, R. A., Mahon, M. F., Warren, M. R., Nowell, H., Paradowski, M., & Spencer, J. (2013). Incorporation by coordination and release of the iron chelator drug deferiprone from zinc-based metal–organic frameworks. *Chemical Communications (Cambridge)*, *49*(96), 11260. doi:10.1039/c3cc45689g

Cai, WWang, JChu, CChen, WWu, CLiu, G. (2019). Metal–Organic Framework-Based Stimuli-Responsive Systems for Drug Delivery. *Advancement of Science*, *6*(1801526), 1–20.

Cavka, J. H., Jakobsen, S., Olsbye, U., Guillou, N., Lamberti, C., Bordiga, S., & Lillerud, K. P. (2008). A New Zirconium Inorganic Building Brick Forming Metal Organic Frameworks with Exceptional Stability. *Journal of the American Chemical Society*, *130*(42), 13850–13851. doi:10.1021/ja8057953 PMID:18817383

Chalat, T., Horcajada, P., Couvreur, P., Serre, C., & Yahia, M. B. (2011). Article. *Nanomedicine (London)*, *6*(10), 1683–1695. doi:10.2217/nnm.11.69

Chatterjee, K., Zhang, J., Honbo, N. S., & Karliner, J. (2010). Doxorubicin Cardiomyopathy. *Cardiology*, *115*(2), 155–162. doi:10.1159/000265166 PMID:20016174

Chen, Y., Wan, Y., Wang, Y., Zhang, H., Jiao, Z., & International, J. (2011). Nanofabrication techniques for controlled drug-release devices. *Nanomedicine (London)*, *6*(1), 2321–2326. doi:10.2217/nnm.10.140

Chowdhuri, A., Bhattacharya, D., & Sahu, S. K. (2016). Article. *Journal of the Chemical Society, Dalton Transactions: Inorganic Chemistry*, *45*(7), 2963. doi:10.1039/C5DT03736K

Chowdhuri, A. R., Singh, T., Ghosh, S. K., & Sahu, S. K. (2016). Carbon Dots Embedded Magnetic Nanoparticles @Chitosan @Metal Organic Framework as a Nanoprobe for pH Sensitive Targeted Anticancer Drug Delivery. *ACS Applied Materials & Interfaces*, *8*(26), 16573–16583. doi:10.1021/acsami.6b03988 PMID:27305490

Chowdhury, M. A. (2016). The Silica-based Formulations for Drug Delivery, Bone Treatment, and Bone Regeneration. *ChemBioEng Rev*, *3*(5), 229–246. doi:10.1002/cben.201500026

Davis, M. E., Chen, Z., & Shin, D. M. (2008). Nanoparticle therapeutics: An emerging treatment modality for cancer. *Nature Reviews. Drug Discovery*, *7*(9), 771–782. doi:10.1038/nrd2614 PMID:18758474

Du, X., Fan, R. Q., Qiang, L. S., Xing, K., Ye, H. X., Ran, X. Y., Song, Y., Wang, P., & Yang, Y. L. (2017). Controlled Zn^{2+} -Triggered Drug Release by Preferred Coordination of Open Active Sites within Functionalization Indium Metal Organic Frameworks. *ACS Applied Materials & Interfaces*, *9*(34), 28939–28948. doi:10.1021/acsami.7b09227 PMID:28776972

Duan, F., Feng, X. C., Yang, X. J., Sun, W. T., Jin, Y., Liu, H. F., Ge, K., Li, Z. H., & Zhang, J. C. (2017). A simple and powerful co-delivery system based on pH-responsive metal-organic frameworks for enhanced cancer immunotherapy. *Biomaterials*, *122*, 23–33. doi:10.1016/j.biomaterials.2017.01.017 PMID:28107662

Dyson, P. J., & Sava, G. (2006). Metal-based antitumour drugs in the post genomic era. *Dalton Transactions (Cambridge, England)*, *16*(16), 1929–1933. doi:10.1039/b601840h PMID:16609762

Esfahanian, M., Ghasemzadeh, M. A., Mohammad, S., Razavian, H., & International, J. (2019). Synthesis, identification and application of the novel metal-organic framework Fe_3O_4 @PAA@ZIF-8 for the drug delivery of ciprofloxacin and investigation of antibacterial activity. *Artificial Cells, Nanomedicine, and Biotechnology*, *47*(1), 2024–2030. doi:10.1080/21691401.2019.1617729 PMID:31112049

Fang, J., Yang, Y., Xiao, W., Zheng, B., Lv, Y. B., Liu, X. L., & Ding, J. (2016). Extremely low frequency alternating magnetic field–triggered and MRI–traced drug delivery by optimized magnetic zeolitic imidazolate framework-90 nanoparticles. *Nanoscale*, *8*(6), 3259–3263. doi:10.1039/C5NR08086J PMID:26809987

Fathabadi, E. G., Shelling, A. N., & Al-Kassas, R. (2012). Nanocarrier systems for delivery of siRNA to ovarian cancer tissues. *Expert Opinion on Drug Delivery*, *9*(7), 743–754. doi:10.1517/17425247.2012.683173 PMID:22657869

Fletcher, A. J., Thomas, K. M., & Rosseinsky, M. J. (2005). Flexibility in metal-organic framework materials: Impact on sorption properties. *Journal of Solid State Chemistry*, *178*(8), 2491–2510. doi:10.1016/j.jssc.2005.05.019

Florea, A. M., & Büsselberg, D. (2011). Cisplatin as an Anti-Tumor Drug: Cellular Mechanisms of Activity, Drug Resistance and Induced Side Effects. *Cancers (Basel)*, *3*(1), 1351–1371. doi:10.3390/cancers3011351 PMID:24212665

Freeman, M. W., Arrott, A., & Watson, J. H. L. (1960). Magnetism in Medicine. *Journal of Applied Physics*, *31*(5), S404–S405. doi:10.1063/1.1984765

Furukawa, H., Gandara, F., Zhang, Y. B., Jiang, J., Queen, W. L., Hudson, M. R., & Yaghi, O. M. (2014). Water Adsorption in Porous Metal–Organic Frameworks and Related Materials. *Journal of the American Chemical Society*, *136*(11), 4369–4381. doi:10.1021/ja500330a PMID:24588307

Gao, P. F., Zheng, L. L., Liang, L. J., Yang, X. X., Li, Y. F., & Huang, C. Z. (2013). A new type of pH-responsive coordination polymer sphere as a vehicle for targeted anticancer drug delivery and sustained release. *Journal of Materials Chemistry. B, Materials for Biology and Medicine*, *1*(25), 3202. doi:10.1039/c3tb00026e PMID:32260920

Graf, N., & Lippard, S. J. (2012). Redox activation of metal-based prodrugs as a strategy for drug delivery. *Advanced Drug Delivery Reviews*, *64*(11), 993–1004. doi:10.1016/j.addr.2012.01.007 PMID:22289471

Gul, R., Ahmed, N., Ullah, N., Khan, M. I., Elaissari, A., & Rehman, A. (2018). Biodegradable Ingredient-Based Emulgel Loaded with Ketoprofen Nanoparticles. *AAPS PharmSciTech*, *19*(4), 1869–1881. doi:10.120812249-018-0997-0 PMID:29651679

Gulbay, B. E., Gurkan, O. U., Yildiz, O. A., Onen, Z. P., Erkekol, F. O., Baccioglu, A., & Acıcan, T. (2006). Side effects due to primary antituberculosis drugs during the initial phase of therapy in 1149 hospitalized patients for tuberculosis. *Respiratory Medicine*, *100*(10), 1834–1842. doi:10.1016/j.rmed.2006.01.014 PMID:16517138

Gupta, M., Agrawal, U., & Vyas, S. P. (2012). Nanocarrier-based topical drug delivery for the treatment of skin diseases. *Expert Opinion on Drug Delivery*, *9*(7), 783–804. doi:10.1517/17425247.2012.686490 PMID:22559240

Haydar, M. A., Abid, H. R., Sunderland, B., & Wang, S. (2017).. . *Drug Design, Development and Therapy*, *11*, 2685. doi:10.2147/DDDT.S145716 PMID:28979098

Horcajada, P., Chalati, T., Serre, C., Gillet, B., Sebr, C., Baati, T., Eubank, J. F., Heurtaux, D., Clayette, C., Kreuz, C., Chang, J. S., Hwang, Y. K., Marsaud, V., Bories, P. N., Cynober, L., Gil, S., Férey, G., Couvreur, P., & Gref, R. (2010). Porous metal–organic-framework nanoscale carriers as a potential platform for drug delivery and imaging. *Nature Materials*, *9*(2), 172–178. doi:10.1038/nmat2608 PMID:20010827

Horcajada, P., Gref, R., Baati, T., Allan, P. K., Maurin, G., Couvreur, P., Férey, G., Morris, R. E., & Serre, C. (2012). Metal—Organic frameworks in biomedicine. *Chemical Reviews*, *112*(2), 1232–1268. doi:10.1021/cr200256v PMID:22168547

Horcajada, P., Serre, C., Maurin, G., Ramsahye, N. A., Balas, F., Vallet-Regi, M., Sebban, M., Taulelle, F., & Férey, G. (2008). Flexible Porous Metal-Organic Frameworks for a Controlled Drug Delivery. *Journal of the American Chemical Society*, *130*(21), 6774–6780. doi:10.1021/ja710973k PMID:18454528

Horcajada, P., Serre, C., Vallet-Regi, M., Sebban, M., Taulelle, F., & Ferey, G. (2006). Metal–Organic Frameworks as Efficient Materials for Drug Delivery. *Angewandte Chemie International Edition*, *45*(36), 5974–5978. doi:10.1002/anie.200601878 PMID:16897793

Hu, Q., Yu, J., Liu, M., Liu, A., Dou, Z., & Yang, Y. (2014). A Low Cytotoxic Cationic Metal–Organic Framework Carrier for Controllable Drug Release. *Journal of Medicinal Chemistry*, *57*(13), 5679–5685. doi:10.1021/jm5004107 PMID:24922463

Huxford, R. C., Della Rocca, J., & Lin, W. (2010). Metal–organic frameworks as potential drug carriers. *Current Opinion in Chemical Biology*, *14*(2), 262–268. doi:10.1016/j.cbpa.2009.12.012 PMID:20071210

Jiang, K., Zhang, L., Hu, Q., Zhao, D., Xia, T., Lin, W., Yang, Y., Cui, Y., Yang, Y., & Qian, G. (2016). Pressure controlled drug release in a Zr-cluster-based MOF. *Journal of Materials Chemistry. B, Materials for Biology and Medicine*, *4*(39), 6398–6401. doi:10.1039/C6TB01756H PMID:32263448

Kalidoss, M., Kumar, T. S. S., Rubaiya, Y., & Doble, M. (2017). Article. *Biomater Tissue Technol*, *1*(1), 1–5.

Kamaly, N. D., Miller, A. D., & Bell, J. (2010).. . *Current Topics in Medicinal Chemistry*, *10*(12), 1158. doi:10.2174/156802610791384199 PMID:20388110

Ke, F., Yuan, Y. P., Qiu, L. G., Shen, Y. H., Xie, A. J., Zhu, J. F., Tian, X. Y., & Zhang, L. D. (2011). Facile fabrication of magnetic metal–organic framework nanocomposites for potential targeted drug delivery. *Journal of Materials Chemistry*, *21*(11), 3843. doi:10.1039/c0jm01770a

Keskin, S., & Kizilel, S. (2011). Biomedical Applications of Metal Organic Frameworks. *Industrial & Engineering Chemistry Research*, *50*(4), 1799–1812. doi:10.1021/ie101312k

Kim, B. J., Cheong, H., Hwang, B. H., & Cha, H. J. (2015). Mussel-Inspired Protein Nanoparticles Containing Iron(III)-DOPA Complexes for pH-Responsive Drug Delivery. *Angewandte Chemie International Edition*, *54*(25), 7318–7322. doi:10.1002/anie.201501748 PMID:25968933

Krishna, R., & Mayer, L. D. (2000). Multidrug resistance (MDR) in cancer. *European Journal of Pharmaceutical Sciences*, *11*(4), 265–283. doi:10.1016/S0928-0987(00)00114-7 PMID:11033070

Kumar, C. S., & Mohammad, F. (2011). Magnetic nanomaterials for hyperthermia-based therapy and controlled drug delivery. *Advanced Drug Delivery Reviews*, *63*(9), 789–808. doi:10.1016/j.addr.2011.03.008 PMID:21447363

Kundu, T., Mitra, S., Patra, P., Goswami, A., Diaz Diaz, D., & Banerjee, R. (2014). Mechanical Downsizing of a Gadolinium(III)-based Metal-Organic Framework for Anticancer Drug Delivery. *Chemistry (Weinheim an der Bergstrasse, Germany)*, *20*(33), 10514–10518. doi:10.1002/chem.201402244 PMID:25044210

Li, L., Wang, J. Q., Kong, H. R., Zeng, Y., & Liu, G. (2018). Functional biomimetic nanoparticles for drug delivery and theranostic applications in cancer treatment. *Science and Technology of Advanced Materials*, *19*(1), 771–790. doi:10.1080/14686996.2018.1528850 PMID:30815042

Li, Z., Peng, Y., Xia, X., Cao, Z., Deng, Y., & Tang, B. (2019). Article. *Scientific Reports*, *9*(17570), 1-7.

Liang, K., Ricco, R., Doherty, C. M., Styles, M. J., Bell, S., Kirby, N., Mudie, S., Haylock, D., Hill, A. J., Doonan, C. J., & Falcaro, P. (2015). Biomimetic mineralization of metal-organic frameworks as protective coatings for biomacromolecules. *Nature Communications*, *6*(1), 7240. doi:10.1038/ncomms8240 PMID:26041070

Liedana, N., Galve, A., Rubio, C., Tellez, C., & Coronas, J. (2012). CAF@ZIF-8: One-Step Encapsulation of Caffeine in MOF. *ACS Applied Materials & Interfaces*, *4*(9), 5016–5021. doi:10.1021/am301365h PMID:22834763

Lin, W., Hu, Q., Jiang, K., Yang, Y., Cui, Y., & Qian, G. (2016). A porphyrin-based metal–organic framework as a pH-responsive drug carrier. *Journal of Solid State Chemistry*, *237*, 307–312. doi:10.1016/j.jssc.2016.02.040

Lin, W. X., Hu, Q., Yu, J. C., Jiang, K., Yang, Y. Y., Xiang, S. C., Cui, Y. J., Yang, Y., Wang, Z. Y., & Qian, G. D. (2016). Low Cytotoxic Metal-Organic Frameworks as Temperature-Responsive Drug Carriers. *ChemPlusChem*, *81*(8), 804–810. doi:10.1002/cplu.201600142 PMID:31968821

Liu, J. J., Wang, H. R., Yi, X., Chao, Y., Geng, Y. H., Xu, L. G., Yang, K., & Liu, Z. (2017). pH-Sensitive Dissociable Nanoscale Coordination Polymers with Drug Loading for Synergistically Enhanced Chemoradiotherapy. *Advanced Functional Materials*, *27*(44), 1703832. doi:10.1002/adfm.201703832

Liu, Z., Cheung, R., Wu, X.Y., Ballinger, J.R., Bendayan, R., & Rauth, A.M.. (2001). Article. *Controlled Release, 77*(3), 213-224.

Lucena, F. R. S., & de Araujo, L. C. C. (2013). Rodrigues MdD, da Silva TG, Pereira VRA, Militao GCG. *Biomedicine and Pharmacotherapy, 67*(8), 707. doi:10.1016/j.biopha.2013.06.003 PMID:24095176

Lzaro, I. A., Haddad, S., Sacca, S., Orellana-Tavra, C., Fairen-Jimenez, D., & Forgan, R. S. (2017). Article. *Chem, 2*, 561. doi:10.1016/j.chempr.2017.02.005 PMID:28516168

McGuire, C. V., & Forgan, R. S. (2015). The surface chemistry of metal–organic frameworks. *Chemical Communications (Cambridge), 51*(25), 5199–5217. doi:10.1039/C4CC04458D

McKinlay, A. C., Morris, R. E., Horcajada, P., Ferey, G., Gref, R., Couvreur, P., & Serre, C. (2010). BioMOFs: Metal-Organic Frameworks for Biological and Medical Applications. *Angewandte Chemie International Edition, 49*(36), 6260–6266. doi:10.1002/anie.201000048 PMID:20652915

Meng, X. S., Gui, B., Yuan, D. Q., Zeller, M., & Wang, C. (2016). Mechanized azobenzene-functionalized zirconium metal-organic framework for on-command cargo release. *Science Advances, 2*(8), e1600480. doi:10.1126ciadv.1600480 PMID:27493996

Miller, S. R., Alvarez, E., Fradcourt, L., Devic, T., Wuttke, S., Wheatley, P. S., Steunou, N., Bonhomme, C., Gervais, C., Laurencin, D., Morris, R. E., Vimont, A., Daturi, M., Horcajada, P., & Serre, C. (2013). A rare example of a porous Ca-MOF for the controlled release of biologically active NO. *Chemical Communications (Cambridge), 49*(71), 7773. doi:10.1039/c3cc41987h

Mourtas, S., Lazar, A. N., Markoutsa, E., Duyckaerts, C., & Antimisiaris, S. G. (2014). Multifunctional nanoliposomes with curcumin–lipid derivative and brain targeting functionality with potential applications for Alzheimer disease. *European Journal of Medicinal Chemistry, 80*, 175–183. doi:10.1016/j.ejmech.2014.04.050 PMID:24780594

Nagata, S., Kokado, K., & Sada, K. (2015). Metal–organic framework tethering PNIPAM for ON–OFF controlled release in solution. *Chemical Communications (Cambridge), 51*(41), 8614–8617. doi:10.1039/C5CC02339D

Nunzio, M., & Agostoni, V. (2014). Article. *Journal of Medicinal Chemistry, 57*(2), 411–420. PMID:24345217

Ogoshi, T., Kanai, S., Fujinami, S., Yamagishi, T., & Nakamoto, Y. (2008). *para* -Bridged Symmetrical Pillar[5]arenes: Their Lewis Acid Catalyzed Synthesis and Host–Guest Property. *Journal of the American Chemical Society, 130*(15), 5022–5023. doi:10.1021/ja711260m PMID:18357989

Ormerod, L. P., & Horsfield, N. (1996). Frequency and type of reactions to antituberculosis drugs: Observations in routine treatment. *Tubercle and Lung Disease, 77*(1), 37–42. doi:10.1016/S0962-8479(96)90073-8 PMID:8733412

Peller, M., Boll, K., Zimpel, A., & Wuttke, S. (2018). Metal–organic framework nanoparticles for magnetic resonance imaging. *Inorganic Chemistry Frontiers, 5*(8), 1760–1779. doi:10.1039/C8QI00149A

Rainsford, K. D. (2005). *Nimesulide – Actions and Uses*. Birkhäuser Basel. doi:10.1007/3-7643-7410-1

Ren, H., Zhang, L., An, J., Wang, T., Li, L., Si, X., He, L., Wu, X., Wang, C., & Su, Z. (2014). Polyacrylic acid@zeolitic imidazolate framework-8 nanoparticles with ultrahigh drug loading capability for pH-sensitive drug release. *Chemical Communications (Cambridge)*, *50*(8), 1000–1002. doi:10.1039/C3CC47666A

Reschiglian, P., Gref, R., Monti, S., Manet, I., Agostoni, V., Reschigl, P., Gref, R., & Monti, S. (2014). Article. *Physical Chemistry*, *118*, 8532–8539.

Rieter, W. J., Pott, K. M., Taylor, K. M. L., & Lin, W. (2008). Nanoscale Coordination Polymers for Platinum-Based Anticancer Drug Delivery. *Journal of the American Chemical Society*, *130*(35), 11584–11585. doi:10.1021/ja803383k PMID:18686947

Rowe, M. D., Thamm, D. H., Kraft, S. L., & Boyes, S. G. (2009). Polymer-Modified Gadolinium Metal-Organic Framework Nanoparticles Used as Multifunctional Nanomedicines for the Targeted Imaging and Treatment of Cancer. *Biomacromolecules*, *10*(4), 983–993. doi:10.1021/bm900043e PMID:19290624

Sahoo, S. C., Kundu, T., & Banerjee, R. (2011). Helical Water Chain Mediated Proton Conductivity in Homochiral Metal–Organic Frameworks with Unprecedented Zeolitic *unh* -Topology. *Journal of the American Chemical Society*, *133*(44), 17950–17958. doi:10.1021/ja2078637 PMID:21919488

Sahu, A., Kasoju, N., & Bora, U. (2008). Fluorescence Study of the Curcumin−Casein Micelle Complexation and Its Application as a Drug Nanocarrier to Cancer Cells. *Biomacromolecules*, *9*(10), 2905–2912. doi:10.1021/bm800683f PMID:18785706

Seo, Y. K., Yoon, J. W., Lee, J. S., Hwang, Y. K., Jun, C. H., Chang, J. S., Wuttke, S., Bazin, P., Vimont, A., Daturi, M., Bourrelly, S., Llewellyn, P. L., Horcajada, P., Serre, C., & Férey, G. (2012). Energy-Efficient Dehumidification over Hierachically Porous Metal-Organic Frameworks as Advanced Water Adsorbents. *Advanced Materials*, *24*(6), 806–810. doi:10.1002/adma.201104084 PMID:22162212

Sethi, K., Sharma, S., & Roy, I. (2016). Article. *RSC Adv.*, *6*, 76861.

Shanmugam, M. K., Rane, G., Kanchi, M. M., Arfuso, F., Chinnathambi, A., Zayed, M. E., Tan, B. K., Kumar, A. P., & Sethi, G. (2015). The Multifaceted Role of Curcumin in Cancer Prevention and Treatment. *Molecules (Basel, Switzerland)*, *20*(2), 2728–2769. doi:10.3390/molecules20022728 PMID:25665066

Sharma, S., Sethi, K., & Roy, I. (2017). Magnetic nanoscale metal–organic frameworks for magnetically aided drug delivery and photodynamic therapy. *New Journal of Chemistry*, *41*(20), 11860–11866. doi:10.1039/C7NJ02032E

Si, W., Xin, P., Li, Z. T., & Hou, J. L. (2015). Tubular Unimolecular Transmembrane Channels: Construction Strategy and Transport Activities. *Accounts of Chemical Research*, *48*(6), 1612–1619. doi:10.1021/acs.accounts.5b00143 PMID:26017272

Simon-Yarza, T., Giménez-Marqués, M., Mrimi, R., Mielcarek, A., Gref, R., Horcajada, P., Serre, C., & Couvreur, P. (2017). A Smart Metal–Organic Framework Nanomaterial for Lung Targeting. *Angewandte Chemie International Edition*, *56*(49), 15565–15569. doi:10.1002/anie.201707346 PMID:28960750

Slowing, I. I., Vivero-Escoto, J. L., Wu, C. W., & Lin, V. (2008). Article. *Advanced Drug Delivery Reviews*, *60*(11), 1278–1288. doi:10.1016/j.addr.2008.03.012 PMID:18514969

Smaldone, R. A., Forgan, R. S., Furukawa, H., Gassensmith, J. J., Slawin, A. M. Z., Yaghi, O. M., & Stoddart, J. F. (2010). Metalorganic frameworks from edible natural products. *Angewandte Chemie International Edition*, *49*(46), 8630–8634. doi:10.1002/anie.201002343 PMID:20715239

Stettner, M., Steinberger, D., Hartmann, C.J., Pabst, T., Konta, L., & Hartung, H.P. (2015). Article. *Brain Behav.*, *5*(8).

Strutt, N. L., Zhang, H., Schneebeli, S. T., & Stoddart, J. F. (2014). Functionalizing Pillar[*n*]arenes. *Accounts of Chemical Research*, *47*(8), 2631–2642. doi:10.1021/ar500177d PMID:24999824

Sun, C. Y., Qin, C., Wang, C. G., Su, Z. M., Wang, S., Wang, X. L., Yang, G.-S., Shao, K.-Z., Lan, Y.-Q., & Wang, E.-B. (2011). Chiral Nanoporous Metal-Organic Frameworks with High Porosity as Materials for Drug Delivery. *Advanced Materials*, *23*(47), 5629–5632. doi:10.1002/adma.201102538 PMID:22095878

Sun, C. Y., Qin, C., Wang, X. L., Yang, G. S., Shao, K. Z., Lan, Y. Q., Su, Z. M., Huang, P., Wang, C. G., & Wang, E. B. (2012). Zeolitic imidazolate framework-8 as efficient pH-sensitive drug delivery vehicle. *Dalton Transactions (Cambridge, England)*, *41*(23), 6906–6909. doi:10.1039/c2dt30357d

Takemura, G., & Fujiwara, H. (2007). Doxorubicin-Induced Cardiomyopathy. *Progress in Cardiovascular Diseases*, *49*(5), 330–352. doi:10.1016/j.pcad.2006.10.002 PMID:17329180

Tan, L. L., Li, H., Qiu, Y. C., Chen, D. X., Wang, X., Pan, R. Y., Wang, Y., Zhang, S. X. A., Wang, B., & Yang, Y. W. (2015). Stimuli-responsive metal–organic frameworks gated by pillar[5]arene supramolecular switches. *Chemical Science (Cambridge)*, *6*(3), 1640–1644. doi:10.1039/C4SC03749A

Tan, L. L., Song, N., Zhang, S. X. A., Li, H., Wang, B., & Yang, Y. W. (2016). Ca^{2+}, pH and thermo triple-responsive mechanized Zr-based MOFs for on-command drug release in bone diseases. *Journal of Materials Chemistry. B, Materials for Biology and Medicine*, *4*(1), 135–140. doi:10.1039/C5TB01789K PMID:32262817

Tat, S. K., Pelletier, J. P., Mineau, F., Caron, J., & Martel-Pelletier, J. (2011). Strontium ranelate inhibits key factors affecting bone remodeling in human osteoarthritic subchondral bone osteoblasts. *Bone*, *49*(3), 559–567. doi:10.1016/j.bone.2011.06.005 PMID:21700005

Taylor-Pashow, K. M. L., Rocca, J. D., Xie, Z., Tran, S., & Lin, W. (2009). Postsynthetic Modifications of Iron-Carboxylate Nanoscale Metal–Organic Frameworks for Imaging and Drug Delivery. *Journal of the American Chemical Society*, *131*(40), 14261–14263. doi:10.1021/ja906198y PMID:19807179

Thommes, M., & Cychosz, K. A. (2014). Physical adsorption characterization of nanoporous materials: Progress and challenges. *Adsorption*, *2*(20), 233–250. doi:10.100710450-014-9606-z

Torchilin, V. P. (2005). Recent advances with liposomes as pharmaceutical carriers. *Nature Reviews. Drug Discovery*, *4*(2), 145–160. doi:10.1038/nrd1632 PMID:15688077

Tran, L. A., & Wilson, L. J. (2011). Nanomedicine: Making controllable magnetic drug delivery possible for the treatment of breast cancer. *Breast Cancer Research*, *13*(2), 303. doi:10.1186/bcr2830 PMID:21457524

Vallet-Regí, M., Balas, F., & Arcos, D. (2007). Vallet-Reg_ M, Balas F, Arcos D.Mesoporous materials for drug delivery. *Angewandte Chemie International Edition*, *46*(40), 7548–7558. doi:10.1002/anie.200604488

Wang, D., Zhou, J., Chen, R., Shi, R., Zhao, G., Xia, G., Li, R., Liu, Z., Tian, J., Wang, H., Guo, Z., Wang, H., & Chen, Q. (2016). Controllable synthesis of dual-MOFs nanostructures for pH-responsive artemisinin delivery, magnetic resonance and optical dual-model imaging-guided chemo/photothermal combinational cancer therapy. *Biomaterials*, *100*, 27–40. doi:10.1016/j.biomaterials.2016.05.027 PMID:27240160

Wang, D. D., Zhou, J. J., Chen, R. H., Shi, R. H., Xia, G. L., Zhou, S., Liu, Z. B., Zhang, N. Q., Wang, H. B., Guo, Z., & Chen, Q. W. (2016). Magnetically guided delivery of DHA and Fe ions for enhanced cancer therapy based on pH-responsive degradation of DHA-loaded Fe 3 O 4 @C@MIL-100(Fe) nanoparticles. *Biomaterials*, *107*, 88–101. doi:10.1016/j.biomaterials.2016.08.039 PMID:27614161

Wang, H. N., Yang, G. S., Wang, X. L., & Su, Z. M. (2013). pH-induced different crystalline behaviors in extended metal–organic frameworks based on the same reactants. *Journal of the Chemical Society, Dalton Transactions: Inorganic Chemistry*, *42*(18), 6294. doi:10.1039/c3dt32958e

Wang, J. Q. (2018). Article. *Angewandte Chemie International Edition*, *57*, 3008. doi:10.1002/anie.201711705 PMID:29450950

Wang, X. G., Dong, Z. Y., Cheng, H., Wan, S. S., Chen, W. H., Zou, M. Z., Huo, J. W., Deng, H. X., & Zhang, X. Z. (2015). A multifunctional metal–organic framework based tumor targeting drug delivery system for cancer therapy. *Nanoscale*, *7*(38), 16061–16070. doi:10.1039/C5NR04045K PMID:26372069

Wu, Y. N., Zhou, M., Li, S., Li, Z., Li, J., Wu, B., Li, G., Li, F., & Guan, X. (2014). Magnetic Metal-Organic Frameworks: γ-Fe $_2$ O $_3$ @MOFs via Confined In Situ Pyrolysis Method for Drug Delivery. *Small*, *10*(14), 2927–2936. doi:10.1002mll.201400362 PMID:24644065

Xue, M., Chi, X. D., Zhang, Z. B., & Huang, F. H. (2012). Pillararenes, A New Class of Macrocycles for Supramolecular Chemistry. *Accounts of Chemical Research*, *45*(8), 1294–1308. doi:10.1021/ar2003418 PMID:22551015

Yang, Q., Zhu, Y., Luo, B., Lan, F., Wu, Y., & Gu, Z. W. (2017). pH-Responsive magnetic metal–organic framework nanocomposites for selective capture and release of glycoproteins. *Nanoscale*, *9*(2), 527–532. doi:10.1039/C6NR08071E PMID:27966715

Yang, Y., Hu, Q., Zhang, Q., Jiang, K., Lin, W., Yang, Y., Cui, Y., & Qian, G. (2016). A Large Capacity Cationic Metal–Organic Framework Nanocarrier for Physiological pH Responsive Drug Delivery. *Molecular Pharmaceutics*, *13*(8), 2782–2786. doi:10.1021/acs.molpharmaceut.6b00374 PMID:27414996

Yoo, J. W., & Lee, C. H. (2006). Drug delivery systems for hormone therapy. *Journal of Controlled Release*, *112*(1), 1–14. doi:10.1016/j.jconrel.2006.01.021 PMID:16530874

Zeng, M. H., Wang, Q. X., Tan, Y. X., Hu, S., Zhao, H. X., Long, L. S., & Kurmoo, M. (2010). Rigid Pillars and Double Walls in a Porous Metal-Organic Framework: Single-Crystal to Single-Crystal, Controlled Uptake and Release of Iodine and Electrical Conductivity. *Journal of the American Chemical Society*, *132*(8), 2561–2563. doi:10.1021/ja908293n PMID:20131766

Zhanet, G., Fontaine, S., & Schurek, K. (2006). Article. *Treatments in Respiratory Medicine*, *5*, 437–465. doi:10.2165/00151829-200605060-00009 PMID:17154673

Zhang, H., & Zhao, Y. (2013). Pillararene-Based Assemblies: Design Principle, Preparation and Applications. *Chemistry (Weinheim an der Bergstrasse, Germany)*, *19*(50), 16862–16879. doi:10.1002/chem.201301635 PMID:24249708

Zhang, P. F., Zhang, L., Qin, Z. N., Hua, S. H., Guo, Z. D., Chu, C. C., Lin, H. R., Zhang, Y., Li, W. G., Zhang, X. Z., Chen, X. Y., & Liu, G. (2018). Genetically Engineered Liposome-like Nanovesicles as Active Targeted Transport Platform. *Advanced Materials*, *30*(7), 1705350. doi:10.1002/adma.201705350

Zheng, H., Zhang, Y., Liu, L., Wan, W., Guo, P., Nystrom, A. M., & Zou, X. (2016). One-pot Synthesis of Metal–Organic Frameworks with Encapsulated Target Molecules and Their Applications for Controlled Drug Delivery. *Journal of the American Chemical Society*, *138*(3), 962–968. doi:10.1021/jacs.5b11720 PMID:26710234

Zheng, M., Liu, S., Guan, X., & Xie, Z. (2015). One-Step Synthesis of Nanoscale Zeolitic Imidazolate Frameworks with High Curcumin Loading for Treatment of Cervical Cancer. *ACS Applied Materials & Interfaces*, *7*(40), 22181–22187. doi:10.1021/acsami.5b04315 PMID:26403308

Zhu, X., Gu, J., Wang, Y., Li, B., Li, Y., Zhao, W., & Shi, J. (2014). Inherent anchorages in UiO-66 nanoparticles for efficient capture of alendronate and its mediated release. *Chemical Communications (Cambridge)*, *50*(63), 8779–8782. doi:10.1039/C4CC02570A

Zhu, Y. D., Chen, S. P., Zhao, H., Yang, Y., Chen, X. Q., Sun, J., Fan, H. S., & Zhang, X. D. (2016). PPy@MIL-100 Nanoparticles as a pH- and Near-IR-Irradiation-Responsive Drug Carrier for Simultaneous Photothermal Therapy and Chemotherapy of Cancer Cells. *ACS Applied Materials & Interfaces*, *8*(50), 34209–34217. doi:10.1021/acsami.6b11378 PMID:27998104

Zhuang, J., Kuo, C. H., Chou, L. Y., Liu, D. Y., Weerapana, E., & Tsung, C. K. (2014). Optimized Metal–Organic-Framework Nanospheres for Drug Delivery: Evaluation of Small-Molecule Encapsulation. *ACS Nano*, *8*(3), 2812–2819. doi:10.1021/nn406590q PMID:24506773

Zou, Z., Li, S. Q., He, D. G., He, X. X., Wang, K. M., Li, L. L., Yang, X., & Li, H. F. (2017). A versatile stimulus-responsive metal–organic framework for size/morphology tunable hollow mesoporous silica and pH-triggered drug delivery. *Journal of Materials Chemistry. B, Materials for Biology and Medicine*, *5*(11), 2126–2132. doi:10.1039/C6TB03379B PMID:32263685

Chapter 9
Metal-Organic Frameworks: A Pharmaceutically-Adapted Drug Delivery System?

Marcus Vinícius Dias-Souza
Faculdade Pitágoras, Brazil

Cleiton Gonçalves Dias
Universidade Vale do Rio Doce, Brazil

ABSTRACT

In the field of bioinorganic chemistry, porosity has been explored for the development of metal-organic frameworks (MOFs), innovative polymers, or networks composed of metal centers that stabilize organic molecules with high contact surface and well-defined crystalline structures. MOFs have also been investigated for drug therapies due to their high loading capacity, thermal stability, low electrical conductivity, biodegradability, biocompatibility, easy functionalization, and size considered adequate to control drug release in vivo. MOFs can be sensitive to variations in pH, magnetism, temperature, pressure, light, and humidity, making them desirable for vectorizing extended release systems. Among the most investigated pharmacological groups for association with MOFs are anti-inflammatory and antineoplasic drugs. Here the authors explore the recent advances of MOFs-based drug delivery systems, their current limitations, and discuss the evidences towards their potential clinical use.

INTRODUCTION

Nanotechnologies have been widely used for several purposes in the fields of engineering, physics and chemistry, which include: computing, internet and telephony, aerospatiale advancements (from airplanes to drones and satellites), liquid and gas separation, improved use of fuels, development of cheaper, lighter, durable and efficient materials for vehicles, houses, generation and conduction of electrical energy (including the green-concept alternatives), improvement of the quality of electronics in general, and even the development of weapons (Quan et al., 2018; Rieth et al., 2019; Burtch et al., 2020). Following the

DOI: 10.4018/978-1-7998-4760-1.ch009

primary advancements in these fields, researchers in pharmaceutical and biological sciences started to investigate nanotechnologies for health issues. Drug delivery is the main theme in this context.

Clinically relevant drugs are generally hydrophobic molecules presented as weak acids or bases. This is adopted in order to avoid pharmacokinetic problems: absorption, the first step, might be hampered is the molecules are too hydrophilic (Patrick, 2009). Thus, the drug will not reach adequate levels on the blood stream, and no therapeutic activity would be observed. Common strategies to overcome this situation include latentiation of the drug and modification of pharmaceutical dosage forms. In latentiation, the drug is synthesized as a hydrophobic inactive form (pro-drug), and the hepatic metabolism converts the drug in a hydrophilic form able to bind to its target, such as a receptor or an enzyme (DiPiro 2012). Modification of pharmaceutical dosage forms often allows that some drugs, which would be poorly absorbed in the dosage form they were prepared, may now be properly absorbed (Patrick, 2009; DiPiro 2012). This is the case of scopolamine, which is quite poorly orally absorbed, but is fully absorbed when in intravenous form (Guay et al., 2003).

In general, the arguments for using the currently available nanotechnologies in drug delivery research (mostly liposomes and nanoparticles) rely on comparisons to the pharmaceutical product not associated to nanotechnology, and include: controlled and vectorized drug delivery in small volumes, controlled clearance, possibility of co-loading drugs, extended duration of clinical benefits, improved distribution, decreased immunogenicity, minimum or no local or systemic toxicity, decreased frequency of side-effects, and also, possibility of adaptations with physicochemical features to activate drug delivery (such as sensibility to light, pH, ions, and magnetism) (Dias-Souza and Dos Santos, 2018; Dias Souza et al., 2019). A number of products have reached successful results in clinical trials and can be found at the international pharmaceutical market (tables 1 and 2), but their availability and authorization for clinical use by sanitary authorities is variable worldwide.

However, the costs and the stability of most of the available nanotechnologies are critical obstacles limiting both the therapeutic use and the industrial production. Some lyophilized formulations for injectable use, for instance, are stable for a few hours after hydration, even if kept in low temperatures. Other formulations are not thermostable and might not be exposed to the light. Aggregation might happen in some cases in physiological conditions, and bioaccumulation in soft tissues may impair the delivery by the nanosystems (Dias-Souza et al., 2019).

Although there are studies are being conducted to overcome the aforementioned problems, metal-organic frameworks (MOF in singular and MOFs in plural) have been investigated as promising, stable and low-cost nanotechnologies for drug delivery (see schematic representations in figures 1 and 2). By the time of preparation of this chapter (early months of 2020), more than 8000 papers could be found in PubMed – being the very first ones published in 1999. Thus, we are talking about a yet recent technology for drug delivery, and for general chemistry and physics as well.

More than 20 types of MOFs have been investigated so far for controlled drug release systems, albeit the transference of the technologies for industrial manufacturing, like several nanotechnologies, is a complex endeavor (Wang et al., 2019). An ideal MOF should present structural stability, high loading capacity, low-cost synthesis processes and flexible topology, such that the MOF might be loaded with different drugs. In general, loading drugs in MOFs requires dissolution in organic solvents, long time stirring (often with heat) and vacuum filtration or freeze-drying (Zheng et al., 2016). As in structures like liposomes and micelles, the self-assembly of MOFs is predetermined by the geometries and connections of its components (metals and organic ligands). Their characterization is generally performed using techniques including zeta-potential, X-ray diffraction, fourier-transformed infrared and nuclear magnetic resonance spectroscopies (Huxford et al., 2010).

Table 1. Examples of liposome-based formulations available at the pharmaceutical market

Loaded drug	Liposomal Formulations[Δ]	Clinical use[♦]
Amphotericin B	AmBisome (Gilead Sciences) Abelcet (Sigma-tau)	Fungal infections
Cytarabine	DepoCyt (Sigma-Tau)	Lymphomatous meningitis
Daunorubicin	DaunoXome (Galen)	Karposi's sarcoma
Daunorubicin combined to cytarabine	Vyxeos (Jazz Pharma)	Acute myeloid leukemia with/without myelodysplasia-related changes
Doxorubicin	Doxil/Caely (Janssen) Myocet (Teva)	Karposi's sarcoma, ovarian cancer, multiple myeloma
Irinotecan	Onivyde (Merrimack)	Pancreatic cancer
Estradiol	Estrasorb (Novavax)	Menopausal therapy
Morphine	DepoDur (Pacira Pharmaceuticals)	Pain management
Proteins SP-B and SP-C	Curosurf/Poractant alpha (Chiesei farmaceutic	Pulmonary surfactant for respiratory distress syndrome
Verteporfin	Visudyne (Bausch and Lomb)	Macular degeneration, myopia, ocular histoplasmosis
Vincristine	Marqibo (Onco TCS)	Acute lymphoblastic leukemia

Δ - All products are presented in commercial names and are registered trademarks from their respective manufacturers.
♦ Clinical recommendation described by the manufacturer.

Here, we discuss some of the recent findings on the development of MOF-based drug delivery systems, focusing in non-steroidal anti-inflammatory drugs, anticancer drugs, and antimicrobials. We shall not discuss methodological aspects of the studies in great detail, as the synthesis, purification, an analysis methods are widely varied amidst the studies referenced here. However, we shed light some of these aspects in order to provide possible explanations to the fact that, so far, tests with humans and even animal models remain scarce.

General Overview of MOFs

MOFs are defined as a class of porous materials, which has grown within the research field of drug-delivery over the past few years. Because of their high porosity, they can be interesting substitutes for materials used today in different areas as well, such as catalysis, separation, and as chemical sensors. MOFs are open structure coordination polymers with potentially empty cavities (Fig.2). The units that form their structure are called construction units, composed of inorganic nodes (or nuclei) connected by organic linkers through covalent binding. Inorganic nodes are usually metallic clusters or metallic ions while organic linkers are usually carboxylates ($R\text{-}COO^-$) or heterocyclic compounds, which can be ditopic (with two types of coordinative functions), tritopic (three types) or polytopic (more than three types). Over the years, a wide variety of MOFs have been produced and used for varied purposes. MOFs have attracted the attention of scientists due to their large surface area, porosity, thermal stability, structure and functionality of the projectable pores.

Table 2. Examples of nanoparticle-based formulations available at the pharmaceutical market

Loaded drug	Nanoparticle formulations[Δ]	Clinical Indication[♦]
Albumin-bound paclitaxel nanoparticles	Abraxane/ABI-007 (Celgene)	Breast and pancreatic cancer;
Calcium phosphate	Vitoss (Stryker)	Bone substitute
Dantrolene sodium	Ryanodex (Eagle Pharmaceuticals)	Malignant hypothermia
Dexmethylphenidate HCl	Focalin XR (Novartis)	Psychostimulant
Engineered protein combining IL-2 and diphtheria toxin	Ontak (Eisai Inc)	Cutaneous T-cell lymphoma
Fenofibrate	Tricor (Lupin Atlantis)	Hyperlipidemia
Hydroxyapatite	Ostim (Heraseus Kulzer)	Bone substitute
	OsSatura (IsoTis Orthobiologics)	
	NanOss (Rti surgical)	
	EquivaBone (Zimmer Biomet)	
Iron dextran	INFeD (Sanofi Avertis) DexIron/Dexferrum (Sanofi Avertis)	Iron deficiency in chronic kidney disease
Iron oxide	Nanotherm (MagForce)	Glioblastoma
Iron sucrose	Venofer (Luitpold pharmaceuticals)	Iron deficiency in chronic kidney disease (CKD)
Megestrol	Megace ES (Par Pharmaceuticals)	Anti-anorexic
Methylphenidate	Ritalin LA (Novartis)	Psychostimulant
Morphine sulphate	Avinza (Pfizer)	Pain Management
Paliperidone palmitate	Invega Sustenna (Janssen Pharms)	Schizophrenia;
Sirolimus	Rapamune (Wyeth Pharmaceuticals)	Immunosuppressant
Sodium ferric gluconate	Ferrlecit (Sanofi Avertis)	Iron deficiency in chronic kidney disease (CKD)
SPION coated with dextran	Feridex/Endorem (AMAG pharmaceuticals)	Imaging agent
SPION coated with silicone	GastroMAR Umirem (AMAG pharmaceuticals)	
Tizanidine	Zanaflex (Acorda)	Muscle relaxant

Δ - All products are presented in commercial names and are registered trademarks from their respective manufacturers.

♦ - Clinical recommendation described by the manufacturer.

Figure 1. Simplified representation of a basic MOF organization comprising organic chains bounded to a metallic nucleus

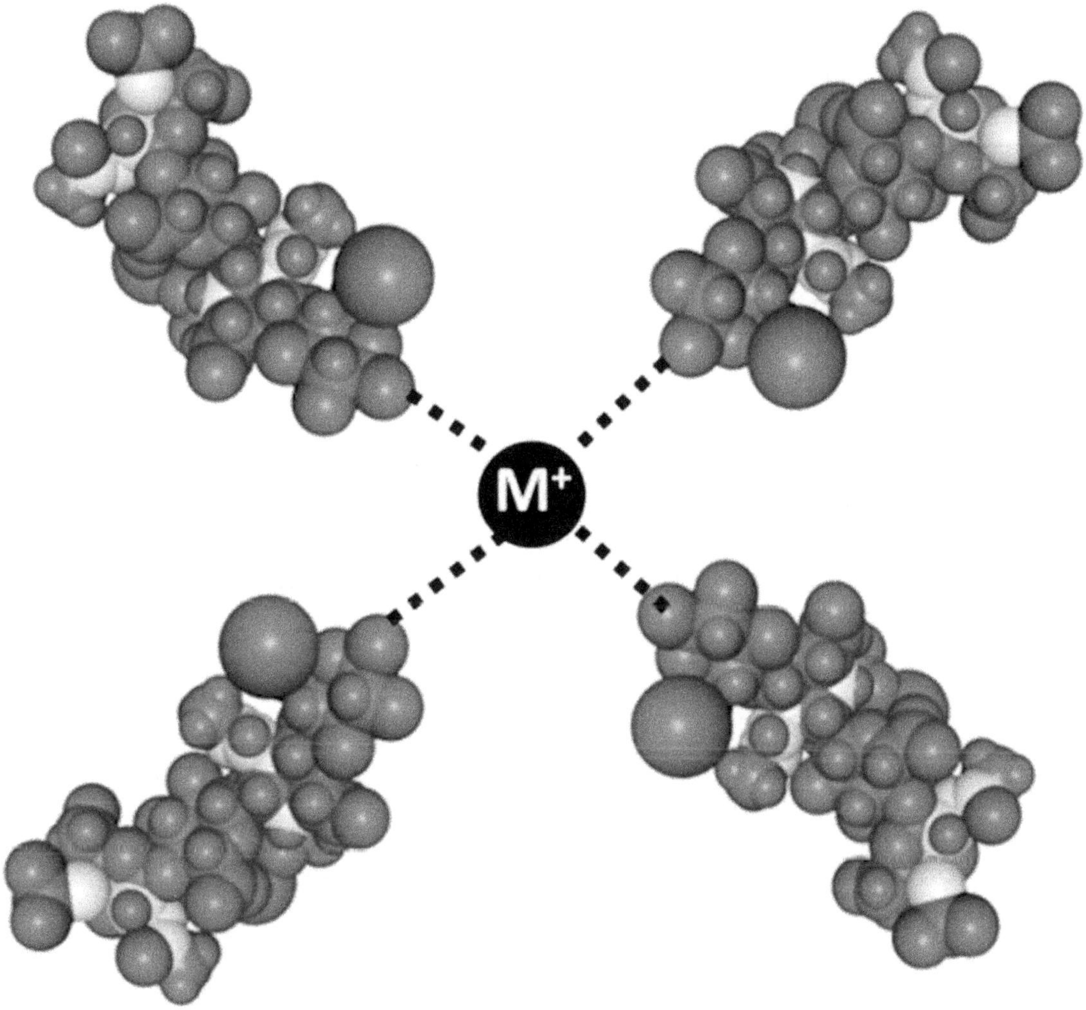

The study of porous metal-organic materials refers to the first ideas for the architecture of extensive structures from well-defined construction units (Kinoshita et al., 1959). The synthesis and determination of the crystalline structure of the complex $[Cu(ADI)_2](NO_3)$ resulted in a diamond-like polymer network. Since then, varied hybrid structures were described, giving rise to the term coordinating polymer. In 1995, Yaghi and Li provided evidence of a porous, robust and crystalline material, suggesting the term "Metal-Organic Framework", which replaced the term coordination polymer (Yaghi and Li, 1995). As a result of their unique characteristics, such as: high porosity, high surface area, presence of coordinated unsaturated sites, existence of strong organic metal-ligand interaction, construction flexibility and high functional capacity, MOFs can be considered promising in different applications, including capture and storage of toxic gases, removal of heavy metals from effluents, and extraction of atmospheric water at desert environments.

Figure 2. Common topology of MOFs. A: primitive cube, B: square lattice, C: Honeycomb

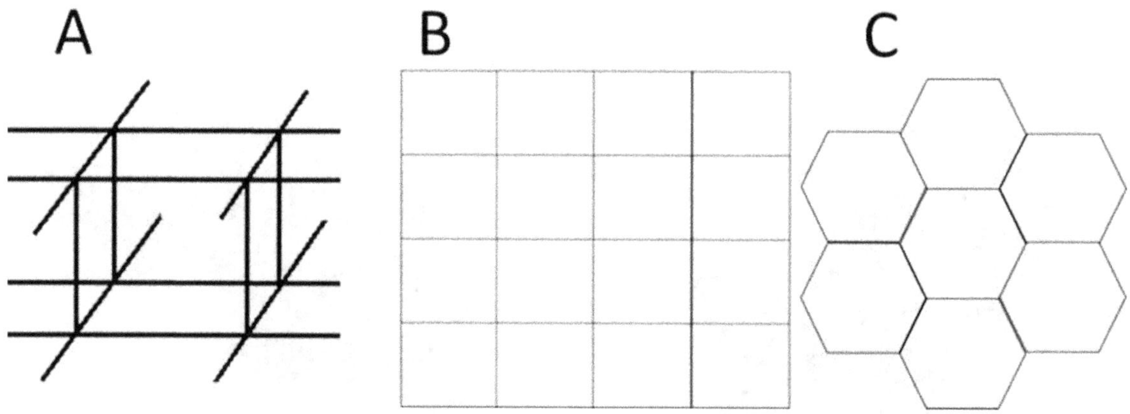

Experimentally, the synthesis of MOFs is usually conducted below 200 °C. Even threaded test tubes, depending on the temperature of the synthesis, can be used under ambient or solvothermal conditions, with pure solvent or mixtures of them. If higher temperatures are required, its preferred to use teflon liners inserted in stainless steel vessels. The process usually occurs in a single recipient, with filtration and washing steps using appropriate solvents.

From the reticular chemistry point of view, MOFs can be rationally designed and synthesized. The topologically different vertices and the connectivity amidst them are numerous, thus, the number of possible structures would be, technically, unlimited. Nevertheless, few basic topologies could be formed, given the principles of geometric design. Depending on the geometry and connectivity of the building blocks, the synthesized MOFs will present simple basic topologies.

MOFs and Non-Steroidal Anti-inflammatory Drugs (NSAID)

Most of the metals used for MOF design are transition metals, but calcium (as Ca^{+2}) was also explored. Technically, it would not be the best metal for preparing MOFs due to its large coordination number, big atom radii, and hard acid properties. However, calcium is part of several biochemical processes of eukaryotic cells, thus, high biocompatibility and minimum toxicity would be expected. A calcium-based MOF was then designed to deliver guaiacol, used as expectorant and anesthetic, and ibuprofen, a non-steroidal anti-inflammatory drug (NSAID). After tests including thermogravimetry, single and powder X-ray diffraction, and molecular storage and release experiments, it was found that guaiacol was loaded on the Ca-based MOF, whereas ibuprofen was not (Wei et al., 2017).

Green chemistry strategies were used to develop a zinc-based MOF to load ibuprofen. Electrosynthesis and sonochemical synthesis were used to prepare the MOF. Although the MOF developed by sonochemical synthesis was smaller than the electrosynthesized one, and with variations on the final particle size as the time of synthesis increased, the loading efficiency was similar and over 94%. Drug release by both MOFs was of about 29% after 24 h (Lestari et al., 2018).

A MOF based in Materials of Institute Lavoisier (MIL) was also developed. MIL are metalloids and nonmetals with *p*-orbital valence electrons, and MIL-based MOFs have been considered more efficient for drug loading than conventional MOFs. A MIL-based MOF was developed with chromium {NH2-MIL-101(Cr)} and functionalized with amino moieties for controlled release of ibuprofen and nimesulide. The COO^- moiety of the deprotonated form of ibuprofen was hypothesized as an explanation to its better incorporation to the NH_2 and chromium groups of the MOF, when compared to nimesulide (Silva et al., 2016).

One calcium-based and three iron-based MOFs were prepared to deliver flurbiprofen. The MOFs were considered biocompatible, and were prepared with the aid of acetonitrile. Fe-based MOFs were prepared with MIL. The centrifuged supernatant was analyzed by high-performance liquid chromatography to indirectly estimate the quantity of loaded drug. Iron-based MOFs presented higher loading efficiency values (up to 46%), and release stoped after 2 days. None of the MOFs could fully deliver flubirprofen, but one of them reached 90% of efficiency (Al Haydar et al., 2017).

An anionic zinc-Adeninate MOF was described as the first example of cation-triggered drug release, and was used to incorporate procainamide, reaching 22% of loading efficiency (An et al., 2009). Cationic zinc-based MOFs were also used to load ibuprofen (Hu et al., 2014).

MOFs and Anticancer Drugs

A zinc-based MOF was prepared using the freeze-drying method to release 5-fluorouracil. *In vitro* tests were conducted using the dialysis method to determine the drug release profile. Total release was reached after 10h, but more than 96% of the drug was already released after 6h (Zhang and Wang, 2018). Anionic Zn-based MOFs were also used for 5-Fluorouracil delivery (Wang et al., 2017).

Light-activated drug delivery strategies were used to design a photo-responsive MOF, namely UiO-AZB. This MOF was prepared to deliver 5-fluorouracil. Aminated polyethylene glycol coating was post-synthetically attached to the MOF to increase its stability, biocompatibility, and half-life. Curiously, in spite of the inaltered pore size distribution at the MOF, pore volume decreased about 20% upon loading, what may help to explain the low loading efficiency of the drug, limited to 15% after a three days process (Roth et al., 2018).

Notwithstanding, pH-responsive MOFs have been investigated as promising alternatives to cancer treatments, as the pH of the neoplasic molecular environment is lower than the one of healthy tissues. ZIF-8 (Zeolitic imidazolate framework-8), a 12 Å pores sodalite-type structure, is a good example of MOF that can cross the nuclear envelope and effectively deliver drugs by protonation-induced breakage of coordination bonds (Ren et al., 2014).

The anticancer drug camptothecin was encapsulated in ZIF-8, generating a 70 nm particle for cellular uptake and intracellular release of the drug; iron oxide nanoparticles were encapsulated in ZIF-8 MOFs, making possible to vectorize drug delivery throughout magnetic features. The viability of MCF-7 (breast cancer) was successfully decreased after 12 h of exposure to the MOF (Zhuang et al., 2014). The strategy of functionalizing ZIF-8 MOFs with polyacrylic acid was used to deliver doxorubicin, a large, antimitotic and cytotoxic molecule used to treat tumors in lungs, breast, bones and other types. The efficiency of encapsulation reached 95%, representing 190 mg of the drug to 100 mg of the MOF. MCF-7 cells were successfully treated after 24h of exposure to the MOF (Ren et al., 2014).

ZIF-8 was also functionalized with bovine serum albumin to encapsulate doxorubicin, which was described as an efficient pH-responsive drug delivery system with higher efficacy than the free drug MCF-7 cells and more biocompatible than the MOF alone, with a biophysical advantage of the presence of positive charges on its surface (Liang et al., 2018). Another ZIF-8 functionalization strategy was conducted Cas9 protein and single guide RNA (sgRNA), which were effective to deliver the CRISPR system in chinese hamster ovary (CHO) cells and decrease gene expression, a very relevant approach to manage malignant neoplasias (Alsaiari et al., 2018).

An elegant system was developed by functionalizing zirconium-based MOFs with nucleic acids, in order to deliver doxorubicin. When the MOFs were exposed to Mg^{+2} ions, these would trigger the activity of a metal ion-dependent DNAzyme, which would then catalyze the cleavage of nucleic acids acting like locks in the MOFs, releasing the loaded drug. This mechanism would only work in the presence of the mentioned ions, which is an interesting strategy to enhance vectorized delivery of drugs. The MOF demonstrated selective cytotoxicity against epithelial human breast MDA-MB-231 cancer cells via receptor-mediated endocytosis (Chen et al., 2017).

MOFs and Antimicrobials, and Considerations on MOF Safety

Silver-based MOFs have been described as effective on Gram-positive and Gram-negative bacteria. As the Ag+ cations are released in the aqueous media, pathogens such as *Staphylococcus aureus, Escherichia coli* and *Pseudomonas aeruginosa* were eradicated (Liu et al., 2010; Berchel et al., 2011). Another Ag-based MOF was effective against bacterial strains of *S. aureus* (including MRSA), *E. coli* and *P. aeruginosa:* when compared to kanamycin and ampicillin, the MOF presented antibacterial activity against all tested species in the same concentrations of the drugs, which were not as effective as the MOF. Cytotoxicity was not detected on human red blood cells (Pelgrift and Friedman, 2013).

Silver nanoparticles and silver-based MOFs were compared for their antimicrobial activity on *E. coli* and *S. aureus*. The MOFs were composed of multinuclear asymmetric silver clusters linked to rod-shaped organic groups. The MIC values of the MOFs varied from 5 to 20 ppm against the strains, but silver nanoparticles were not effective in up to 40 ppm. It was hypothesized that the antimicrobial mechanism would involve interaction of silver ions with proteins and DNA, interference with ion balance, hamper ion channels and interference on membrane integrity. The MOFs were not toxic to red and white blood cells (Lu et al., 2014).

Porphyrin units were loaded into zinc-based MOF against *E.coli* K12 wildtype strain. The system was effective to generate reactive oxygen species upon exposure to visible light. Thus, it could be used in antimicrobial photodynamic therapy (Zhou et al., 2018).

In the previous sections we mentioned the zeolitic imidazolate framework-8 (ZIF-8) as a promissing MOF structure for drug delivery with the ability to cross the nuclear envelope. ZIF-8 delivery of drugs is related to a protonation-induced breakage of its coordination bonds, and a recent study has explored it for antimicrobial purposes. Ceftazidime was loaded in ZIF-8 for systemic and localized antimicrobial therapy, including the use in nebulization. The MOF was tested against *E. coli* K12 T7 strain, and was effective at the concentration of 100 μg/mL (Sava Gallis et al., 2019).

Beyond the clinical issues of antimicrobials, environmental issues concerning the presence of bacteria also deserved the attention of MOF research. A study developed an antimicrobial Cu-based MOF for marine coatings. An interesting mechanism of action was described: *Cobetia marina* strains, commonly detected in colonized submerged surfaces, would form biofilms over the MOF network, resulting in its disruption and release of cupper ions. The free ions damage the bacterial cell wall and eradicate the biofilm, making it a promising alternative to prevent biofouling (Sancet et al., 2013).

Most of the available data on cytotoxicity and general safety of MOFs are related to their isolated parts, and not of the entire structures. One of the rare studies in this context was conducted in Wistar female rats, and used iron-based MOFs (MIL-88A, 100, 88Bt) to load busulfan, azidothymidine, doxorubicin or cidofovir. No inflammatory reactions were detected, but the treated organs were darker than the ones from untreated animals, and the liver and the spleen of treated animals were different from the control rats. The liver and spleen returned to a similar appearance to that of the untreated animals three months after the administration of the MOFs (Horcajada et al., 2010). An increase in the spleen, liver and lung weights was found in another study, beyond ALT and AST increased activities (Baati et al., 2013).

Most of the metals used in the cited MOF studies, although often chemically compatible to varied drugs, are not compatible in the same proportion with some pharmaceutical and biomedical materials and applications. Inorganic nanoparticles are often not biodegradable, thus, more toxicity studies are of paramount importance.

CONCLUSION

For pharmaceutical purposes, further MOF studies should be focused on the preparation of responsive drug delivery systems lacking toxicity and adequate colloidal stability, as chemical compatibility problems are more easily solved. The green synthesis strategy and the use of endogenous organic components as binders can be good possibilities to make MOFs biocompatible and functional. Further investigations of their toxicity after prolonged use and *in vivo* research with these systems are necessary as well.

REFERENCES

Al Haydar, M., Abid, H. R., Sunderland, B., & Wang, S. (2017). Metal organic frameworks as a drug delivery system for flurbiprofen. *Drug Design, Development and Therapy*, *11*, 2685–2695. doi:10.2147/DDDT.S145716 PMID:28979098

Alsaiari, S. K., Patil, S., Alyami, M., Alamoudi, K. O., Aleisa, F. A., Merzaban, J. S., Li, M., & Khashab, N. M. (2018, January 10). Endosomal Escape and Delivery of CRISPR/Cas9 Genome Editing Machinery Enabled by Nanoscale Zeolitic Imidazolate Framework. *Journal of the American Chemical Society*, *140*(1), 143–146. doi:10.1021/jacs.7b11754 PMID:29272114

An, J., Geib, S. J., & Rosi, N. L. (2009). Cation-Triggered Drug Release from a Porous Zinc−Adeninate Metal−Organic Framework. *Journal of the American Chemical Society*, *131*(24), 8376–8377. doi:10.1021/ja902972w PMID:19489551

Baati, T., Njim, L., Neffati, F., Kerkeni, A., Bouttemi, M., Gref, R., & Horcajada, P. (2013). In depth analysis of the in vivo toxicity of nanoparticles of porous iron(iii) metal–organic frameworks. *Chemical Science (Cambridge)*, *4*(4), 1597. doi:10.1039/c3sc22116d

Banerjee, D., Chen, X., Lobanov, S. S., Plonka, A. M., Chan, X., Daly, J. A., Kim, T., Thallapally, P. K., & Parise, J. B. (2018). Iodine adsorption in metal organic frameworks in the presence of humidity. *ACS Applied Materials & Interfaces*, *10*(13), 10622–10626. doi:10.1021/acsami.8b02651 PMID:29547256

Berchel, M., Le Gall, T., Denis, C., Le Hir, S., Quentel, F., Elleouet, C., Montier, T., Rueff, J. M., Salauen, J. Y., Haelters, J. P., Hix, G. B., Lehn, P., & Jaffres, P. A. (2011). A silverbased metal–organic framework material as a 'reservoir' of bactericidal metal ions. *New Journal of Chemistry*, *35*(5), 1000–1003. doi:10.1039/c1nj20202b

Burtch, N. C., Heinen, J., Bennett, T. D., Dubbeldam, D., & Allendorf, M. D. (2018). Mechanical properties in metal–organic frameworks: Emerging opportunities and challenges for device functionality and technological applications. *Advanced Materials*, *30*(37), 1704124. doi:10.1002/adma.201704124 PMID:29149545

Burtch, N. C., Walton, I. M., Hungerford, J. T., Morelock, C. R., Jiao, Y., Heinen, J., Chen, Y.-S., Yakovenko, A. A., Xu, W., Dubbeldam, D., & Walton, K. S. (2020). In situ visualization of loading-dependent water effects in a stable metal–organic framework. *Nature Chemistry*, *12*(2), 186–192. doi:10.103841557-019-0374-y PMID:31792386

Chen, W.-H., Yu, X., Cecconello, A., Sohn, Y. S., Nechushtai, R., & Willner, I. (2017, August 1). Stimuli-responsive nucleic acid-functionalized metal-organic framework nanoparticles using pH- and metal-ion-dependent DNAzymes as locks. *Chemical Science (Cambridge)*, *8*(8), 5769–5780. doi:10.1039/C7SC01765K

Dias-Souza, M. V., & Dos Santos, R. M. (2018). Drug and food applications of liposomes and nanoparticles: from benchmark to bedside? Drug Delivery Approaches and Nanosystems, 1, 299-326.

Dias-Souza, M. V., Nascimento, A. A., & Vasconcelos, T. Y. L. (2019). Immunomodulatory and Antimicrobial effects of nanoemulsions: pharmaceutical development aspects and perspectives on clinical treatments. In Nanodispersons for Drug delivery. Toronto: Apple Academic Press/CRC Press/Taylor and Francis.

DiPiro, J. T., Talbert, R. L., Yee, G. C., Matzke, G. R., Wells, B. G., & Posey, L. M. (Eds.). (2012). *Pharmacotherapy: a pathophysiologic approach* (8th ed.). McGraw-Hill.

Gonzalez, M. I., Turkiewicz, A. B., Darago, L. E., Oktawiec, J., Bustillo, K., Grandjean, F., Long, G. J., & Long, J. R. (2020). Confinement of atomically defined metal halide sheets in a metal–organic framework. *Nature*, *577*(7788), 64–68. doi:10.103841586-019-1776-0 PMID:31739311

Guay, D. R. (2003). Clinical pharmacokinetics of drugs used to treat urge incontinence. *Clinical Pharmacokinetics*, *42*(14), 1243–1285. doi:10.2165/00003088-200342140-00004 PMID:14606931

Han, Z., Wang, K., Guo, Y., Chen, W., Zhang, J., Zhang, X., Siligardi, G., Yang, S., Zhou, Z., Sun, P., Shi, W., & Cheng, P. (2019). Cation-induced chirality in a bifunctional metal-organic framework for quantitative enantioselective recognition. *Nature Communications*, *10*(1), 5117. doi:10.103841467-019-13090-9 PMID:31712651

Horcajada, P., Chalati, T., Serre, C., Gillet, B., Sebrie, C., Baati, T., Eubank, J. F., Heurtaux, D., Clayette, P., Kreuz, C., Chang, J. S., Hwang, Y. K., Marsaud, V., Bories, P. N., Cynober, L., Gil, S., Férey, G., Couvreur, P., & Gref, R. (2010, February). Porous metal-organic-framework nanoscale carriers as a potential platform for drug delivery and imaging. *Nature Materials*, *9*(2), 172–178. doi:10.1038/nmat2608 PMID:20010827

Hu, Q., Yu, J., Liu, M., Liu, A., Dou, Z., & Yang, Y. (2014). A Low Cytotoxic Cationic Metal–Organic Framework Carrier for Controllable Drug Release. *Journal of Medicinal Chemistry, 57*(13), 5679–5685. doi:10.1021/jm5004107 PMID:24922463

Huxford, R. C., Della Rocca, J., & Lin, W. (2010, April). Metal-organic frameworks as potential drug carriers. *Current Opinion in Chemical Biology, 14*(2), 262–268. doi:10.1016/j.cbpa.2009.12.012 PMID:20071210

Jaramillo, D. E., Reed, D. A., Jiang, H. Z. H., Oktawiec, J., Mara, M. W., Forse, A. C., Lussier, D. J., Murphy, R. A., Cunningham, M., Colombo, V., Shuh, D. K., Reimer, J. A., & Long, J. R. (2020). Selective nitrogen adsorption via backbonding in a metal–organic framework with exposed vanadium sites. *Nature Materials, 19*(5), 517–521. Advance online publication. doi:10.103841563-019-0597-8 PMID:32015534

Kinoshita, Y., Matsubara, I., & Saito, Y. (1959). The crystal structure of bis(succinonitrilo)copper(I) nitrate. *Bulletin of the Chemical Society of Japan, 32*(7), 741–747. doi:10.1246/bcsj.32.741

Liang, Z., Yang, Z., Yuan, H., Wang, C., Qi, J., Liu, K., Cao, R., & Zheng, H. (2018, July 31). A protein@ metal-organic framework nanocomposite for pH-triggered anticancer drug delivery. *Dalton Transactions (Cambridge, England), 47*(30), 10223–10228. doi:10.1039/C8DT01789A PMID:30014058

Liu, Y., Xu, X., Xia, Q., Yuan, G., He, Q., & Cui, Y. (2010). Multiple topological isomerism of three-connected networks in silver-based metal–organoboron frameworks. *Chemical Communications (Cambridge), 46*(15), 2608–2610. doi:10.1039/b923365b PMID:20449323

Lu, X. Y., Ye, J., Zhang, D., Xie, R., Bogale, R. F., Sun, Y., Zhao, L., Zhao, Q., & Ning, G. (2014). Silver carboxylate metal-organic frameworks with highly antibacterial activity and biocompatibility. *Journal of Inorganic Biochemistry, 138*, 114–121. doi:10.1016/j.jinorgbio.2014.05.005 PMID:24935093

Milner, P. J., Siegelman, R. L., Forse, A. C., Gonzalez, M. I., Runčevski, T., Martell, J. D., Reimer, J. A., & Long, J. R. (2017). A diaminopropane-appended metal organic framework enabling efficient CO2 capture from coal flue gas via a mixed adsorption mechanism. *Journal of the American Chemical Society, 139*(38), 13541–13553. doi:10.1021/jacs.7b07612 PMID:28906108

Patrick, G. L. (2009). *Introduction to medicinal chemistry. 4.* Oxford University.

Pelgrift, R. Y., & Friedman, A. J. (2013). Nanotechnology as a therapeutic tool to combat microbial resistance. *Advanced Drug Delivery Reviews, 65*(13-14), 1803–1815. doi:10.1016/j.addr.2013.07.011 PMID:23892192

Quan, B., Jin, A., Yu, S.-H., Kang, S. M., Jeong, J., Abruña, H. D., Jin, L., Piao, Y., & Sung, Y.-E. (2018). Solvothermal-derived S-doped graphene as an anode material for sodium-ion batteries. *Advancement of Science, 5*(5), 1700880. doi:10.1002/advs.201700880 PMID:29876213

Ren, H., Zhang, L., An, J., Wang, T., Li, L., Si, X., He, L., Wu, X., Wang, C., & Su, Z. (2014, January 28). Polyacrylic acid@zeolitic imidazolate framework-8 nanoparticles with ultrahigh drug loading capability for pH-sensitive drug release. *Chemical Communications (Cambridge), 50*(8), 1000–1002. doi:10.1039/C3CC47666A

Rieth, A. J., Wright, A. M., & Dincă, M. (2019). Kinetic stability of metal–organic frameworks for corrosive and coordinating gas capture. *Nature Reviews. Materials*, *4*(11), 708–725. doi:10.103841578-019-0140-1

Sancet, M. P. A. (2013). Surface anchored metal-organic frameworks as stimulus responsive antifouling coatings. *Biointerphases*, *8*. PMID:24706148

Sava Gallis, D. F., Butler, K. S., Agola, J. O., Pearce, C., & McBride, A. (2019). Antibacterial Countermeasures via MOF – Supported Sustained Therapeutic Release. *ACS Applied Materials & Interfaces*. Advance online publication. doi:10.1021/acsami.8b21698 PMID:30682243

Silva, I. M. P., Carvalho, M. A., Oliveira, C. S., Profirio, D. M., Ferreira, R. B., Corbi, P. P., & Formiga, A. L. B. (2016). Enhanced performance of a metal-organic framework analogue to MIL-101(Cr) containing amine groups for ibuprofen and nimesulide controlled release. *Inorganic Chemistry Communications*, *70*, 47–50. doi:10.1016/j.inoche.2016.05.020

Wang, J., Ma, D., Liao, W., Li, S., Huang, M., Liu, H., Wang, Y., Xie, R., & Xu, J. (2017). A hydrostable anionic zinc-organic framework carrier with a **bcu** topology for drug delivery. *CrystEngComm*, *19*(35), 5244–5250. doi:10.1039/C7CE01238A

Wang, Y., Yan, J., Wen, N., Xiong, H., Cai, S., He, Q., Hu, Y., Peng, D., Liu, Z., & Liu, Y. (2020, February). Metal-organic frameworks for stimuli-responsive drug delivery. *Biomaterials*, *230*, 119619. Advance online publication. doi:10.1016/j.biomaterials.2019.119619 PMID:31757529

Wei, L. Q., Lu, J. Y., Li, Q. Q., Zhou, Y., Tang, L.-L., & Li, F.-Y. (2017). *, Tang LL, Li FY. A porous Ca-MOF with nano-sized {Ca11} as building unit: Structure, drug loading and release properties. *Inorganic Chemistry Communications*, *78*, 43–47. doi:10.1016/j.inoche.2017.02.010

Yaghi, O. M., & Li, H. (1995). Hydrothermal synthesis of a metal–organic framework containing large rectangular channels. *Journal of the American Chemical Society*, *117*(41), 10401–10402. doi:10.1021/ja00146a033

Yu, D., Shao, Q., Song, Q., Cui, J., Zhang, Y., Wu, B., Ge, L., Wang, Y., Zhang, Y., Qin, Y., Vajtai, R., Ajayan, P. M., Wang, H., Xu, T., & Wu, Y. (2020). A solvent-assisted ligand exchange approach enables metal-organic frameworks with diverse and complex architectures. *Nature Communications*, *11*(1), 927. doi:10.103841467-020-14671-9 PMID:32066754

Zhang, Y., & Wang, J. (2018). A water-stable and biofriendly 2D anionic zinc(II) metal–organic framework for drug delivery. *Inorganica Chimica Acta*, *477*, 8–14. doi:10.1016/j.ica.2018.01.015

Zheng, H., Zhang, Y., Liu, L., Wan, W., Guo, P., Nyström, A. M., & Zou, X. (2016). One-pot synthesis of metal–organic frameworks with encapsulated targetmolecules and their applications for controlled drug delivery. *Journal of the American Chemical Society*, *3*(3), 962–968. doi:10.1021/jacs.5b11720 PMID:26710234

Zhou, W., Begum, S., Wang, Z., Krolla, P., Wagner, D., Bräse, S., & Tsotsalas, M. (2018). High Antimicrobial Activity of Metal–Organic Framework-Templated Porphyrin Polymer Thin Films. *ACS Applied Materials & Interfaces*, *10*(2), 1528–1533. doi:10.1021/acsami.7b14866 PMID:29243917

Zhuang, J., Kuo, C.-H., Chou, L.-Y., Liu, D.-Y., Weerapana, E., & Tsung, C.-K. (2014, March 25). Optimized metal-organic-framework nanospheres for drug delivery: Evaluation of small-molecule encapsulation. *ACS Nano*, *8*(3), 2812–2819. doi:10.1021/nn406590q PMID:24506773

Compilation of References

Zheng, M., Liu, S., Guan, X., & Xie, Z. (2015). One-Step Synthesis of Nanoscale Zeolitic Imidazolate Frameworks with High Curcumin Loading for Treatment of Cervical Cancer. *ACS Applied Materials & Interfaces*, *7*(40), 22181–22187. doi:10.1021/acsami.5b04315 PMID:26403308

Shanmugam, M. K., Rane, G., Kanchi, M. M., Arfuso, F., Chinnathambi, A., Zayed, M. E., Tan, B. K., Kumar, A. P., & Sethi, G. (2015). The Multifaceted Role of Curcumin in Cancer Prevention and Treatment. *Molecules (Basel, Switzerland)*, *20*(2), 2728–2769. doi:10.3390/molecules20022728 PMID:25665066

Anand, P., Kunnumakkara, A. B., Newman, R. A., & Aggarwal, B. B. (2007). Bioavailability of Curcumin: Problems and Promises. *Molecular Pharmaceutics*, *4*(6), 807–818. doi:10.1021/mp700113r PMID:17999464

Mourtas, S., Lazar, A. N., Markoutsa, E., Duyckaerts, C., & Antimisiaris, S. G. (2014). Multifunctional nanoliposomes with curcumin–lipid derivative and brain targeting functionality with potential applications for Alzheimer disease. *European Journal of Medicinal Chemistry*, *80*, 175–183. doi:10.1016/j.ejmech.2014.04.050 PMID:24780594

Sahu, A., Kasoju, N., & Bora, U. (2008). Fluorescence Study of the Curcumin−Casein Micelle Complexation and Its Application as a Drug Nanocarrier to Cancer Cells. *Biomacromolecules*, *9*(10), 2905–2912. doi:10.1021/bm800683f PMID:18785706

Zhanet, G., Fontaine, S., & Schurek, K. (2006). Article. *Treatments in Respiratory Medicine*, *5*, 437–465. doi:10.2165/00151829-200605060-00009 PMID:17154673

Esfahanian, M., Ghasemzadeh, M. A., Mohammad, S., Razavian, H., & International, J. (2019). Synthesis, identification and application of the novel metal-organic framework Fe_3O_4@PAA@ZIF-8 for the drug delivery of ciprofloxacin and investigation of antibacterial activity. *Artificial Cells, Nanomedicine, and Biotechnology*, *47*(1), 2024–2030. doi:10.1080/21691401.2019.1617729 PMID:31112049

Gul, R., Ahmed, N., Ullah, N., Khan, M. I., Elaissari, A., & Rehman, A. (2018). Biodegradable Ingredient-Based Emulgel Loaded with Ketoprofen Nanoparticles. *AAPS PharmSciTech*, *19*(4), 1869–1881. doi:10.120812249-018-0997-0 PMID:29651679

Tat, S. K., Pelletier, J. P., Mineau, F., Caron, J., & Martel-Pelletier, J. (2011). Strontium ranelate inhibits key factors affecting bone remodeling in human osteoarthritic subchondral bone osteoblasts. *Bone*, *49*(3), 559–567. doi:10.1016/j.bone.2011.06.005 PMID:21700005

Bertini, I. (2007). *Biological inorganic chemistry:structure and reactivity*. University Science Books.

Li, Z., Peng, Y., Xia, X., Cao, Z., Deng, Y., & Tang, B. (2019). Article. *Scientific Reports, 9*(17570), 1-7.

Ormerod, L. P., & Horsfield, N. (1996). Frequency and type of reactions to antituberculosis drugs: Observations in routine treatment. *Tubercle and Lung Disease*, *77*(1), 37–42. doi:10.1016/S0962-8479(96)90073-8 PMID:8733412

Gulbay, B. E., Gurkan, O. U., Yildiz, O. A., Onen, Z. P., Erkekol, F. O., Baccioglu, A., & Acıcan, T. (2006). Side effects due to primary antituberculosis drugs during the initial phase of therapy in 1149 hospitalized patients for tuberculosis. *Respiratory Medicine*, *100*(10), 1834–1842. doi:10.1016/j.rmed.2006.01.014 PMID:16517138

Bray, P. F. (1984). Isoniazid-induced acute toxic encephalopathy. *Neurology*, *34*(5), 703. doi:10.1212/WNL.34.5.703-a PMID:6538670

Stettner, M., Steinberger, D., Hartmann, C.J., Pabst, T., Konta, L., & Hartung, H.P. (2015). Article. *Brain Behav.*, *5*(8).

Simon-Yarza, T., Giménez-Marqués, M., Mrimi, R., Mielcarek, A., Gref, R., Horcajada, P., Serre, C., & Couvreur, P. (2017). A Smart Metal–Organic Framework Nanomaterial for Lung Targeting. *Angewandte Chemie International Edition*, *56*(49), 15565–15569. doi:10.1002/anie.201707346 PMID:28960750

Thommes, M., & Cychosz, K. A. (2014). Physical adsorption characterization of nanoporous materials: Progress and challenges. *Adsorption*, *2*(20), 233–250. doi:10.100710450-014-9606-z

Kalidoss, M., Kumar, T. S. S., Rubaiya, Y., & Doble, M. (2017). Article. *Biomater Tissue Technol*, *1*(1), 1–5.

Chowdhury, M. A. (2016). The Silica-based Formulations for Drug Delivery, Bone Treatment, and Bone Regeneration. *ChemBioEng Rev*, *3*(5), 229–246. doi:10.1002/cben.201500026

McGuire, C. V., & Forgan, R. S. (2015). The surface chemistry of metal–organic frameworks. *Chemical Communications (Cambridge)*, *51*(25), 5199–5217. doi:10.1039/C4CC04458D

Horcajada, P., Gref, R., Baati, T., Allan, P. K., Maurin, G., Couvreur, P., Férey, G., Morris, R. E., & Serre, C. (2012). Metal—Organic frameworks in biomedicine. *Chemical Reviews*, *112*(2), 1232–1268. doi:10.1021/cr200256v PMID:22168547

Keskin, S., & Kizilel, S. (2011). Biomedical Applications of Metal Organic Frameworks. *Industrial & Engineering Chemistry Research*, *50*(4), 1799–1812. doi:10.1021/ie101312k

Slowing, I. I., Vivero-Escoto, J. L., Wu, C. W., & Lin, V. (2008). Article. *Advanced Drug Delivery Reviews*, *60*(11), 1278–1288. doi:10.1016/j.addr.2008.03.012 PMID:18514969

Wang, J. Q. (2018). Article. *Angewandte Chemie International Edition*, *57*, 3008. doi:10.1002/anie.201711705 PMID:29450950

Zhang, P. F., Zhang, L., Qin, Z. N., Hua, S. H., Guo, Z. D., Chu, C. C., Lin, H. R., Zhang, Y., Li, W. G., Zhang, X. Z., Chen, X. Y., & Liu, G. (2018). Genetically Engineered Liposome-like Nanovesicles as Active Targeted Transport Platform. *Advanced Materials*, *30*(7), 1705350. doi:10.1002/adma.201705350

Li, L., Wang, J. Q., Kong, H. R., Zeng, Y., & Liu, G. (2018). Functional biomimetic nanoparticles for drug delivery and theranostic applications in cancer treatment. *Science and Technology of Advanced Materials*, *19*(1), 771–790. doi:10.1080/14686996.2018.1528850 PMID:30815042

Baati, T., Njim, L., Neffati, F., Kerkeni, A., Bouttemi, M., Gref, R., Najjar, M. F., Zakhama, A., Couvreur, P., Serre, C., & Horcajada, P. (2013). In depth analysis of the in vivo toxicity of nanoparticles of porous iron(iii) metal–organic frameworks. *Chemical Science (Cambridge)*, *4*(4), 1597. doi:10.1039/c3sc22116d

Smaldone, R. A., Forgan, R. S., Furukawa, H., Gassensmith, J. J., Slawin, A. M. Z., Yaghi, O. M., & Stoddart, J. F. (2010). Metalorganic frameworks from edible natural products. *Angewandte Chemie International Edition*, *49*(46), 8630–8634. doi:10.1002/anie.201002343 PMID:20715239

Burrows, A. D., Jurcic, M., Keenan, L. L., Lane, R. A., Mahon, M. F., Warren, M. R., Nowell, H., Paradowski, M., & Spencer, J. (2013). Incorporation by coordination and release of the iron chelator drug deferiprone from zinc-based metal–organic frameworks. *Chemical Communications (Cambridge)*, *49*(96), 11260. doi:10.1039/c3cc45689g

Banerjee, R., Sahoo, S. C., & Kundu, T. (2013). *World Patent WO2013051035*. Academic Press.

Furukawa, H., Gandara, F., Zhang, Y. B., Jiang, J., Queen, W. L., Hudson, M. R., & Yaghi, O. M. (2014). Water Adsorption in Porous Metal–Organic Frameworks and Related Materials. *Journal of the American Chemical Society*, *136*(11), 4369–4381. doi:10.1021/ja500330a PMID:24588307

Seo, Y. K., Yoon, J. W., Lee, J. S., Hwang, Y. K., Jun, C. H., Chang, J. S., Wuttke, S., Bazin, P., Vimont, A., Daturi, M., Bourrelly, S., Llewellyn, P. L., Horcajada, P., Serre, C., & Férey, G. (2012). Energy-Efficient Dehumidification over Hierachically Porous Metal-Organic Frameworks as Advanced Water Adsorbents. *Advanced Materials*, *24*(6), 806–810. doi:10.1002/adma.201104084 PMID:22162212

Sahoo, S. C., Kundu, T., & Banerjee, R. (2011). Helical Water Chain Mediated Proton Conductivity in Homochiral Metal–Organic Frameworks with Unprecedented Zeolitic *unh* -Topology. *Journal of the American Chemical Society*, *133*(44), 17950–17958. doi:10.1021/ja2078637 PMID:21919488

Fletcher, A. J., Thomas, K. M., & Rosseinsky, M. J. (2005). Flexibility in metal-organic framework materials: Impact on sorption properties. *Journal of Solid State Chemistry*, *178*(8), 2491–2510. doi:10.1016/j.jssc.2005.05.019

Huxford, R. C., Della Rocca, J., & Lin, W. (2010). Metal–organic frameworks as potential drug carriers. *Current Opinion in Chemical Biology*, *14*(2), 262–268. doi:10.1016/j.cbpa.2009.12.012 PMID:20071210

Torchilin, V. P. (2005). Recent advances with liposomes as pharmaceutical carriers. *Nature Reviews. Drug Discovery*, *4*(2), 145–160. doi:10.1038/nrd1632 PMID:15688077

An, J., Geib, S. J., & Rosi, N. L. (2009). Cation-Triggered Drug Release from a Porous Zinc–Adeninate Metal–Organic Framework. *Journal of the American Chemical Society*, *131*(24), 8376–8377. doi:10.1021/ja902972w PMID:19489551

McKinlay, A. C., Morris, R. E., Horcajada, P., Ferey, G., Gref, R., Couvreur, P., & Serre, C. (2010). BioMOFs: Metal-Organic Frameworks for Biological and Medical Applications. *Angewandte Chemie International Edition*, *49*(36), 6260–6266. doi:10.1002/anie.201000048 PMID:20652915

Cai, WWang, JChu, CChen, WWu, CLiu, G. (2019). Metal–Organic Framework-Based Stimuli-Responsive Systems for Drug Delivery. *Advancement of Science*, *6*(1801526), 1–20.

Ren, H., Zhang, L., An, J., Wang, T., Li, L., Si, X., He, L., Wu, X., Wang, C., & Su, Z. (2014). Polyacrylic acid@zeolitic imidazolate framework-8 nanoparticles with ultrahigh drug loading capability for pH-sensitive drug release. *Chemical Communications (Cambridge)*, *50*(8), 1000–1002. doi:10.1039/C3CC47666A

Zhuang, J., Kuo, C. H., Chou, L. Y., Liu, D. Y., Weerapana, E., & Tsung, C. K. (2014). Optimized Metal–Organic-Framework Nanospheres for Drug Delivery: Evaluation of Small-Molecule Encapsulation. *ACS Nano*, *8*(3), 2812–2819. doi:10.1021/nn406590q PMID:24506773

Alsaiari, S. K., Patil, S., Alyami, M., Alamoudi, K. O., Aleisa, F. A., Merzaban, J. S., Li, M., & Khashab, N. M. (2018). Endosomal Escape and Delivery of CRISPR/Cas9 Genome Editing Machinery Enabled by Nanoscale Zeolitic Imidazolate Framework. *Journal of the American Chemical Society*, *140*(1), 143–146. doi:10.1021/jacs.7b11754 PMID:29272114

Zheng, H., Zhang, Y., Liu, L., Wan, W., Guo, P., Nystrom, A. M., & Zou, X. (2016). One-pot Synthesis of Metal–Organic Frameworks with Encapsulated Target Molecules and Their Applications for Controlled Drug Delivery. *Journal of the American Chemical Society*, *138*(3), 962–968. doi:10.1021/jacs.5b11720 PMID:26710234

Zou, Z., Li, S. Q., He, D. G., He, X. X., Wang, K. M., Li, L. L., Yang, X., & Li, H. F. (2017). A versatile stimulus-responsive metal–organic framework for size/morphology tunable hollow mesoporous silica and pH-triggered drug delivery. *Journal of Materials Chemistry. B, Materials for Biology and Medicine*, *5*(11), 2126–2132. doi:10.1039/C6TB03379B PMID:32263685

Chowdhuri, A. R., Singh, T., Ghosh, S. K., & Sahu, S. K. (2016). Carbon Dots Embedded Magnetic Nanoparticles @ Chitosan @Metal Organic Framework as a Nanoprobe for pH Sensitive Targeted Anticancer Drug Delivery. *ACS Applied Materials & Interfaces*, *8*(26), 16573–16583. doi:10.1021/acsami.6b03988 PMID:27305490

Gao, P. F., Zheng, L. L., Liang, L. J., Yang, X. X., Li, Y. F., & Huang, C. Z. (2013). A new type of pH-responsive coordination polymer sphere as a vehicle for targeted anticancer drug delivery and sustained release. *Journal of Materials Chemistry. B, Materials for Biology and Medicine*, *1*(25), 3202. doi:10.1039/c3tb00026e PMID:32260920

Yoo, J. W., & Lee, C. H. (2006). Drug delivery systems for hormone therapy. *Journal of Controlled Release*, *112*(1), 1–14. doi:10.1016/j.jconrel.2006.01.021 PMID:16530874

Kim, B. J., Cheong, H., Hwang, B. H., & Cha, H. J. (2015). Mussel-Inspired Protein Nanoparticles Containing Iron(III)-DOPA Complexes for pH-Responsive Drug Delivery. *Angewandte Chemie International Edition*, *54*(25), 7318–7322. doi:10.1002/anie.201501748 PMID:25968933

Wang, D. D., Zhou, J. J., Chen, R. H., Shi, R. H., Xia, G. L., Zhou, S., Liu, Z. B., Zhang, N. Q., Wang, H. B., Guo, Z., & Chen, Q. W. (2016). Magnetically guided delivery of DHA and Fe ions for enhanced cancer therapy based on pH-responsive degradation of DHA-loaded Fe 3 O 4 @C@MIL-100(Fe) nanoparticles. *Biomaterials*, *107*, 88–101. doi:10.1016/j.biomaterials.2016.08.039 PMID:27614161

Yang, Q., Zhu, Y., Luo, B., Lan, F., Wu, Y., & Gu, Z. W. (2017). pH-Responsive magnetic metal–organic framework nanocomposites for selective capture and release of glycoproteins. *Nanoscale*, *9*(2), 527–532. doi:10.1039/C6NR08071E PMID:27966715

Zhu, Y. D., Chen, S. P., Zhao, H., Yang, Y., Chen, X. Q., Sun, J., Fan, H. S., & Zhang, X. D. (2016). PPy@MIL-100 Nanoparticles as a pH- and Near-IR-Irradiation-Responsive Drug Carrier for Simultaneous Photothermal Therapy and Chemotherapy of Cancer Cells. *ACS Applied Materials & Interfaces*, *8*(50), 34209–34217. doi:10.1021/acsami.6b11378 PMID:27998104

Zhu, X., Gu, J., Wang, Y., Li, B., Li, Y., Zhao, W., & Shi, J. (2014). Inherent anchorages in UiO-66 nanoparticles for efficient capture of alendronate and its mediated release. *Chemical Communications (Cambridge)*, *50*(63), 8779–8782. doi:10.1039/C4CC02570A

Cavka, J. H., Jakobsen, S., Olsbye, U., Guillou, N., Lamberti, C., Bordiga, S., & Lillerud, K. P. (2008). A New Zirconium Inorganic Building Brick Forming Metal Organic Frameworks with Exceptional Stability. *Journal of the American Chemical Society*, *130*(42), 13850–13851. doi:10.1021/ja8057953 PMID:18817383

Yang, Y., Hu, Q., Zhang, Q., Jiang, K., Lin, W., Yang, Y., Cui, Y., & Qian, G. (2016). A Large Capacity Cationic Metal–Organic Framework Nanocarrier for Physiological pH Responsive Drug Delivery. *Molecular Pharmaceutics*, *13*(8), 2782–2786. doi:10.1021/acs.molpharmaceut.6b00374 PMID:27414996

Lin, W., Hu, Q., Jiang, K., Yang, Y., Cui, Y., & Qian, G. (2016). A porphyrin-based metal–organic framework as a pH-responsive drug carrier. *Journal of Solid State Chemistry*, *237*, 307–312. doi:10.1016/j.jssc.2016.02.040

Lzaro, I. A., Haddad, S., Sacca, S., Orellana-Tavra, C., Fairen-Jimenez, D., & Forgan, R. S. (2017). Article. *Chem*, *2*, 561. doi:10.1016/j.chempr.2017.02.005 PMID:28516168

Liu, J. J., Wang, H. R., Yi, X., Chao, Y., Geng, Y. H., Xu, L. G., Yang, K., & Liu, Z. (2017). pH-Sensitive Dissociable Nanoscale Coordination Polymers with Drug Loading for Synergistically Enhanced Chemoradiotherapy. *Advanced Functional Materials*, *27*(44), 1703832. doi:10.1002/adfm.201703832

Gupta, M., Agrawal, U., & Vyas, S. P. (2012). Nanocarrier-based topical drug delivery for the treatment of skin diseases. *Expert Opinion on Drug Delivery*, *9*(7), 783–804. doi:10.1517/17425247.2012.686490 PMID:22559240

Duan, F., Feng, X. C., Yang, X. J., Sun, W. T., Jin, Y., Liu, H. F., Ge, K., Li, Z. H., & Zhang, J. C. (2017). A simple and powerful co-delivery system based on pH-responsive metal-organic frameworks for enhanced cancer immunotherapy. *Biomaterials*, *122*, 23–33. doi:10.1016/j.biomaterials.2017.01.017 PMID:28107662

Kundu, T., Mitra, S., Patra, P., Goswami, A., Diaz Diaz, D., & Banerjee, R. (2014). Mechanical Downsizing of a Gadolinium(III)-based Metal-Organic Framework for Anticancer Drug Delivery. *Chemistry (Weinheim an der Bergstrasse, Germany)*, *20*(33), 10514–10518. doi:10.1002/chem.201402244 PMID:25044210

Kumar, C. S., & Mohammad, F. (2011). Magnetic nanomaterials for hyperthermia-based therapy and controlled drug delivery. *Advanced Drug Delivery Reviews*, *63*(9), 789–808. doi:10.1016/j.addr.2011.03.008 PMID:21447363

Freeman, M. W., Arrott, A., & Watson, J. H. L. (1960). Magnetism in Medicine. *Journal of Applied Physics*, *31*(5), S404–S405. doi:10.1063/1.1984765

Peller, M., Boll, K., Zimpel, A., & Wuttke, S. (2018). Metal–organic framework nanoparticles for magnetic resonance imaging. *Inorganic Chemistry Frontiers*, *5*(8), 1760–1779. doi:10.1039/C8QI00149A

Ke, F., Yuan, Y. P., Qiu, L. G., Shen, Y. H., Xie, A. J., Zhu, J. F., Tian, X. Y., & Zhang, L. D. (2011). Facile fabrication of magnetic metal–organic framework nanocomposites for potential targeted drug delivery. *Journal of Materials Chemistry*, *21*(11), 3843. doi:10.1039/c0jm01770a

Wu, Y. N., Zhou, M., Li, S., Li, Z., Li, J., Wu, B., Li, G., Li, F., & Guan, X. (2014). Magnetic Metal-Organic Frameworks: γ-Fe $_2$ O $_3$ @MOFs via Confined In Situ Pyrolysis Method for Drug Delivery. *Small*, *10*(14), 2927–2936. doi:10.1002mll.201400362 PMID.24644065

Sethi, K., Sharma, S., & Roy, I. (2016). Article. *RSC Adv.*, *6*, 76861.

Sharma, S., Sethi, K., & Roy, I. (2017). Magnetic nanoscale metal–organic frameworks for magnetically aided drug delivery and photodynamic therapy. *New Journal of Chemistry*, *41*(20), 11860–11866. doi:10.1039/C7NJ02032E

Hu, Q., Yu, J., Liu, M., Liu, A., Dou, Z., & Yang, Y. (2014). A Low Cytotoxic Cationic Metal–Organic Framework Carrier for Controllable Drug Release. *Journal of Medicinal Chemistry*, *57*(13), 5679–5685. doi:10.1021/jm5004107 PMID:24922463

Vallet-Regí, M., Balas, F., & Arcos, D. (2007). Vallet-Reg_ M, Balas F, Arcos D.Mesoporous materials for drug delivery. *Angewandte Chemie International Edition*, *46*(40), 7548–7558. doi:10.1002/anie.200604488

Du, X., Fan, R. Q., Qiang, L. S., Xing, K., Ye, H. X., Ran, X. Y., Song, Y., Wang, P., & Yang, Y. L. (2017). Controlled Zn $^{2+}$ -Triggered Drug Release by Preferred Coordination of Open Active Sites within Functionalization Indium Metal Organic Frameworks. *ACS Applied Materials & Interfaces*, *9*(34), 28939–28948. doi:10.1021/acsami.7b09227 PMID:28776972

Nagata, S., Kokado, K., & Sada, K. (2015). Metal–organic framework tethering PNIPAM for ON–OFF controlled release in solution. *Chemical Communications (Cambridge)*, *51*(41), 8614–8617. doi:10.1039/C5CC02339D

Lin, W. X., Hu, Q., Yu, J. C., Jiang, K., Yang, Y. Y., Xiang, S. C., Cui, Y. J., Yang, Y., Wang, Z. Y., & Qian, G. D. (2016). Low Cytotoxic Metal-Organic Frameworks as Temperature-Responsive Drug Carriers. *ChemPlusChem*, *81*(8), 804–810. doi:10.1002/cplu.201600142 PMID:31968821

Jiang, K., Zhang, L., Hu, Q., Zhao, D., Xia, T., Lin, W., Yang, Y., Cui, Y., Yang, Y., & Qian, G. (2016). Pressure controlled drug release in a Zr-cluster-based MOF. *Journal of Materials Chemistry. B, Materials for Biology and Medicine*, *4*(39), 6398–6401. doi:10.1039/C6TB01756H PMID:32263448

Ogoshi, T., Kanai, S., Fujinami, S., Yamagishi, T., & Nakamoto, Y. (2008). *para* -Bridged Symmetrical Pillar[5]arenes: Their Lewis Acid Catalyzed Synthesis and Host–Guest Property. *Journal of the American Chemical Society*, *130*(15), 5022–5023. doi:10.1021/ja711260m PMID:18357989

Strutt, N. L., Zhang, H., Schneebeli, S. T., & Stoddart, J. F. (2014). Functionalizing Pillar[*n*]arenes. *Accounts of Chemical Research*, *47*(8), 2631–2642. doi:10.1021/ar500177d PMID:24999824

Zhang, H., & Zhao, Y. (2013). Pillararene-Based Assemblies: Design Principle, Preparation and Applications. *Chemistry (Weinheim an der Bergstrasse, Germany)*, *19*(50), 16862–16879. doi:10.1002/chem.201301635 PMID:24249708

Si, W., Xin, P., Li, Z. T., & Hou, J. L. (2015). Tubular Unimolecular Transmembrane Channels: Construction Strategy and Transport Activities. *Accounts of Chemical Research*, *48*(6), 1612–1619. doi:10.1021/acs.accounts.5b00143 PMID:26017272

Xue, M., Chi, X. D., Zhang, Z. B., & Huang, F. H. (2012). Pillararenes, A New Class of Macrocycles for Supramolecular Chemistry. *Accounts of Chemical Research*, *45*(8), 1294–1308. doi:10.1021/ar2003418 PMID:22551015

Aznar, E., Oroval, M., Pascual, L., Murguia, J. R., Manez, R. M., & Sancenon, F. (2016). Gated Materials for On-Command Release of Guest Molecules. *Chemical Reviews*, *116*(2), 561–718. doi:10.1021/acs.chemrev.5b00456 PMID:26730615

Davis, M. E., Chen, Z., & Shin, D. M. (2008). Nanoparticle therapeutics: An emerging treatment modality for cancer. *Nature Reviews. Drug Discovery*, *7*(9), 771–782. doi:10.1038/nrd2614 PMID:18758474

Tan, L. L., Li, H., Qiu, Y. C., Chen, D. X., Wang, X., Pan, R. Y., Wang, Y., Zhang, S. X. A., Wang, B., & Yang, Y. W. (2015). Stimuli-responsive metal–organic frameworks gated by pillar[5]arene supramolecular switches. *Chemical Science (Cambridge)*, *6*(3), 1640–1644. doi:10.1039/C4SC03749A

Tan, L. L., Song, N., Zhang, S. X. A., Li, H., Wang, B., & Yang, Y. W. (2016). Ca^{2+}, pH and thermo triple-responsive mechanized Zr-based MOFs for on-command drug release in bone diseases. *Journal of Materials Chemistry. B, Materi als for Biology and Medicine*, *4*(1), 135–140. doi:10.1039/C5TB01789K PMID:32262817

Wang, X. G., Dong, Z. Y., Cheng, H., Wan, S. S., Chen, W. H., Zou, M. Z., Huo, J. W., Deng, H. X., & Zhang, X. Z. (2015). A multifunctional metal–organic framework based tumor targeting drug delivery system for cancer therapy. *Nanoscale*, *7*(38), 16061–16070. doi:10.1039/C5NR04045K PMID:26372069

Meng, X. S., Gui, B., Yuan, D. Q., Zeller, M., & Wang, C. (2016). Mechanized azobenzene-functionalized zirconium metal-organic framework for on-command cargo release. *Science Advances*, *2*(8), e1600480. doi:10.1126ciadv.1600480 PMID:27493996

Sun, C. Y., Qin, C., Wang, C. G., Su, Z. M., Wang, S., Wang, X. L., Yang, G.-S., Shao, K.-Z., Lan, Y.-Q., & Wang, E.-B. (2011). Chiral Nanoporous Metal-Organic Frameworks with High Porosity as Materials for Drug Delivery. *Advanced Materials*, *23*(47), 5629–5632. doi:10.1002/adma.201102538 PMID:22095878

Liedana, N., Galve, A., Rubio, C., Tellez, C., & Coronas, J. (2012). CAF@ZIF-8: One-Step Encapsulation of Caffeine in MOF. *ACS Applied Materials & Interfaces*, *4*(9), 5016–5021. doi:10.1021/am301365h PMID:22834763

Miller, S. R., Alvarez, E., Fradcourt, L., Devic, T., Wuttke, S., Wheatley, P. S., Steunou, N., Bonhomme, C., Gervais, C., Laurencin, D., Morris, R. E., Vimont, A., Daturi, M., Horcajada, P., & Serre, C. (2013). A rare example of a porous Ca-MOF for the controlled release of biologically active NO. *Chemical Communications (Cambridge)*, *49*(71), 7773. doi:10.1039/c3cc41987h

Zeng, M. H., Wang, Q. X., Tan, Y. X., Hu, S., Zhao, H. X., Long, L. S., & Kurmoo, M. (2010). Rigid Pillars and Double Walls in a Porous Metal-Organic Framework: Single-Crystal to Single-Crystal, Controlled Uptake and Release of Iodine and Electrical Conductivity. *Journal of the American Chemical Society*, *132*(8), 2561–2563. doi:10.1021/ja908293n PMID:20131766

Lucena, F. R. S., & de Araujo, L. C. C. (2013). Rodrigues MdD, da Silva TG, Pereira VRA, Militao GCG. *Biomedicine and Pharmacotherapy*, *67*(8), 707. doi:10.1016/j.biopha.2013.06.003 PMID:24095176

Horcajada, P., Serre, C., Maurin, G., Ramsahye, N. A., Balas, F., Vallet-Regi, M., Sebban, M., Taulelle, F., & Férey, G. (2008). Flexible Porous Metal-Organic Frameworks for a Controlled Drug Delivery. *Journal of the American Chemical Society*, *130*(21), 6774–6780. doi:10.1021/ja710973k PMID:18454528

Dyson, P. J., & Sava, G. (2006). Metal-based antitumour drugs in the post genomic era. *Dalton Transactions (Cambridge, England)*, *16*(16), 1929–1933. doi:10.1039/b601840h PMID:16609762

Taylor-Pashow, K. M. L., Rocca, J. D., Xie, Z., Tran, S., & Lin, W. (2009). Postsynthetic Modifications of Iron-Carboxylate Nanoscale Metal–Organic Frameworks for Imaging and Drug Delivery. *Journal of the American Chemical Society*, *131*(40), 14261–14263. doi:10.1021/ja906198y PMID:19807179

Wang, H. N., Yang, G. S., Wang, X. L., & Su, Z. M. (2013). pH-induced different crystalline behaviors in extended metal–organic frameworks based on the same reactants. *Journal of the Chemical Society, Dalton Transactions: Inorganic Chemistry*, *42*(18), 6294. doi:10.1039/c3dt32958e

Horcajada, P., Serre, C., Vallet-Regi, M., Sebban, M., Taulelle, F., & Ferey, G. (2006). Metal–Organic Frameworks as Efficient Materials for Drug Delivery. *Angewandte Chemie International Edition*, *45*(36), 5974–5978. doi:10.1002/anie.200601878 PMID:16897793

Rieter, W. J., Pott, K. M., Taylor, K. M. L., & Lin, W. (2008). Nanoscale Coordination Polymers for Platinum-Based Anticancer Drug Delivery. *Journal of the American Chemical Society*, *130*(35), 11584–11585. doi:10.1021/ja803383k PMID:18686947

Rowe, M. D., Thamm, D. H., Kraft, S. L., & Boyes, S. G. (2009). Polymer-Modified Gadolinium Metal-Organic Framework Nanoparticles Used as Multifunctional Nanomedicines for the Targeted Imaging and Treatment of Cancer. *Biomacromolecules*, *10*(4), 983–993. doi:10.1021/bm900043e PMID:19290624

Liang, K., Ricco, R., Doherty, C. M., Styles, M. J., Bell, S., Kirby, N., Mudie, S., Haylock, D., Hill, A. J., Doonan, C. J., & Falcaro, P. (2015). Biomimetic mineralization of metal-organic frameworks as protective coatings for biomacromolecules. *Nature Communications*, *6*(1), 7240. doi:10.1038/ncomms8240 PMID:26041070

Chowdhuri, A., Bhattacharya, D., & Sahu, S. K. (2016). Article. *Journal of the Chemical Society, Dalton Transactions: Inorganic Chemistry*, *45*(7), 2963. doi:10.1039/C5DT03736K

Fang, J., Yang, Y., Xiao, W., Zheng, B., Lv, Y. B., Liu, X. L., & Ding, J. (2016). Extremely low frequency alternating magnetic field–triggered and MRI–traced drug delivery by optimized magnetic zeolitic imidazolate framework-90 nanoparticles. *Nanoscale*, *8*(6), 3259–3263. doi:10.1039/C5NR08086J PMID:26809987

Wang, D., Zhou, J., Chen, R., Shi, R., Zhao, G., Xia, G., Li, R., Liu, Z., Tian, J., Wang, H., Guo, Z., Wang, H., & Chen, Q. (2016). Controllable synthesis of dual-MOFs nanostructures for pH-responsive artemisinin delivery, magnetic resonance and optical dual-model imaging-guided chemo/photothermal combinational cancer therapy. *Biomaterials*, *100*, 27–40. doi:10.1016/j.biomaterials.2016.05.027 PMID:27240160

Kamaly, N. D., Miller, A. D., & Bell, J. (2010).. . *Current Topics in Medicinal Chemistry*, *10*(12), 1158. doi:10.2174/156802610791384199 PMID:20388110

Fathabadi, E. G., Shelling, A. N., & Al-Kassas, R. (2012). Nanocarrier systems for delivery of siRNA to ovarian cancer tissues. *Expert Opinion on Drug Delivery*, *9*(7), 743–754. doi:10.1517/17425247.2012.683173 PMID:22657869

Graf, N., & Lippard, S. J. (2012). Redox activation of metal-based prodrugs as a strategy for drug delivery. *Advanced Drug Delivery Reviews*, *64*(11), 993–1004. doi:10.1016/j.addr.2012.01.007 PMID:22289471

Florea, A. M., & Büsselberg, D. (2011). Cisplatin as an Anti-Tumor Drug: Cellular Mechanisms of Activity, Drug Resistance and Induced Side Effects. *Cancers (Basel)*, *3*(1), 1351–1371. doi:10.3390/cancers3011351 PMID:24212665

Andersson, B. S. (2008). *Biology of Blood and Marrow Transplantation*. doi:10.1016/j.bbmt.2008.03.009 PMID:18489993

Chalat, T., Horcajada, P., Couvreur, P., Serre, C., & Yahia, M. B. (2011). Article. *Nanomedicine (London)*, *6*(10), 1683–1695. doi:10.2217/nnm.11.69

Reschiglian, P., Gref, R., Monti, S., Manet, I., Agostoni, V., Reschigl, P., Gref, R., & Monti, S. (2014). Article. *Physical Chemistry*, *118*, 8532–8539.

Takemura, G., & Fujiwara, H. (2007). Doxorubicin-Induced Cardiomyopathy. *Progress in Cardiovascular Diseases*, *49*(5), 330–352. doi:10.1016/j.pcad.2006.10.002 PMID:17329180

Liu, Z., Cheung, R., Wu, X.Y., Ballinger, J.R., Bendayan, R., & Rauth, A.M.. (2001). Article. *Controlled Release, 77*(3), 213-224.

Chatterjee, K., Zhang, J., Honbo, N. S., & Karliner, J. (2010). Doxorubicin Cardiomyopathy. *Cardiology*, *115*(2), 155–162. doi:10.1159/000265166 PMID:20016174

Chen, Y., Wan, Y., Wang, Y., Zhang, H., Jiao, Z., & International, J. (2011). Nanofabrication techniques for controlled drug-release devices. *Nanomedicine (London)*, *6*(1), 2321–2326. doi:10.2217/nnm.10.140

Haydar, M. A., Abid, H. R., Sunderland, B., & Wang, S. (2017).. . *Drug Design, Development and Therapy*, *11*, 2685. doi:10.2147/DDDT.S145716 PMID:28979098

Krishna, R., & Mayer, L. D. (2000). Multidrug resistance (MDR) in cancer. *European Journal of Pharmaceutical Sciences*, *11*(4), 265–283. doi:10.1016/S0928-0987(00)00114-7 PMID:11033070

Adhikar, C., Das, A., & Chakrabort, A. (2015). Article. *Molecular Pharmaceutics*, *12*(9), 3158–3166. doi:10.1021/acs.molpharmaceut.5b00043 PMID:26196058

Nunzio, M., & Agostoni, V. (2014). Article. *Journal of Medicinal Chemistry*, *57*(2), 411–420. PMID:24345217

Sun, C. Y., Qin, C., Wang, X. L., Yang, G. S., Shao, K. Z., Lan, Y. Q., Su, Z. M., Huang, P., Wang, C. G., & Wang, E. B. (2012). Zeolitic imidazolate framework-8 as efficient pH-sensitive drug delivery vehicle. *Dalton Transactions (Cambridge, England)*, *41*(23), 6906–6909. doi:10.1039/c2dt30357d

Arias, J. L. (2008). Novel Strategies to Improve the Anticancer Action of 5-Fluorouracil by Using Drug Delivery Systems. *Molecules (Basel, Switzerland)*, *13*(10), 2340–2369. doi:10.3390/molecules13102340 PMID:18830159

Tran, L. A., & Wilson, L. J. (2011). Nanomedicine: Making controllable magnetic drug delivery possible for the treatment of breast cancer. *Breast Cancer Research, 13*(2), 303. doi:10.1186/bcr2830 PMID:21457524

Rainsford, K. D. (2005). *Nimesulide – Actions and Uses*. Birkhäuser Basel. doi:10.1007/3-7643-7410-1

Abd El Salam, H. M., & Zaki, T. (2019). A Novel Microwave Synthesis of Manganese Based MOF for Adsorptive of Cd(II), Pb(II) and Hg(II) Ions from Aqua Medium. *Egyptian Journal of Chemistry, 62*(5), 837–851. doi:10.21608/ejchem.2019.6524.1550

Abdi, H. R., Tian, H., Ang, H. M., Tade, M. O., Buckley, C. E., & Wang, S. (2012). Nanosize Zr-metal-organic framework (UiO-66) for hydrogen and carbon dioxide storage. *Chemical Engineering Journal, 187*, 415–420. doi:10.1016/j.cej.2012.01.104

Abdulgader, H. A., Kochkodan, V., & Hilal, N. (2013). Hybrid Ion Exchange - Pressure Driven Membrane Processes in Water Treatment: A Review. *Separation and Purification Technology, 116*, 253–264. doi:10.1016/j.seppur.2013.05.052

Abdullah, N., Rahman, M. A., Dzarfan Othman, M. H., Jaafar, J., & Aziz, A. A. (2018). Preparation, Characterizations and Performance Evaluations of Alumina Hollow Fiber Membrane Incorporated with UiO-66 Particles for Humic Acid Removal. *Journal of Membrane Science, 563*(May), 162–174. doi:10.1016/j.memsci.2018.05.059

Aboutorabi, L., Morsali, A., Tahmasebi, E., & Büyükgüngor, O. (2016). Metal-Organic Framework based on isonicotinate N-oxide for fast and highly efficient aqueous phase Cr(VI) adsorption. *Inorganic Chemistry, 55*(11), 5507–5513. doi:10.1021/acs.inorgchem.6b00522 PMID:27195982

Abu Seman, M. N., Khayet, M., & Hilal, N. (2010). Nanofiltration Thin-Film Composite Polyester Polyethersulfone-Based Membranes Prepared by Interfacial Polymerization. *Journal of Membrane Science, 348*(1–2), 109–116. doi:10.1016/j.memsci.2009.10.047

Adhikari, A. K., & Lin, K. S. (2016). Improving CO2 adsorption capacities and CO2/N2 separation efficiencies of MOF-74(Ni, Co) by doping palladium-containing activated carbon. *Chemical Engineering Journal, 284*(January), 1348–1360. doi:10.1016/j.cej.2015.09.086

Adriano, D. C. (2001). *Trace Elements in Terrestrial Environments*. Springer SBM., doi:10.1007/978-0-387-21510-5

Afshariazara, F., Morsalia, A., Wangb, J., & Junk, P. C. (2020). Highest and Fastest Removal Rate of Pb(II) Ions through Rational Functionalized decoration of a MOF cavity. *Chemistry (Weinheim an der Bergstrasse, Germany), 26*(6), 1355–1362. doi:10.1002/chem.201904436 PMID:31756257

Aghakhani, A., Mohamadi, F., & Ghadimi, J. (2019). Novel alcohol vapour sensor based on the mixed-ligand modified MOF-199 coated quartz crystal microbalance. *International Journal of Environmental Analytical Chemistry*, 1–18. do i:10.1080/03067319.2019.1689968

Ahmed, I., & Jhung, S. H. (2014). Composites of metal-organic frameworks: Preparation and application in adsorption. *Materials Today, 17*(3), 136–146. doi:10.1016/j.mattod.2014.03.002

Aijaz, A., & Xu, Q. (2014). *Catalysis with Metal Nanoparticles Immobilized within the Pores of Metal – Organic Frameworks*. Academic Press.

Akbarzadeh, F., Motaghi, M., Chauhan, N. P. S., & Sargazi, G. (2020). A novel synthesis of new antibacterial nanostructures based on Zn-MOF compound: Design, characterization and a high performance application. *Heliyon, 6*(1), e03231. doi:10.1016/j.heliyon.2020.e03231 PMID:32021929

Al-Amshawee, S., Yunus, M. Y. B. M., Azoddein, A. A. M., Hassell, D. G., Dakhil, I. H., & Hasan, H. A. (2020). Electrodialysis Desalination for Water and Wastewater: A Review. *Chemical Engineering Journal*, *380*, 122231. doi:10.1016/j.cej.2019.122231

Alizadeh, S., & Nematollahi, D. (2019). Convergent and Divergent Paired Electrodeposition of Metal-Organic Framework Thin Films. *Scientific Reports*, *9*(1), 1–13. doi:10.103841598-019-50390-y PMID:31586078

Aljundi, I. H. (2016, December). Desalination Characteristics of TFN-RO Membrane Incorporated with ZIF-8 Nanoparticles. *Desalination*, *420*, 12–20. doi:10.1016/j.desal.2017.06.020

Allendorf, M. D., Bauer, C. A., Bhakta, R. K., & Houk, R. J. T. (2009). Luminescent metal-organic frameworks. *Chemical Society Reviews*, *38*(5), 1330–1352. doi:10.1039/b802352m PMID:19384441

Alloway, B. J. (1995). *Heavy Metals in Soils*. Blackie Academic and Professional. doi:10.1007/978-94-011-1344-1

Alvisi, M., Aversa, P., Cassano, G., Serra, E., Tagliente, M. A., Schioppa, M., & Penza, M. (2012). Organic vapor detection by QCM sensors using CNT-composite films. In *Sensors and Microsystems* (pp. 79–85). Springer. doi:10.1007/978-1-4614-0935-9_14

Ameloot, R., Stappers, L., Fransaer, J., Alaerts, L., Sels, B. F., & De Vos, D. E. (2009). Patterned growth of metal-organic framework coatings by electrochemical synthesis. *Chemistry of Materials*, *21*(13), 2580–2582. doi:10.1021/cm900069f

Ameloot, R., Vermoortele, F., Vanhove, W., Roeffaers, M. B. J., Sels, B. F., & De Vos, D. E. (2011). Interfacial Synthesis of Hollow Metalâ€"organic Framework Capsules Demonstrating Selective Permeability. *Nature Chemistry*, *3*(5), 382–387. doi:10.1038/nchem.1026 PMID:21505497

Ang, H., & Hong, L. (2017). Polycationic Polymer-Regulated Assembling of 2D MOF Nanosheets for High-Performance Nanofiltration. *ACS Applied Materials & Interfaces*, *9*(33), 28079–28088. doi:10.1021/acsami.7b08383 PMID:28752999

Arjmandi, M., Peyravi, M., Chenar, M. P., & Jahanshahi, M. (2019). A New Concept of MOF-Based PMM by Modification of Conventional Dense Film Casting Method: Significant Impact on the Performance of FO Process. *Journal of Membrane Science*, *579*(February), 253–265. doi:10.1016/j.memsci.2019.02.020

Ash, D. C., Joyce, M. J., Barnes, C., Booth, C. J., & Jefferies, A. C. (2003). Viscosity measurement of industrial oils using the droplet quartz crystal microbalance. *Measurement Science & Technology*, *14*(11), 1955–1962. doi:10.1088/0957-0233/14/11/013

Babazadeh, M., Hosseinzadeh-Khanmiri, R., Abolhasani, J., Ghorbani-Kalhor, E., & Hassanpour, A. (2015). Solid phase extraction of heavy metal ions from agricultural samples with the aid of a novel functionalized magnetic metal-organic framework. *RSC Advances*, *5*(26), 19884–19892. doi:10.1039/C4RA15532G

Babel, S., Kurniawan, T.A. (2004). Cr (VI) removal from synthetic wastewater using coconut shell charcoal and commercial activated carbon modified with oxidizing agents and/or chitosan, *Chemosphere*, 54, 951-967.

Bai, Y., He, G., Zhao, Y., Duan, C., Dang, D., & Meng, Q. (2006). Porous material for absorption and luminescent detection of aromatic molecules in water. *Chemical Communications (Cambridge)*, *43*(14), 1530–1532. doi:10.1039/b517002h PMID:16575450

Bai, Z.-Q., Yuan, L.-Y., Zhu, L., Liu, Z.-R., Chu, S.-Q., Zheng, L.-R., Zhang, J., Chai, Z.-F., & Shi, W.-Q. (2015). Introduction of amino groups into acid-resistant MOFs for enhanced U(VI) sorption. *Journal of Materials Chemistry. A, Materials for Energy and Sustainability*, *3*(2), 525–534. doi:10.1039/C4TA04878D

Bakhtiari, N., & Azizian, S. (2015). Adsorption of copper ion from aqueous solution by nanoporous MOF-5: a kinetic and equilibrium study, *J. Mol. Liq.*, 206, 114-118.

Bakker, K. (2012). Water security: Research challenges and opportunities. *Science*, *337*, 914–915.

Banerjee, D., Chen, X., Lobanov, S. S., Plonka, A. M., Chan, X., Daly, J. A., Kim, T., Thallapally, P. K., & Parise, J. B. (2018). Iodine adsorption in metal organic frameworks in the presence of humidity. *ACS Applied Materials & Interfaces*, *10*(13), 10622–10626. doi:10.1021/acsami.8b02651 PMID:29547256

Barakat, M. A. (2011). New trends in removing heavy metals from industrial wastewater. *Arabian Journal of Chemistry*, *4*(4), 361–377. doi:10.1016/j.arabjc.2010.07.019

Barthelet, K., Marrot, J., Riou, D., & Férey, G. (2002). A breathing hybrid organic–inorganic solid with very large pores and high magnetic characteristics. *Angewandte Chemie International Edition*, *41*(2), 281–284. doi:10.1002/1521-3773(20020118)41:2<281::AID-ANIE281>3.0.CO;2-Y PMID:12491409

Basu, S., & Balakrishnan, M. (2017). Polyamide Thin Film Composite Membranes Containing ZIF-8 for the Separation of Pharmaceutical Compounds from Aqueous Streams. *Separation and Purification Technology*, *179*, 118–125. doi:10.1016/j.seppur.2017.01.061

Bataille, T., Bracco, S., Comotti, A., Costantino, F., Guerri, A., Ienco, A., & Marmottini, F. (2012). Solvent dependent synthesis of micro-and nano-crystalline phosphinate based 1D tubular MOF: Structure and CO_2 adsorption selectivity. *CrystEngComm*, *14*(21), 7170–7173. doi:10.1039/c2ce26138c

Batten, S. R., Champness, N. R., Chen, X. M., Garcia-Martinez, J., Kitagawa, S., Öhrström, L., O'Keeffe, M., Suh, M. P., & Reedijk, J. (2013). Terminology of metal–organic frameworks and coordination polymers (IUPAC Recommendations 2013). *Pure and Applied Chemistry*, *85*(8), 1715–1724. doi:10.1351/PAC-REC-12-11-20

Baumann, A. E., Burns, D. A., Liu, B., & Thoi, S. (2019). Metal-Organic Framework Functionalization and Design Strategies for Advanced Electrochemical Energy Storage Devices. *Communications Chemistry*, *2*(1), 86. doi:10.103842004-019-0184-6

Bechelany, M., Drobek, M., Vallicari, C., Abou Chaaya, A., Julbe, A., & Miele, P. (2015). Highly crystalline MOF-based materials grown on electrospun nanofibers†. *Nanoscale*, *7*(13), 5794–5802. Advance online publication. doi:10.1039/C4NR06640E PMID:25759092

Benito, J., Sorribas, S., Lucas, I., Coronas, J., & Gascon, I. (2016). Langmuir–Blodgett Films of the Metal–Organic Framework MIL-101 (Cr): Preparation, Characterization, and CO2 Adsorption Study Using a QCM-Based Setup. *ACS Applied Materials & Interfaces*, *8*(25), 16486–16492. doi:10.1021/acsami.6b04272 PMID:27268426

Ben, T., Lu, C., Pei, C., Xu, S., & Qiu, S. (2012). Polymer-Supported and Free-Standing Metal-Organic Framework Membrane. *Polymer-Supported and Free-Standing Metal–Organic Framework Membrane*, *18*(3), 1–5. doi:10.1002/chem.201201574 PMID:22807003

Berchel, M., Le Gall, T., Denis, C., Le Hir, S., Quentel, F., Elleouet, C., Montier, T., Rueff, J. M., Salauen, J. Y., Haelters, J. P., Hix, G. B., Lehn, P., & Jaffres, P. A. (2011). A silverbased metal–organic framework material as a 'reservoir' of bactericidal metal ions. *New Journal of Chemistry*, *35*(5), 1000–1003. doi:10.1039/c1nj20202b

Bétard, A., & Fischer, R. A. (2012). Metal-Organic Framework Thin Films: From Fundamentals to Applications. *Chemical Reviews*, *112*(2), 1055–1083. doi:10.1021/cr200167v PMID:21928861

Bethi, Sonawane, Bhanvase, & Gumfekar. (2016). Nanomaterials-Based Advanced Oxidation Processes for Wastewater Treatment: A Review. *Chemical Engineering and Processing - Process Intensification*, *109*, 178–89. https://linkinghub.elsevier.com/retrieve/pii/S0255270116303312

Bhanoth, S., Chappanda, K. N., Teja, V. M., Majhi, S. M., Surya, S. G., & More, Y. D. (2019). *A silver nanoparticle-anchored UiO-66 (Zr) metal-organic framework (MOF)-based capacitive H_2S gas sensor*. Academic Press.

Bhattacharjee, S., Lee, Y.-R., & Ahn, W.-S. (2015). Post-synthesis functionalization of a zeolitic imidazolate structure ZIF-90: A study on removal of Hg(II) from water and epoxidation of alkenes. *CrystEngComm*, *17*(12), 2575–2582. doi:10.1039/C4CE02555E

Bhattacharyya, P., Chakrabarti, K., Chakraborty, A., Tripathy, S., & Powell, M. A. (2008). Fractionation and bioavailability of Pb in municipal solid waste compost and Pb uptake by rice straw and grain under submerged condition in amended soil. *Geosciences Journal*, *12*(1), 41–45. doi:10.100712303-008-0006-9

Boix, G., Troyano, J., Tovar, L. C. G., Camur, C., Bermejo, N., Yazdi, A., Piella, J., Bastús, N. G., Puntes, V., Imaz, I., & Maspoch, D. (2020). MOF-Beads containing inorganic nanoparticles for the simultaneous removal of multiple heavy metals from water. *ACS Applied Materials & Interfaces*, *12*(9), 10554–10562. doi:10.1021/acsami.9b23206 PMID:32026677

Bonnett, B. L., Smith, E. D., De La Garza, M., Cai, M., Haag, J. V. IV, Serrano, J. M., Cornell, H. D., Gibbons, B., Martin, S. M., & Morris, A. J. (2020). PCN-222 Metal-Organic Framework Nanoparticles with Tunable Pore Size for Nanocomposite Reverse Osmosis Membranes. *ACS Applied Materials & Interfaces*, *12*(13), 15765–15773. doi:10.1021/acsami.0c04349 PMID:32163702

Bradl, H. B. (2005). *Heavy Metals in the Environment: Origin, Interaction and Remediation* (1st ed.). Elsevier Academic Press.

Budyanto, S., Kuo, Y. L., & Liu, J. C. (2015). Adsorption and precipitation of fluoride on calcite nanoparticles: A spectroscopic study. *Separation and Purification Technology*, *150*, 325–331. doi:10.1016/j.seppur.2015.07.016

Burtch, N. C., Heinen, J., Bennett, T. D., Dubbeldam, D., & Allendorf, M. D. (2018). Mechanical properties in metal–organic frameworks: Emerging opportunities and challenges for device functionality and technological applications. *Advanced Materials*, *30*(37), 1704124. doi:10.1002/adma.201704124 PMID:29149545

Burtch, N. C., Walton, I. M., Hungerford, J. T., Morelock, C. R., Jiao, Y., Heinen, J., Chen, Y.-S., Yakovenko, A. A., Xu, W., Dubbeldam, D., & Walton, K. S. (2020). In situ visualization of loading-dependent water effects in a stable metal–organic framework. *Nature Chemistry*, *12*(2), 186–192. doi:10.103841557-019-0374-y PMID:31792386

Butova, V. V. E., Soldatov, M. A., Guda, A. A., Lomachenko, K. A., & Lamberti, C. (2016). Metal-organic frameworks: Structure, properties, methods of synthesis and characterization. *Russian Chemical Reviews*, *85*(3), 280–307. doi:10.1070/RCR4554

Bux, H., Liang, F., Li, Y., Cravillon, J., Wiebcke, M., & Caro, J. (2009). Zeolitic imidazolate framework membrane with molecular sieving properties by microwave-assisted solvothermal synthesis. *Journal of the American Chemical Society*, *131*(44), 16000–16001. doi:10.1021/ja907359t PMID:19842668

Campbell, J., Davies, R. P., Braddock, D. C., & Livingston, A. G. (2015). Improving the Permeance of Hybrid Polymer/Metal-Organic Framework (MOF) Membranes for Organic Solvent Nanofiltration (OSN)-Development of MOF Thin Films via Interfacial Synthesis. *Journal of Materials Chemistry. A, Materials for Energy and Sustainability*, *3*(18), 9668–9674. doi:10.1039/C5TA01315A

Campbell, J., Székely, G., Davies, R. P., Braddock, D. C., & Livingston, A. G. (2014). Fabrication of Hybrid Polymer/Metal Organic Framework Membranes: Mixed Matrix Membranes versus in Situ Growth. *Journal of Materials Chemistry. A, Materials for Energy and Sustainability*, *2*(24), 9260–9271. doi:10.1039/C4TA00628C

Canivet, J., Fateeva, A., Guo, Y., Coasne, B., & Farrusseng, D. (2014). Water Adsorption in MOFs: Fundamentals and Applications. *Chemical Society Reviews*, *43*(16), 5594–5617. doi:10.1039/C4CS00078A PMID:24875439

Cao, J., Su, Y., Liu, Y., Guan, J., He, M., Zhang, R., & Jiang, Z. (2018). Self-Assembled MOF Membranes with Underwater Superoleophobicity for Oil/Water Separation. *Journal of Membrane Science*, *566*, 268–277. doi:10.1016/j.memsci.2018.08.068

Cao, X., Zheng, B., Shi, W., Yang, J., Fan, Z., Luo, Z., Rui, X., Chen, B., Yan, Q., & Zhang, H. (2015). Reduced Graphene Oxide-Wrapped MoO_3 Composites prepared by using Metal-Organic Framework as Precursor for All-Solid-State Flexible Supercapacitors. *Advanced Materials*, *27*(32), 4695–4701. doi:10.1002/adma.201501310 PMID:26178419

Cao, Z., Chen, L., Li, S., Yu, M., Li, Z., Zhou, K., Liu, C., Jiang, F., & Hong, M. (2019). A Flexible Two-Fold Interpenetrated Indium MOF Exhibiting Dynamic Response to Gas Adsorption and High-Sensitivity Detection of Nitroaromatic Explosives. *Chemistry, an Asian Journal*, *14*(20), 3597–3602. doi:10.1002/asia.201900458 PMID:31069916

Carter, K. P., Young, A. M., & Palmer, A. E. (2014). Fluorescent sensors for measuring metal ions in living systems. *Chemical Reviews*, *114*(8), 4564–4601. doi:10.1021/cr400546e PMID:24588137

Cassiède, M., Pauly, J., Milhet, M., Rivaletto, M., Marrucho, I. M., Coutinho, J. A. P., & Daridon, J. L. (2008). A quartz crystal microbalance technique to study wax crystallization in the presence of gas. *Measurement Science & Technology*, *19*(6), 065704. doi:10.1088/0957-0233/19/6/065704

Chappanda, K. N., Shekhah, O., Yassine, O., Patole, S. P., Eddaoudi, M., & Salama, K. N. (2018). The quest for highly sensitive QCM humidity sensors: The coating of CNT/MOF composite sensing films as case study. *Sensors and Actuators. B, Chemical*, *257*, 609–619. doi:10.1016/j.snb.2017.10.189

Chem, J. M. (2012). *Electrospinning technology applied in zeolitic imidazolate framework membrane synthesis*. doi:10.1039/c2jm35401b

Chena, S., Fenga, F., Lia, S., Lia, X.-X., & Shu, L. (2018). Metal-organic framework DUT-67 (Zr) for adsorptive removal of trace Hg^{2+} and CH_3Hg^+ in water. *Chemical Speciation and Bioavailability*, *30*(1), 99–106. doi:10.1080/09542299.2018.1509020

Chen, B., Wang, L., Zapata, F., Qian, G., & Lobkovsky, E. B. (2008). A luminescent microporous metal-organic framework for the recognition and sensing of anions. *Journal of the American Chemical Society*, *130*(21), 6718–6719. doi:10.1021/ja802035e PMID:18452294

Chen, D. M., Zhang, N. N., Liu, C. S., & Du, M. (2017). Dual-emitting dye-MOF composite as a self-calibrating sensor for 2, 4, 6-trinitrophenol. *ACS Applied Materials & Interfaces*, *9*(29), 24671–24677. doi:10.1021/acsami.7b07901 PMID:28682050

Chen, D. M., Zhang, N. N., Liu, C. S., & Du, M. (2017). Template-directed synthesis of a luminescent Tb-MOF material for highly selective Fe^{3+} and Al^{3+} ion detection and VOC vapor sensing. *Journal of Materials Chemistry. C, Materials for Optical and Electronic Devices*, *5*(9), 2311–2317. doi:10.1039/C6TC05349A

Cheng, J., Liang, J., Dong, L., Chai, J., Zhao, N., Ullah, S., Wang, H., Zhang, D., Imtiaz, S., Shan, G., & Zheng, G. (2018). Self-Assembly of 2D-Metal-Organic Framework/Graphene Oxide Membranes as Highly Efficient Adsorbents for the Removal of Cs+ from Aqueous Solutions. *RSC Advances*, *8*(71), 40813–40822. doi:10.1039/C8RA08410F

Cheng, X., Jiang, X., Zhang, Y., Lau, C. H., Xie, Z., Ng, D., Smith, S. J. D., Hill, M. R., & Shao, L. (2017). Building Additional Passageways in Polyamide Membranes with Hydrostable Metal Organic Frameworks to Recycle and Remove Organic Solutes from Various Solvents. *ACS Applied Materials & Interfaces*, *9*(44), 38877–38886. doi:10.1021/acsami.7b07373 PMID:29022696

Chen, H. W., Wu, R. J., Chan, K. H., Sun, Y. L., & Su, P. G. (2005). The application of CNT/Nafion composite material to low humidity sensing measurement. *Sensors and Actuators. B, Chemical*, *104*(1), 80–84. doi:10.1016/j.snb.2004.04.105

Chen, L., & Xu, Q. (2019). Metal-Organic Framework Composites for Catalysis. *Matter*, *1*(1), 57–89. doi:10.1016/j.matt.2019.05.018

Chen, N., Li, M. X., Yang, P., He, X., Shao, M., & Zhu, S. R. (2013). Chiral coordination polymers with SHG-active and luminescence: An unusual homochiral 3D MOF constructed from achiral components. *Crystal Growth & Design*, *13*(6), 2650–2660. doi:10.1021/cg400426m

Chen, Q., Yu, P., Huang, W., Yu, S., Liu, M., & Gao, C. (2015). High-Flux Composite Hollow Fiber Nanofiltration Membranes Fabricated through Layer-by-Layer Deposition of Oppositely Charged Crosslinked Polyelectrolytes for Dye Removal. *Journal of Membrane Science*, *492*, 312–321. doi:10.1016/j.memsci.2015.05.068

Chen, R., Zhao, T., Tian, T., Cao, S., Coxon, P. R., Xi, K., Fairen-Jimenez, D., Kumar, R. V., & Cheetham, A. K. (2014). Graphene-Wrapped Sulfur/Metal Organic Framework-derived Microporous Carbon Composite for Lithium Sulfur Batteries. *APL Materials*, *2*(12), 124109. doi:10.1063/1.4901751

Chen, S. G., Shi, Z. Z., Qin, L., Jia, H. L., & Zheng, H. G. (2017). Two New Luminescent Cd (II)-Metal-Organic Frameworks as Bifunctional Chemosensors for Detection of Cations Fe^{3+}, Anions CrO_4^{2-} and $Cr_2O_7^{2-}$ in Aqueous Solution. *Crystal Growth & Design*, *17*(1), 67–72. doi:10.1021/acs.cgd.6b01197

Chen, W. M., Meng, X. L., Zhuang, G. L., Wang, Z., Kurmoo, M., Zhao, Q. Q., Wang, X. P., Shan, B., Tung, C. H., & Sun, D. (2017). A superior fluorescent sensor for Al^{3+} and UO_2^{2+} based on a Co (II) metal-organic framework with exposed pyrimidyl Lewis base sites. *Journal of Materials Chemistry. A, Materials for Energy and Sustainability*, *5*(25), 13079–13085. doi:10.1039/C7TA01546A

Chen, W.-H., Yu, X., Cecconello, A., Sohn, Y. S., Nechushtai, R., & Willner, I. (2017, August 1). Stimuli-responsive nucleic acid-functionalized metal-organic framework nanoparticles using pH- and metal-ion-dependent DNAzymes as locks. *Chemical Science (Cambridge)*, *8*(8), 5769–5780. doi:10.1039/C7SC01765K

Chen, W., Wang, J. Y., Chen, C., Yue, Q., Yuan, H. M., Chen, J. S., & Wang, S. N. (2003). Photoluminescent metal-organic polymer constructed from trimetallic clusters and mixed carboxylates. *Inorganic Chemistry*, *42*(4), 944–946. doi:10.1021/ic025871j PMID:12588123

Chernikova, V., Yassine, O., Shekhah, O., Eddaoudi, M., & Salama, K. N. (2018). Highly sensitive and selective SO2 MOF sensor: The integration of MFM-300 MOF as a sensitive layer on a capacitive interdigitated electrode. *Journal of Materials Chemistry. A, Materials for Energy and Sustainability*, *6*(14), 5550–5554. doi:10.1039/C7TA10538J

Cho, H. Y., Yang, D. A., Kim, J., Jeong, S. Y., & Ahn, W. S. (2012). CO_2 adsorption and catalytic application of Co-MOF-74 synthesized by microwave heating. *Catalysis Today*, *185*(1), 35–40. doi:10.1016/j.cattod.2011.08.019

Chui, S. S. Y., Lo, S. M. F., Charmant, J. P., Orpen, A. G., & Williams, I. D. (1999). A chemically functionalizable nanoporous material [Cu3 (TMA) 2 (H2O) 3] n. *Science*, *283*(5405), 1148–1150. doi:10.1126cience.283.5405.1148 PMID:10024237

Collins, D. J., & Zhou, H. C. (2007). Hydrogen storage in metal-organic frameworks. *Journal of Materials Chemistry*, *17*(30), 3154–3160. doi:10.1039/b702858j

Consales, M., Campopiano, S., Cutolo, A., Penza, M., Aversa, P., Cassano, G., Giordano, M., & Cusano, A. (2006). Sensing properties of buffered and not buffered carbon nanotubes by fibre optic and acoustic sensors. *Measurement Science & Technology*, *17*(5), 1220–1228. doi:10.1088/0957-0233/17/5/S46

Cook, T. R., Yao, R. Z., & Stang, P. J. (2013). Metal-Organic Frameworks and Self-Assembled Supramolecular Coordination Complexes: Comparing and Contrasting the Design, Synthesis, and Functionality of Metal-Organic Materials. *Chemical Reviews*, *113*(1), 734–777. doi:10.1021/cr3002824 PMID:23121121

Czaja, A. U., Trukhan, N., & Müller, U. (2009). Industrial applications of metal–organic frameworks. *Chemical Society Reviews*, *38*(5), 1284–1293. doi:10.1039/b804680h PMID:19384438

D'Ans, P., Courbon, E., Permyakova, A., Nouar, F., Simonnet-Jégat, C., Bourdreux, F., Malet, L., Serre, C., Frère, M., & Steunou, N. (2019). A new strontium bromide MOF composite with improved performance for solar energy storage application. *Journal of Energy Storage*, *25*, 100881. doi:10.1016/j.est.2019.100881

De Decker, J., Folens, K., De Clercq, J., Meledina, M., Van Tendeloo, G., Du Laing, G., & Van Der Voort, P. (2017). Ship-in-a-bottle CMPO in MIL-101(Cr) for selective uranium recovery from aqueous streams through adsorption. *Journal of Hazardous Materials*, *335*, 1–9. doi:10.1016/j.jhazmat.2017.04.029 PMID:28414943

Dechnik, J., Sumby, C. J., & Janiak, C. (2017). Enhancing Mixed-Matrix Membrane Performance with Metal-Organic Framework Additives. *Crystal Growth & Design*, *17*(8), 4467–4488. doi:10.1021/acs.cgd.7b00595

Deng, Y., Chen, N., Li, Q., Wu, X., Huang, X., Lin, Z., & Zhao, Y. (2017). Highly fluorescent metal-organic frameworks based on a benzene-cored tetraphenylethene derivative with the ability to detect 2, 4, 6-trinitrophenol in water. *Crystal Growth & Design*, *17*(6), 3170–3177. doi:10.1021/acs.cgd.7b00131

Denny, M. S. Jr, & Cohen, S. M. (2015). In Situ Modification of Metal-Organic Frameworks in Mixed-Matrix Membranes. *Angewandte Chemie International Edition*, *54*(31), 9029–9032. doi:10.1002/anie.201504077 PMID:26073065

Dias-Souza, M. V., & Dos Santos, R. M. (2018). Drug and food applications of liposomes and nanoparticles: from benchmark to bedside? Drug Delivery Approaches and Nanosystems, 1, 299-326.

Dias-Souza, M. V., Nascimento, A. A., & Vasconcelos, T. Y. L. (2019). Immunomodulatory and Antimicrobial effects of nanoemulsions: pharmaceutical development aspects and perspectives on clinical treatments. In Nanodispersons for Drug delivery. Toronto: Apple Academic Press/CRC Press/Taylor and Francis.

Ding, B., Liu, S. X., Cheng, Y., Guo, C., Wu, X. X., Guo, J. H., Liu, Y. Y., & Li, Y. (2016). Heterometallic Alkaline Earth-Lanthanide Ba^{II}-La^{II} Microporous Metal-Organic Framework as Bifunctional Luminescent probes of Al^{3+} and MnO_4. *Inorganic Chemistry*, *55*(9), 4391–4402. doi:10.1021/acs.inorgchem.6b00111 PMID:27088966

DiPiro, J. T., Talbert, R. L., Yee, G. C., Matzke, G. R., Wells, B. G., & Posey, L. M. (Eds.). (2012). *Pharmacotherapy: a pathophysiologic approach* (8th ed.). McGraw-Hill.

Doherty, C. M., Buso, D., Hill, A. J., Furukawa, S., & Kitagawa, S. (2014). Using Functional Nano- and Microparticles for the Preparation of Metal À Organic Framework. *Accounts of Chemical Research*, *47*(2). Advance online publication. doi:10.1021/ar400130a PMID:24205847

Donato, L., Garofalo, A., Drioli, E., Alharbi, O., Aljlil, S. A., Criscuoli, A., & Algieri, C. (2020). Improved Performance of Vacuum Membrane Distillation in Desalination with Zeolite Membranes. *Separation and Purification Technology*, *237*, 116376. doi:10.1016/j.seppur.2019.116376

Drever, J. I. (1997). *The Geochemistry of Natural Waters: Surface and Groundwater Environments*. Prentice Hall.

Drisdell, W. S., Poloni, R., McDonald, T. M., Long, J. R., Smit, B., Neaton, J. B., & Kortright, J. B. (2013). Probing Adsorption Interactions in Metal–Organic Frameworks using X-ray Spectroscopy. *Journal of the American Chemical Society*, *135*(48), 18183–18190. doi:10.1021/ja408972f PMID:24224556

Duan, J., Higuchi, M., Horike, S., Foo, M. L., Rao, K. P., Inubushi, Y., Fukushima, T., & Kitagawa, S. (2013). Methane Separation: High CO2 /CH4 and C2 Hydrocarbons/CH4 Selectivity in a Chemically Robust Porous Coordination Polymer (Adv. Funct. Mater. 28/2013). *Advanced Functional Materials*, *23*(28), 3481. doi:10.1002/adfm.201370138

Duan, J., Higuchi, M., Zheng, J., Noro, S., Chang, I.-Y., Hyeon-Deuk, K., Mathew, S., Kusaka, S., Sivaniah, E., Matsuda, R., Sakaki, S., & Kitagawa, S. (2017). Density Gradation of Open Metal Sites in the Mesospace of Porous Coordination Polymers. *Journal of the American Chemical Society*, *139*(33), 11576–11583. doi:10.1021/jacs.7b05702 PMID:28747050

Duan, J., Pan, Y., Liu, G., & Jin, W. (2018). Metal-Organic Framework Adsorbents and Membranes for Separation Applications. *Current Opinion in Chemical Engineering*, *20*, 122–131. doi:10.1016/j.coche.2018.04.005

Duan, J., Pan, Y., Pacheco, F., Litwiller, E., Lai, Z., & Pinnau, I. (2015). High-Performance Polyamide Thin-Film-Nanocomposite Reverse Osmosis Membranes Containing Hydrophobic Zeolitic Imidazolate Framework-8. *Journal of Membrane Science*, *476*, 303–310. doi:10.1016/j.memsci.2014.11.038

Du, N., Song, J., Li, S., Chi, Y. X., Bai, F. Y., & Xing, Y. H. (2016). A Highly Stable 3D Luminescent Indium-Polycarboxylic Framework of UO_2^{2+}, Ru^{3+} and Biomolecules Thiamines. *ACS Applied Materials & Interfaces*, *8*(42), 28718–28726. doi:10.1021/acsami.6b09456 PMID:27748584

Duruibe, J. O., Ogwuegbu, M. O. C., & Egwurugwu, J. N. (2007). Heavy metal pollution and human biotoxic effects. *International Journal of Physical Sciences*, *2*, 112–118.

Du, Z., Deng, S., Chen, Y., Wang, B., Huang, J., Wang, Y., & Yu, G. (2015). Removal of perfluorinated carboxylates from washing wastewater of perfluorooctanesulfonyl fluoride using activated carbons and resins. *Journal of Hazardous Materials*, *286*, 136–143. doi:10.1016/j.jhazmat.2014.12.037 PMID:25585266

Efome, J. E., Rana, D., Matsuura, T., & Lan, C. Q. (2018). Insight Studies on Metal-Organic Framework Nanofibrous Membrane Adsorption and Activation for Heavy Metal Ions Removal from Aqueous Solution. *ACS Applied Materials & Interfaces*, *10*(22), 18619–18629. doi:10.1021/acsami.8b01454 PMID:29763287

Efome, J. E., Rana, D., Matsuura, T., & Lan, C. Q. (2018). Metal-Organic frameworks supported on nanofibers to remove heavy metals. *Journal of Materials Chemistry. A, Materials for Energy and Sustainability*, *6*(10), 4550–4555. doi:10.1039/C7TA10428F

Elrasheedy, A., Nady, N., Bassyouni, M., & El-Shazly, A. (2019). Metal Organic Framework Based Polymer Mixed Matrix Membranes: Review on Applications in Water Purification. *Membranes*, *9*(7), 88. doi:10.3390/membranes9070088 PMID:31330993

Escobar-Ferrand, L., Li, D., Lee, D., & Durning, C. J. (2014). All-Nanoparticle Layer-by-Layer Surface Modification of Micro- and Ultrafiltration Membranes. *Langmuir*, *30*(19), 5545–5556. doi:10.1021/la5002489 PMID:24568094

Evans, J. D., Sumby, C. J., & Doonan, C. J. (2014). Post-synthetic metalation of metal-organic frameworks. *Chemical Society Reviews*, *43*(16), 5933–5951. doi:10.1039/C4CS00076E PMID:24736674

Ezugbe, E. O., & Rathilal, S. (2020). Membrane Technologies in Wastewater Treatment: A Review. *Membranes*, *10*(5). PMID:32365810

Fei, H., Bresler, M. R., & Oliver, S. R. J. (2011). A new paradigm for anion trapping in high capacity and selectivity: Crystal-to-crystal transformation of cationic materials. *Journal of the American Chemical Society*, *133*(29), 11110–11113. doi:10.1021/ja204577p PMID:21707072

Feng, D., Chung, W. C., Wei, Z., Gu, Z. Y., Jiang, H. L., Chen, Y. P., Darensbourg, D. J., & Zhou, H. C. (2013). Construction of ultrastable porphyrin Zr metal-organic frameworks through linker elimination. *Journal of the American Chemical Society*, *135*(45), 17105–17110. doi:10.1021/ja408084j PMID:24125517

Feng, M., Zhang, P., Zhou, H.-C., & Sharma, V. K. (2018). Water-stable metal-organic frameworks for aqueous removal of heavy metals and radionuclides: A review. *Chemosphere*, *209*, 783–800. doi:10.1016/j.chemosphere.2018.06.114 PMID:29960946

Feng, X., Song, Y., Li, Z., Kaufmann, M., Pi, Y., Chen, J. S., Xu, Z., Li, Z., Wang, C., & Lin, W. (2019). Metal–Organic Framework Stabilizes a Low-Coordinate Iridium Complex for Catalytic Methane Borylation. *Journal of the American Chemical Society*, *141*(28), 11196–11203. doi:10.1021/jacs.9b04285 PMID:31265280

Filippov, A. P., Strizhak, P. E., Vlasenko, N. V., Kochkin, Y. N., & Serebrii, T. G. (2014). Adsorption—Desorption Dynamics of Alcohols on H-Beta and H-CMK Zeolites Nanocrystallites Studied by Quartz Crystal Microbalance Method. *Adsorption Science and Technology*, *32*(10), 807–820. doi:10.1260/0263-6174.32.10.807

Folens, K., Leus, K., Nicomel, N. R., Meledina, M., Turner, S., Tendeloo, G. V., Laing, G. D., & Voort, P. V. D. (2016). Fe_3O_4@MIL-101-A Selective and regenerable adsorbent for the removal of As species from water. *European Journal of Inorganic Chemistry*, *2016*(27), 4395–4401. doi:10.1002/ejic.201600160

Freedman, A., Kebabian, P. L., Li, Z., Robinson, W. A., & Wormhoudt, J. C. (2008). Apparatus for determination of vapor pressures at ambient temperatures employing a Knudsen effusion cell and quartz crystal microbalance. *Measurement Science & Technology*, *19*(12), 125102. doi:10.1088/0957-0233/19/12/125102

Frerking, M. E. (1978). Temperature Compensation. In *Crystal Oscillator Design and Temperature Compensation* (pp. 130–176). Springer. doi:10.1007/978-94-011-6056-8_10

Fu, F., & Wang, Q. (2011). Removal of heavy metal ions from wastewaters: a review. Journal of Environmental Management. *J. Environ. Manage.*, *92*, 407-418.

Gao, J., He, P., Yang, T., Wang, X., Zhou, L., He, Q., Jia, L., Deng, H., Zhang, H., Jia, B., & He, X. (2020). Short rod-like Ni-MOF anchored on graphene oxide nanosheets: A promising voltammetric platform for highly sensitive determination of p-chloronitrobenzene. *Journal of Electroanalytical Chemistry (Lausanne, Switzerland)*, *861*, 113954. doi:10.1016/j.jelechem.2020.113954

Garrido, L., Mar, L., & Castillo, L. F. (2010). *CO 2 Transport in Polysulfone Membranes Containing Zeolitic Imidazolate Frameworks As Determined by Permeation and PFG NMR Techniques*. doi:10.1021/ma902303e

Gascon, J., & Kapteijn, F. (2010). Metal-organic framework membranes-high potential, bright future? *Angewandte Chemie International Edition*, *49*(9), 1530–1532. doi:10.1002/anie.200906491 PMID:20127924

Ge, K. M., Wang, D., Xu, Z. J., & Chu, R. Q. (2020). A luminescent Eu (III)-MOF for selective sensing of Ag^+ in aqueous solution. *Journal of Molecular Structure*, *1208*, 127862. doi:10.1016/j.molstruc.2020.127862

Gholami, F., Zinadini, S., Zinatizadeh, A. A., & Abbasi, A. R. (2018). TMU-5 Metal-Organic Frameworks (MOFs) as a Novel Nanofiller for Flux Increment and Fouling Mitigation in PES Ultrafiltration Membrane. *Separation and Purification Technology*, *194*, 272–280. doi:10.1016/j.seppur.2017.11.054

Ghorbani-Kalhor, E. (2016). A metal-organic framework nanocomposite made from functionalized magnetite nanoparticles and HKUST-1 (MOF-199) for preconcentration of Cd(II), Pb(II), and Ni(II). *Mikrochimica Acta*, *183*(9), 2639–2647. doi:10.100700604-016-1896-2

Ghorbani-Kalhor, E., Hosseinzadeh-Khanmiri, R., Abolhasani, J., Babazadeh, M., & Hassanpour, A. (2015). Determination of mercury(II) ions in seafood samples after extraction and pre-concentration by a novel functionalized magnetic metal-organic framework nanocomposite. *Journal of Separation Science*, *38*(7), 1179–1186. doi:10.1002/jssc.201401320 PMID:25631191

Ghosh, A., Chakrabarti, S., Biswas, K., & Ghosh, U. C. (2014). Agglomerated nanoparticles of hydrous Ce(IV)+Zr(IV) mixed oxide: Preparation, characterization and physicochemical aspects on fluoride adsorption. *Applied Surface Science*, *307*, 665–676. doi:10.1016/j.apsusc.2014.04.095

Glover, T. G., Peterson, G. W., Schindler, B. J., Britt, D., & Yaghi, O. (2011). MOF-74 building unit has a direct impact on toxic gas adsorption. *Chemical Engineering Science*, *66*(2), 163–170. doi:10.1016/j.ces.2010.10.002

Gnanasekarana, G., Balagurua, S., Gangasalama, A., & Das, D. B. (2018). Removal of hazardous material from wastewater by using Metal-Organic Framework (MOFs) embedded polymeric membranes. *Separation Science and Technology*, *54*(3), 1–13.

Gnanasekaran, G., Balaguru, S., Arthanareeswaran, G., & Das, D. B. (2019). Removal of Hazardous Material from Wastewater by Using Metal Organic Framework (MOF) Embedded Polymeric Membranes. *Separation Science and Technology (Philadelphia)*, *54*(3), 434–446. doi:10.1080/01496395.2018.1508232

Godfrey, H. G. W., da Silva, I., Briggs, L., Carter, J. H., Morris, C. G., Savage, M., & Schröder, M. (2018). Ammonia Storage by Reversible Host–Guest Site Exchange in a Robust Metal–Organic Framework. *Angewandte Chemie International Edition*, *57*(45), 14778–14781. doi:10.1002/anie.201808316 PMID:30098090

Golpour, M., & Pakizeh, M. (2018). Preparation and Characterization of New PA-MOF/PPSU-GO Membrane for the Separation of KHI from Water. *Chemical Engineering Journal*, *345*, 221–232. doi:10.1016/j.cej.2018.03.154

Gonzalez, M. I., Turkiewicz, A. B., Darago, L. E., Oktawiec, J., Bustillo, K., Grandjean, F., Long, G. J., & Long, J. R. (2020). Confinement of atomically defined metal halide sheets in a metal–organic framework. *Nature*, *577*(7788), 64–68. doi:10.103841586-019-1776-0 PMID:31739311

Gorchev, H. G., & Ozolins, G. (2004). *Guidelines for Drinking-Water Quality (3rd ed.)*. WHO.

Górka, J., Fulvio, P. F., Pikus, S., & Jaroniec, M. (2010). Mesoporous metal organic framework-boehmite and silica composites. *Chemical Communications*, *46*(36), 6798–6800. doi:10.1039/c0cc01578d PMID:20717603

Griffith, R. L. (1943). The crystal structure of silver oxalate. *The Journal of Chemical Physics*, *11*(11), 499–505. doi:10.1063/1.1723789

Guala, S. D., Vega, F. A., & Covelo, E. F. (2010). The dynamics of heavy metals in plant-soil interactions. *Ecological Modelling*, *221*(8), 1148–1152. doi:10.1016/j.ecolmodel.2010.01.003

Guay, D. R. (2003). Clinical pharmacokinetics of drugs used to treat urge incontinence. *Clinical Pharmacokinetics*, *42*(14), 1243–1285. doi:10.2165/00003088-200342140-00004 PMID:14606931

Guo, Y., Feng, X., Han, T., Wang, S., Lin, Z., Dong, Y., & Wang, B. (2014). Tuning the luminescence of metal-organic frameworks for detection of energetic heterocyclic compounds. *Journal of the American Chemical Society*, *136*(44), 15485–15488. doi:10.1021/ja508962m PMID:25325884

Guo, Y., Wang, X., Hu, P., & Peng, X. (2016). ZIF-8 Coated Polyvinylidenefluoride (PVDF) Hollow Fiber for Highly Efficient Separation of Small Dye Molecules. *Applied Materials Today*, *5*, 103–110. doi:10.1016/j.apmt.2016.07.007

Gu, Q., Albert Ng, T. C., Sun, Q., Kotb Elshahawy, A. M., Lyu, Z., He, Z., Zhang, L., Ng, H. Y., Zeng, K., & Wang, J. (2019). Heterogeneous ZIF-L Membranes with Improved Hydrophilicity and Anti-Bacterial Adhesion for Potential Application in Water Treatment. *RSC Advances*, *9*(3), 1591–1601. doi:10.1039/C8RA08758J

Gu, Y., Wang, Y., Li, H., Qin, W., Zhang, H., Wang, G., Zhang, Y., & Zhao, H. (2020). Fabrication of Hierarchically Porous NH2-MIL-53/Wood-Carbon Hybrid Membrane for Highly Effective and Selective Sequestration of Pb2+. *Chemical Engineering Journal*, *387*, 124141. doi:10.1016/j.cej.2020.124141

Gu, Z. G., Fu, W. Q., Liu, M., & Zhang, J. (2017). Surface-mounted MOF templated fabrication of homochiral polymer thin film for enantioselective adsorption of drugs. *Chemical Communications*, *53*(9), 1470–1473. doi:10.1039/C6CC08342K PMID:28093585

Gu, Z., Song, W., Yanga, Z., & Zhou, R. (2018). Metal-organic framework as an efficient filter for the removal of heavy metal cations in water. *Physical Chemistry Chemical Physics*, *20*(48), 30384–30391. doi:10.1039/C8CP05129A PMID:30489583

Haghighi, E., & Zeinali, S. (2019). Nanoporous MIL-101 (Cr) as a sensing layer coated on a quartz crystal microbalance (QCM) nanosensor to detect volatile organic compounds (VOCs). *RSC Advances*, *9*(42), 24460–24470. doi:10.1039/C9RA04152D

Haghighi, E., & Zeinali, S. (2020). Formaldehyde detection using quartz crystal microbalance (QCM) nanosensor coated by nanoporous MIL-101 (Cr) film. *Microporous and Mesoporous Materials*, *300*, 110065. doi:10.1016/j.micromeso.2020.110065

Haldar, D., Duarah, P., & Purkait, M. K. (2020). MOFs for the Treatment of Arsenic, Fluoride and Iron Contaminated Drinking Water: A Review. *Chemosphere*, *251*, 126388. doi:10.1016/j.chemosphere.2020.126388 PMID:32443223

Halder, S., Mondal, J., Ortega-Castro, J., Frontera, A., & Roy, P. (2017). A Ni-based MOF for selective detection and removal of Hg^{2+} in aqueous medium: A facile strategy. *Dalton Transactions (Cambridge, England)*, *46*(6), 1943–1950. doi:10.1039/C6DT04722J PMID:28112319

Han, Z., Wang, K., Guo, Y., Chen, W., Zhang, J., Zhang, X., Siligardi, G., Yang, S., Zhou, Z., Sun, P., Shi, W., & Cheng, P. (2019). Cation-induced chirality in a bifunctional metal-organic framework for quantitative enantioselective recognition. *Nature Communications*, *10*(1), 5117. doi:10.103841467-019-13090-9 PMID:31712651

Hao, J. N., & Yan, B. (2016). Ln^{3+} post-functionalized metal-organic frameworks for color tunable emission and highly sensitive sensing of toxic anions and small molecules. *New Journal of Chemistry*, *40*(5), 4654–4661. doi:10.1039/C5NJ03419A

Hartmann, M., Kunz, S., Himsl, D., Tangermann, O., Ernst, S., & Wagener, A. (2008). Adsorptive separation of isobutene and isobutene on $Cu_3(BTC)_2$. *Langmuir*, *24*(16), 8634–8642. doi:10.1021/la8008656 PMID:18611044

Hayden, O., Bindeus, R., & Dickert, F. L. (2003). Combining atomic force microscope and quartz crystal microbalance studies for cell detection. *Measurement Science & Technology*, *14*(11), 1876–1881. doi:10.1088/0957-0233/14/11/004

He, J., Cai, X., Chen, K., Li, Y., Zhang, K., Jin, Z., & Kong, L. (2016). Performance of a novelly-defined zirconium metal-organic frameworks adsorption membrane in fluoride removal. *Journal of Colloid and Interface Science*, *484*, 162–172. doi:10.1016/j.jcis.2016.08.074 PMID:27610471

He, J., Xu, J., Yin, J., Li, N., & Bu, X. H. (2019). Recent advances in luminescent metal-organic frameworks for chemical sensors. *Science China Materials*, *62*(11), 1655–1678. doi:10.100740843-019-1169-9

He, T., Zhang, Y. Z., Kong, X. J., Yu, J., Lv, X. L., Wu, Y., Guo, Z. J., & Li, J. R. (2018). Zr (IV)-based metal-organic framework with T-shaped ligand: Unique structure, high stability, selective detection, and rapid adsorption of $Cr_2O_7^{2-}$ in water. *ACS Applied Materials & Interfaces*, *10*(19), 16650–16659. doi:10.1021/acsami.8b03987 PMID:29733570

He, Y., Yu, P. T., Ma, D., & Chung, T. S. (2017). UiO-66 Incorporated Thin-Film Nanocomposite Membranes for Efficient Selenium and Arsenic Removal. *Journal of Membrane Science*, *541*(March), 262–270. doi:10.1016/j.memsci.2017.06.061

Hindocha, S. (2017). The chemistry of metal-organic frameworks: Synthesis, characterization, and applications. *Johnson Matthey Technology Review*, *61*(2), 138–141. doi:10.1595/205651317X695091

Hinterholzinger, F. M., Rühle, B., Wuttke, S., Karaghiosoff, K., & Bein, T. (2013). Highly sensitive and selective fluoride detection in water through fluorophore release from a metal-organic framework. *Scientific Reports*, *3*(1), 1–7. doi:10.1038rep02562 PMID:24008779

Horcajada, P., Chalati, T., Serre, C., Gillet, B., Sebrie, C., Baati, T., Eubank, J. F., Heurtaux, D., Clayette, P., Kreuz, C., Chang, J.-S., Hwang, Y. K., Marsaud, V., Bories, P.-N., Cynober, L., Gil, S., Férey, G., Couvreur, P., & Gref, R. (2010). Porous metal-organic framework nanoscale carriers as a potential platform for drug delivery and imaging. *Nature Materials*, *9*(2), 172–178. doi:10.1038/nmat2608 PMID:20010827

Hoskins, B. F., & Robson, R. (1990). Design and construction of a new class of scaffolding-like materials comprising infinite polymeric frameworks of 3D-linked molecular rods. A reappraisal of the zinc cyanide and cadmium cyanide structures and the synthesis and structure of the diamond-related frameworks $[N(CH_3)_4][Cu^IZn^{II}(CN)_4]$ and Cu^I[4, 4′, 4″, 4‴-tetracyanotetraphenylmethane] $BF_4.xC_6H_5NO_2$. *Journal of the American Chemical Society*, *112*(4), 1546–1554. doi:10.1021/ja00160a038

Howarth, A. J., Katz, M. J., Wang, T. C., Platero-Prats, A. E., Chapman, K. W., Hupp, J. T., & Farha, O. K. (2015). High efficiency adsorption and removal of selenate and selenite from water using metal-organic frameworks. *Journal of the American Chemical Society*, *137*(23), 7488–7494. doi:10.1021/jacs.5b03904 PMID:26000611

Huang, H., Zhou, J., Chen, S., Zeng, L., & Huang, Y. (2004). A highly sensitive QCM sensor coated with Ag^+-ZSM-5 film for medical diagnosis. *Sensors and Actuators. B, Chemical*, *101*(3), 316–321. doi:10.1016/j.snb.2004.04.001

Huang, X., Xu, Q., Gao, W., Yang, T., Zhan, R., Deng, J., Guo, B., Tao, M., Liu, H., & Xu, M. (2019). Rechargeable K-Se Batteries based on Metal-Organic Frameworks-derived Porous Carbon Matrix confined Selenium as Cathode Materials. *Journal of Colloid and Interface Science*, *539*, 326–331. doi:10.1016/j.jcis.2018.12.083 PMID:30594007

Huang, Y., Zeng, X., Guo, L., Lan, J., Zhang, L., & Cao, D. (2018). Heavy metal ion removal of wastewater by zeolite-imidazolate frameworks. *Separation and Purification Technology*, *194*, 462–469. doi:10.1016/j.seppur.2017.11.068

Hua, Y., Li, X., Chen, C., & Pang, H. (2019). Cobalt based Metal-Organic Frameworks and Their Derivatives for Electrochemical Energy Conversion and Storage. *Chemical Engineering Journal*, *370*, 37–59. doi:10.1016/j.cej.2019.03.163

Huo, J.-B., Xu, L., Yang, J.-C. E., Cui, H.-J., Yuan, B., & Fu, M. L. (2018). Magnetic responsive Fe_3O_4-ZIF-8 core-shell composites for efficient removal of As(III) from water. *Colloids and Surfaces. A, Physicochemical and Engineering Aspects*, *539*, 59–68. doi:10.1016/j.colsurfa.2017.12.010

Hu, R., Zhang, K., Fan, G., Luo, Z., & Li, G. (2015). Development of a high-sensitivity plasticizer sensor based on a quartz crystal microbalance modified with a nanostructured nickel hydroxide film. *Measurement Science & Technology*, *26*(5), 055102. doi:10.1088/0957-0233/26/5/055102

Hu, Z., Chen, Y., & Jiang, J. (2011). Zeolitic Imidazolate Framework-8 as a Reverse Osmosis Membrane for Water Desalination: Insight from Molecular Simulation. *The Journal of Chemical Physics*, *134*(13), 134705. doi:10.1063/1.3573902 PMID:21476767

Hu, Z., Deibert, B. J., & Li, J. (2014). Luminescent metal-organic frameworks for chemical sensing and explosive detection. *Chemical Society Reviews*, *43*(16), 5815–5840. doi:10.1039/C4CS00010B PMID:24577142

Ishida, T., Nagaoka, M., Akita, T., & Haruta, M. (2008). Deposition of gold clusters on porous coordination polymers by solid grinding and their catalytic activity in aerobic oxidation of alcohols. *Chemistry (Weinheim an der Bergstrasse, Germany)*, *14*(28), 8456–8460. doi:10.1002/chem.200800980 PMID:18752232

Jaramillo, D. E., Reed, D. A., Jiang, H. Z. H., Oktawiec, J., Mara, M. W., Forse, A. C., Lussier, D. J., Murphy, R. A., Cunningham, M., Colombo, V., Shuh, D. K., Reimer, J. A., & Long, J. R. (2020). Selective nitrogen adsorption via backbonding in a metal–organic framework with exposed vanadium sites. *Nature Materials*, *19*(5), 517–521. Advance online publication. doi:10.103841563-019-0597-8 PMID:32015534

Jhung, S. H., Lee, J. H., Yoon, J. W., Serre, C., Ferey, G., & Chang, J. S. (2007). Microwave synthesis of chromium terephthalate MIL-101 its benzene sorption. *Advanced Materials*, *19*(1), 121–124. doi:10.1002/adma.200601604

Jiang, K., Zhang, L., Xia, T., Yang, Y., Li, B., Cui, Y., & Qian, G. (2019). A water-stable fcu-MOF material with exposed amino groups for the multi-functional separation of small molecules. *Science China Materials*, *62*(9), 1315–1322. doi:10.100740843-019-9427-5

Jian, M., Liu, B., Zhang, G., Liu, R., & Zhang, X. (2015). Adsorptive removal of arsenic from aqueous solution by zeolitic imidazolate framework-8 (ZIF-8) nanoparticles. *Colloids and Surfaces. A, Physicochemical and Engineering Aspects*, *465*, 67–76. doi:10.1016/j.colsurfa.2014.10.023

Jian, M., Wang, H., Liu, R., Qu, J., Wang, H., & Zhang, X. (2016). Self-assembled one-dimensional MnO_2@zeolitic imidazolate framework-8 nanostructures for highly efficient arsenite removal. *Environmental Science. Nano*, *3*(5), 1186–1194. doi:10.1039/C6EN00246C

Ji, G., Gao, X., Zheng, T., Guan, W., Liu, H., & Liu, Z. (2018). Post synthetic metalation metal-organic framework as a fluorescent probe for the ultrasensitive and reversible detection of PO_4^{3-} ions. *Inorganic Chemistry*, *57*(17), 10525–10532. doi:10.1021/acs.inorgchem.8b00313 PMID:30095895

Jin, F., Zhang, Y., Wang, H. Z., Zhu, H. Z., Yan, Y., Zhang, J., Wu, J. Y., Tian, Y.-P., & Zhou, H.-P. (2013). Diverse structural Ag (I) supramolecular complexes constructed from multidentate dicyanoisophorone-based ligands: Structure and enhanced luminescence. *Crystal Growth & Design*, *13*(5), 1978–1987. doi:10.1021/cg400025j

Jin, W.-W., Li, H.-J., Zou, J.-Z., Iguva, S., Zhang, Q., Zeng, S.-Z., Xu, G.-Z., & Zeng, X.-R. (2020). Metal Organic Framework-derived Carbon Nanosheets with Fish-Scale Surface Morphology as Cathode Materials for Lithium-Selenium Batteries. *Journal of Alloys and Compounds*, *820*, 153084. doi:10.1016/j.jallcom.2019.153084

Ji, X., Yao, W., Peng, J., Ren, N., Zhou, J., & Huang, Y. (2012). Evaluation of Cu-ZSM-5 zeolites as QCM sensor coatings for DMMP detection. *Sensors and Actuators. B, Chemical*, *166*, 50–55. doi:10.1016/j.snb.2011.12.014

Joaristi, A. M., Juan-Alcanpiz, J., Serra-Crespo, P., Kapteijin, F., & Gascon, J. (2012). Electrochemical synthesis of some Zn^{2+}, Cu^{2+} and Al^{3+} metal-organic frameworks. *Crystal Growth & Design*, *12*(7), 3489–3498. doi:10.1021/cg300552w

Johansson, S. (2006, March). New frequency counting principle improves resolution. In *Proceedings of the 20th European Frequency and Time Forum* (pp. 139-146). IEEE.

Kadhom, M., & Deng, B. (2018). Metal-Organic Frameworks (MOFs) in Water Filtration Membranes for Desalination and Other Applications. *Applied Materials Today*, *11*, 219–230. doi:10.1016/j.apmt.2018.02.008

Kadhom, M., Hu, W., & Deng, B. (2017). Thin Film Nanocomposite Membrane Filled with Metal-Organic Frameworks UiO-66 and MIL-125 Nanoparticles for Water Desalination. *Membranes*, *7*(2), 31. doi:10.3390/membranes7020031 PMID:28613247

Kamarehie, B., Noraee, Z., Jafari, A., Ghaderpoori, M., Karami, M. A., & Ghaderpoury, A. (2018). Data on the fluoride adsorption from aqueous solutions by metal-organic frameworks (ZIF-8 and Uio-66). *Data in Brief*, *20*, 799–804. doi:10.1016/j.dib.2018.08.159 PMID:30211278

Kandiah, M., Nilsen, M. H., Usseglio, S., Jakobsen, S., Olsbye, U., Tilset, M., Larabi, C., Quadrelli, E. A., Bonino, F., & Lillerud, K. P. (2010). Synthesis and stability of tagged UiO-66 Zr-MOFs. *Chemistry of Materials*, *22*(24), 6632–6640. doi:10.1021/cm102601v

Karimi, A., Vatanpour, V., Khataee, A., & Safarpour, M. (2019). Contra-Diffusion Synthesis of ZIF-8 Layer on Polyvinylidene Fluoride Ultrafiltration Membranes for Improved Water Purification. *Journal of Industrial and Engineering Chemistry*, *73*, 95–105. doi:10.1016/j.jiec.2019.01.010

Karmakar, A., Kumar, N., Samanta, P., Desai, A. V., & Ghosh, S. K. (2016). A post-synthetically modified MOF for selective and sensitive aqueous-phase detection of highly toxic cyanide ions. *Chemistry (Weinheim an der Bergstrasse, Germany)*, *22*(3), 864–868. doi:10.1002/chem.201503323 PMID:26584835

Karmakar, S., Bhattacharjee, S., & De, S. (2018). Aluminium fumarate metal organic framework incorporated polyacrylonitrile hollow fiber membranes: Spinning, characterization and application in fluoride removal from groundwater. *Chemical Engineering Journal*, *334*, 41–53. doi:10.1016/j.cej.2017.10.021

Karmakar, S., Dechanik, J., Janiak, C., & De, S. (2016). Aluminium fumarate metal-organic framework; A super adsorbent for fluoride removal from water. *Journal of Hazardous Materials*, *303*, 10–20. doi:10.1016/j.jhazmat.2015.10.030 PMID:26513559

Kaur, R., Kim, K. H., & Deep, A. (2017). A convenient electrolytic assembly of graphene-MOF composite thin film and its photoanodic application. *Applied Surface Science*, *396*, 1303–1309. doi:10.1016/j.apsusc.2016.11.150

Kazemipour, M., Ansari, M., Tajrobehkar, S., Majdzadeh, M., & Kermani, H. R. (2008). Removal of lead, cadmium, zinc, and copper from industrial wastewater by carbon developed from walnut, hazelnut, almond, pistachio shell, and apricot stone. *Journal of Hazardous Materials*, *150*(2), 322–327. doi:10.1016/j.jhazmat.2007.04.118 PMID:17548149

Kebria, M. R. S., Rahimpour, A., Bakeri, G., & Abedini, R. (2018, October). 2019. "Experimental and Theoretical Investigation of Thin ZIF-8/Chitosan Coated Layer on Air Gap Membrane Distillation Performance of PVDF Membrane. *Desalination*, *450*, 21–32. doi:10.1016/j.desal.2018.10.023

Keeffe, M., & Yaghi, O. M. (2012). Deconstructing the Crystal Structures of Metal Organic Frameworks and Related Materials into Their Underlying Nets. *Chemical Reviews*, *112*(2), 675–702. doi:10.1021/cr200205j PMID:21916513

Ke, F., Peng, C., Zhang, T., Zhang, M., Zhou, C., Cai, H., Zhu, J., & Wan, X. (2018). Fumarate-based metal-organic frameworks as a new platform for highly selective removal of fluoride from brick tea. *Scientific Reports*, *8*(1), 1–11. doi:10.103841598-018-19277-2 PMID:29343778

Kepert, C. J. (2010). Metal-Organic Framework Materials. In D. W. Bruce, D. O'Hare, R. I. Walton, D. W. Bruce, D. O'Hare, & R. I. Walton (Eds.), *Porous Materials*. doi:10.1002/9780470711385.ch1

Khanjani, S., & Morsali, A. (2014). Ultrasound-Promoted Coating of MOF-5 on Silk Fiber and Study of Adsorptive Removal and Recovery of Hazardous Anionic Dye 'Congo Red.'. *Ultrasonics Sonochemistry*, *21*(4), 1424–1429. doi:10.1016/j.ultsonch.2013.12.012 PMID:24412182

Khan, N. A., Hasan, Z., & Jhung, S. H. (2013). Adsorptive Removal of Hazardous Materials Using Metal-Organic Frameworks (MOFs): A Review. *Journal of Hazardous Materials*, *244–245*, 444–456. doi:10.1016/j.jhazmat.2012.11.011 PMID:23195596

Khatua, S., Goswami, S., Biswas, S., Tomar, K., Jena, S. H., & Konar, S. (2015). Stable multiresponsive luminescent MOF for colorimetric detection of small molecules in selective and reversible manner. *Chemistry of Materials*, *27*(15), 5349–5360. doi:10.1021/acs.chemmater.5b01773

Kim, J., Kim, S. H., Yang, S. T., & Ahn, W. S. (2012). Bench-scale preparation of $Cu_3(BTC)_2$ by ethanol reflux: Synthesis optimization and adsorption/catalytic application. *Microporous and Mesoporous Materials*, *161*, 48–55. doi:10.1016/j.micromeso.2012.05.021

Kinoshita, Y., Matsubara, I., & Saito, Y. (1959). The crystal structure of bis(succinonitrilo)copper(I) nitrate. *Bulletin of the Chemical Society of Japan*, *32*(7), 741–747. doi:10.1246/bcsj.32.741

Kitagawa, S., Kitaura, R., & Shin-ichiro, N. (2004). Functional Porous Coordination Polymers. *Angewandte Chemie International Edition*, *43*(18), 2334–2375. doi:10.1002/anie.200300610 PMID:15114565

Kitagawa, S., Matsuyama, S., Munakata, M., & Emori, T. (1991). Synthesis and crystal structures of novel one-dimensional polymers,[$\{M(bpen)X\}_\infty$][M= Cu^I, X= [{Cu (bpen)(CO)(CH3 CN)(PF6)}∞]. *Journal of the Chemical Society, Dalton Transactions: Inorganic Chemistry*, (11), 2869–2874. doi:10.1039/DT9910002869

Klimakow, M., Klobes, P., Thunemann, A. F., Rademann, K., & Emmerling, F. (2010). Mechanochemical synthesis of metal-organic frameworks: A fast and facile approach towards quantitative yields and high specific surface areas. *Chemistry of Materials*, *22*(18), 5216–5221. doi:10.1021/cm1012119

Kolliopoulos, A. V., Kampouris, D. K., & Banks, C. E. (2015). Rapid and portable electrochemical quantification of phosphorus. *Analytical Chemistry*, *87*(8), 4269–4274. doi:10.1021/ac504602a PMID:25856498

Kosuru, L., Bouchaala, A., Jaber, N., & Younis, M. I. (2016). Humidity detection using metal organic framework coated on QCM. *Journal of Sensors*, *2016*, 2016. doi:10.1155/2016/4902790

Kou, D., Ma, W., Zhang, S., Li, R., & Zhang, Y. (2020). BTEX vapors detection with flexible MOF and functional polymer by means of composite photonic crystal. *ACS Applied Materials & Interfaces*, *12*(10), 11955–11964. doi:10.1021/acsami.9b22033

Kozlowski, H., Janicka-Klos, A., Brasun, J., Gaggelli, E., Valensin, D., & Valensin, G. (2009). Copper, iron, and zinc ions homeostasis and their role in neurodegenerative disorders (metal uptake, transport, distribution and regulation). *Coordination Chemistry Reviews*, *253*(21), 2665–2685. doi:10.1016/j.ccr.2009.05.011

Kumar, B. (2015). *Fluoride in Groundwater and its Health Hazards: A Review in Indian Scenario*. Academic Press.

Kumar, P., Anand, B., Tsang, Y. F., Kim, K.-H., Khullar, S., & Wang, B. (2019). Regeneration, degradation, and toxicity effect of MOFs: Opportunities and challenges. *Environmental Research*, *176*, 108488. doi:10.1016/j.envres.2019.05.019 PMID:31295665

Kung, C.-W., Han, P.-C., Chuang, C.-H., & Wu, K. C.-W. (2019). Electronically Conductive Metal-Organic Framework-based Materials. *APL Materials*, *7*(11), 110902. doi:10.1063/1.5125487

Kurosawa, S., Aizawa, H., Tozuka, M., Nakamura, M., & Park, J. W. (2003). Immunosensors using a quartz crystal microbalance. *Measurement Science & Technology*, *14*(11), 1882–1887. doi:10.1088/0957-0233/14/11/005

Kurwadkar, S. (2019). Occurrence and Distribution of Organic and Inorganic Pollutants in Groundwater. *Water Environment Research*, *91*(10), 1001–1008. doi:10.1002/wer.1166 PMID:31230394

Kwon, S. Y., Kim, J. C., & Choi, B. I. (2008). Accurate dew-point measurement over a wide temperature range using a quartz crystal microbalance dew-point sensor. *Measurement Science & Technology*, *19*(11), 115206. doi:10.1088/0957-0233/19/11/115206

Lakowicz, J. R. (2006). *Principles of Fluorescence Spectroscopy*. Springer. doi:10.1007/978-0-387-46312-4

Lalor, G. C. (2008). Review of cadmium transfers from soil to humans and its health effects in the Jamaican environment. *The Science of the Total Environment*, *400*(1-3), 162–172. doi:10.1016/j.scitotenv.2008.07.011 PMID:18752835

Lan, A., Li, K., Wu, H., Olson, D. H., Emge, T. J., Ki, W., Hong, M., & Li, J. (2009). A luminescent microporous metalorganic framework for the fast and reversible detection of high explosives. *Angewandte Chemie International Edition*, *48*(13), 2334–2338. doi:10.1002/anie.200804853 PMID:19180622

Lang, X., Dong, C., Cai, K., Li, L., & Zhang, Q. (2019). Highly Active Cluster Structure Manganese-based Metal Organic Frameworks (Mn-MOFs) like Corals in the Sea Synthesized by Facile Solvothermal Method as Gas Electrode Catalysts for Lithium-Oxygen Batteries. *International Journal of Energy Research*, *44*(2), 1256–1263. doi:10.1002/er.4954

Lee, Tang, & Huo. (2014). Fabrication of Porous Matrix Membrane (PMM) Using Metal-Organic Framework as Green Template for Water Treatment. *Scientific Reports, 4*, 1–5.

Lee, J. Y., She, Q., Huo, F., & Tang, C. Y. (2015). Metal-Organic Framework-Based Porous Matrix Membranes for Improving Mass Transfer in Forward Osmosis Membranes. *Journal of Membrane Science*, *492*, 392–399. doi:10.1016/j.memsci.2015.06.003

Lee, J., Farha, O. K., Roberts, J., Scheidt, K. A., Nguyen, S. T., & Hupp, J. T. (2009). Metal-organic framework materials as catalysts. *Chemical Society Reviews*, *38*(5), 1450–1459. doi:10.1039/b807080f PMID:19384447

Lee, T. H., Oh, J. Y., Hong, S. P., Lee, J. M., Roh, S. M., Kim, S. H., & Park, H. B. (2019). ZIF-8 Particle Size Effects on Reverse Osmosis Performance of Polyamide Thin-Film Nanocomposite Membranes: Importance of Particle Deposition. *Journal of Membrane Science*, *570–571*, 23–33. doi:10.1016/j.memsci.2018.10.015

Lee, Y. R., & Ahn, W. S. (2013). Synthesis of metal-organic frameworks: A mini review. *Korean Journal of Chemical Engineering*, *30*(9), 1667–1680. doi:10.100711814-013-0140-6

Leung, E., Muller, U., Trukhan, N., Mattenheimer, H., Cox, G., & Blei, S. (2012). *Process preparing porous metal organic frameworks based on aluminium fumarate*. US patent (05.04.12) 2012/0082864A1

Leus, K., Muylaert, I., Van Speybroeck, V., Marin, G. B., & Van Der Voort, P. (2010). *A coordinative saturated vanadium containing metal organic framework that shows a remarkable catalytic activity*. doi:10.1016/S0167-2991(10)75053-9

Li. (2018). Efficient Removal of Nickel(II) from High Salinity Wastewater by a Novel PAA/ZIF-8/PVDF Hybrid Ultrafiltration Membrane. *Water Research, 143*(2), 87–98.

Liang, D., Zhang, J., Govindaiah, M. T., Tanguy, L., Ernst, A., Zengerle, R., & Koltay, P. (2014). Liquid volume measurement method for the picoliter to nanoliter volume range based on quartz crystal microbalance technology. *Measurement Science & Technology*, *25*(9), 095302. doi:10.1088/0957-0233/25/9/095302

Liang, Z., Yang, Z., Yuan, H., Wang, C., Qi, J., Liu, K., Cao, R., & Zheng, H. (2018, July 31). A protein@metal-organic framework nanocomposite for pH-triggered anticancer drug delivery. *Dalton Transactions (Cambridge, England)*, *47*(30), 10223–10228. doi:10.1039/C8DT01789A PMID:30014058

Liao, Z., Fang, X., Xie, J., Li, Q., Wang, D., Sun, X., Wang, L., & Li, J. (2019). Hydrophilic Hollow Nanocube-Functionalized Thin Film Nanocomposite Membrane with Enhanced Nanofiltration Performance. *ACS Applied Materials & Interfaces*, *11*(5), 5344–5352. doi:10.1021/acsami.8b19121 PMID:30638012

Li, C. M., Huang, J. P., Zhu, H., Liu, L., Feng, Y., Hu, G., & Yu, X. (2017). Dual-emitting fluorescence of Eu/Zr-MOF for ratiometric sensing formaldehyde. *Sensors and Actuators. B, Chemical*, *253*, 275–282. doi:10.1016/j.snb.2017.06.064

Li, D., Zhang, S., Feng, X., Yang, H., Nie, F., & Zhang, W. (2019). A novel peroxidase mimetic Co-MOF enhanced luminol chemiluminescence and its application in glucose sensing. *Sensors and Actuators. B, Chemical, 296*, 126631. doi:10.1016/j.snb.2019.126631

Lien, T. T. N., Dai Lam, T., An, V. T. H., Hoang, T. V., Quang, D. T., Khieu, D. Q., ... Kim, J. S. (2010). Multi-wall carbon nanotubes (MWCNTs)-doped polypyrrole DNA biosensor for label-free detection of genetically modified organisms by QCM and EIS. *Talanta, 80*(3), 1164–1169. doi:10.1016/j.talanta.2009.09.002 PMID:20006069

Li, J. Q., Gong, L. L., Feng, X. F., Zhang, L., Wu, H. Q., Yan, C. S., Xiong, Y. Y., Gao, H. Y., & Luo, F. (2017). Direct extraction of U(VI) from alkaline solution and seawater via anion exchange by metal-organic framework. *Chemical Engineering Journal, 316*, 154–159. doi:10.1016/j.cej.2017.01.046

Li, J. R., Ma, Y., McCarthy, M. C., Sculley, J., Yu, J., Jeong, H. K., Balbuena, P. B., & Zhou, H. C. (2011). Carbon dioxide capture-related gas adsorption and separation in metal-organic frameworks. *Coordination Chemistry Reviews, 255*(15), 1791–1823. doi:10.1016/j.ccr.2011.02.012

Li, J. R., Sculley, J., & Hong, C. Z. (2012). Metal-Organic Frameworks for Separations. *Chemical Reviews, 112*(2), 869–932. doi:10.1021/cr200190s PMID:21978134

Li, J., Liu, Y., Wang, X., Zhao, G., Ai, Y., Han, B., Wen, T., Hayat, T., Alsaedi, A., & Wang, X. (2017). Experimental and theoretical study on selenate uptake to zirconium metal-organic frameworks: Effect of defects and ligands. *Chemical Engineering Journal, 330*, 1012–1021. doi:10.1016/j.cej.2017.08.038

Li, J., Wang, H., Yuan, X., Zhang, J., & Chew, J. W. (2020). Metal-Organic Framework Membranes for Wastewater Treatment and Water Regeneration. *Coordination Chemistry Reviews, 404*, 213116. doi:10.1016/j.ccr.2019.213116

Li, L., Chen, Q., Niu, Z. G., Zhou, X. H., Yang, T., & Huang, W. (2016). Lanthanide metal-organic frameworks assembled from a fluorene-based ligand: Selective sensing of Pb^{2+} and Fe^{3+} ions. *Journal of Materials Chemistry. C, Materials for Optical and Electronic Devices, 4*(9), 1900–1905. doi:10.1039/C5TC04320D

Li, L., Feng, X., Han, R., Zang, S., & Yang, G. (2017). Cr(VI) removal via anion exchange on a silvertriazolate MOF. *Journal of Hazardous Materials, 321*, 622–628. doi:10.1016/j.jhazmat.2016.09.029 PMID:27694026

Li, L., Ma, W., Shen, S., Huang, H., Bai, Y., & Liu, H. (2016). A combined experimental and theoretical study on the extraction of uranium by amino-derived metal-organic frameworks through post-synthetic strategy. *ACS Applied Materials & Interfaces, 8*(45), 31032–31041. doi:10.1021/acsami.6b11332 PMID:27788576

Lin, K., Adhikari, A. K., Lin, K., & Tu, M. (2016). Hydrogen storage capacity enhancement of MIL-53 (Cr) by Pd loaded activated carbon doping. *Journal of the Taiwan Institute of Chemical Engineers*. doi:10.1016/j.jtice.2016.02.033

Lin, K.Y.A., Liu, Y.T., & Chen, S.Y. (2016). Adsorption of fluoride to UiO-66-NH_2 in water: stability, kinetic, isotherm and thermodynamic studies. *Journal of Colloid and Interfaces Science, 461*, 79-87.

Lin, K.-Y. A., Liu, Y.-T., & Chen, S.-Y. (2016). Adsorption of fluoride to UiO-66-NH2 in water: Stability, kinetic, isotherm and thermodynamic studies. *Journal of Colloid and Interface Science, 461*, 79–87. doi:10.1016/j.jcis.2015.08.061 PMID:26397913

Lin, X. M., Gao, G. M., Zheng, L. Y., Chi, Y. W., & Chen, G. N. (2014). Encapsulation of strongly fluorescent carbon quantum dots in metal-organic frameworks for enhancing chemical sensing. *Analytical Chemistry, 86*(2), 1223–1228. doi:10.1021/ac403536a PMID:24328057

Li, S., Dong, Y., Zhou, J., Liu, Y., Wang, J., Gao, X., Han, Y., Qi, P., & Wang, B. (2018). Carbon Dioxide in the Cage: Manganese Metal-Organic Frameworks for High Performance CO_2 Electrodes in Li-CO_2 Batteries. *Energy & Environmental Science, 11*(5), 1318–1325. doi:10.1039/C8EE00415C

Li, S., Xie, L., He, M., Hu, X., Luo, G., Chen, C., & Zhu, Z. (2020). Metal-Organic Frameworks-Derived Bamboo-Like CuO/In2O3 Heterostructure for High-Performance H2S Gas Sensor with Low Operating Temperature. *Sensors and Actuators. B, Chemical, 310*, 127828. doi:10.1016/j.snb.2020.127828

Li, T. (2019, July). Integrating Cationic Metal-Organic Frameworks with Ultrafiltration Membrane for Selective Removal of Perchlorate from Water. *Journal of Hazardous Materials*, 381.

Liu, X., Sun, T., Hu, J., & Wang, S. (2016). Composites of metal–organic frameworks and carbon-based materials: preparations, functionalities and applications. *Journal of Materials Chemistry A: Materials for Energy and Sustainability, 4*(October), 3584–3616. doi:10.1039/C5TA09924B

Liu, C., & Yan, B. (2016). A novel photofunctional hybrid materials of pyrene functionalized metal-organic framework with conformation change for fluorescence sensing of Cu^{2+}. *Sensors and Actuators. B, Chemical, 235*, 541–546. doi:10.1016/j.snb.2016.05.127

Liu, F., Song, S., Cheng, G., Xiong, W., Shi, L., & Zhang, Y. (2018). MIL-101(Cr) Metal-Organic Framework functionalized with tetraethylenepentamine for potential removal of Uranium (VI) from waste water. *Adsorption Science and Technology, 36*(7–8), 1550–1567. doi:10.1177/0263617418789516

Liu, G., Chernikova, V., Liu, Y., Zhang, K., Belmabkhout, Y., Shekhah, O., Zhang, C., Yi, S., Eddaoudi, M., & Koros, W. J. (2018). Mixed matrix formulations with MOF molecular sieving for key energy-intensive separations. *Nature Materials, 17*(3), 283–289. doi:10.103841563-017-0013-1 PMID:29434309

Liu, P., Redekop, E., Gao, X., Liu, W. C., Olsbye, U., & Somorjai, G. A. (2019). Oligomerization of Light Olefins Catalyzed by Brønsted-Acidic Metal–Organic Framework-808. *Journal of the American Chemical Society, 141*(29), 11557–11564. doi:10.1021/jacs.9b03867 PMID:31264857

Liu, R., Sui, Y., & Wang, X. (2019). Metal–Organic Framework-Based Ultrafiltration Membrane Separation with Capacitive-Type for Enhanced Phosphate Removal. *Chemical Engineering Journal, 371*(February), 903–913. doi:10.1016/j.cej.2019.04.136

Liu, S., Luo, M., Li, J., Luo, F., Ke, L., & Ma, J. (2016). Adsorption equilibrium and kinetics of uranium onto porous azo-metal–organic frameworks. *Journal of Radioanalytical and Nuclear Chemistry, 310*(1), 353–362. doi:10.100710967-016-4852-z

Liu, T. Y., Yuan, H.-G., Liu, Y.-Y., Ren, D., Su, Y.-C., & Wang, X. (2018). Metal-Organic Framework Nanocomposite Thin Films with Interfacial Bindings and Self-Standing Robustness for High Water Flux and Enhanced Ion Selectivity. *ACS Nano, 12*(9), 9253–9265. doi:10.1021/acsnano.8b03994 PMID:30153418

Liu, W., Dai, X., Bai, Z., Wang, Y., Yang, Z., Zhang, L., Xu, L., Chen, L., Li, Y., Gui, D., Diwu, J., Wang, J., Zhou, R., Chai, Z., & Wang, S. (2017). Highly sensitive and selective uranium detection in natural water systems using a luminescent mesoporous metal-organic framework equipped with abundant Lewis basic sites: A combined batch, X-ray absorption spectroscopy, and first principles simulation investigation. *Environmental Science & Technology, 51*(7), 3911–3921. doi:10.1021/acs.est.6b06305 PMID:28271891

Liu, W., Shao, Q., Ji, G., Liang, X., Cheng, Y., Quan, B., & Du, Y. (2017). Metal–organic-frameworks derived porous carbon-wrapped Ni composites with optimized impedance matching as excellent lightweight electromagnetic wave absorber. *Chemical Engineering Journal, 313*, 734–744. doi:10.1016/j.cej.2016.12.117

Liu, W., Yan, Z., Zhang, Z., Zhang, Y., Cai, G., & Li, Z. (2019). Bioactive and anti-corrosive bio-MOF-1 coating on magnesium alloy for bone repair application. *Journal of Alloys and Compounds*, *788*, 705–711. doi:10.1016/j.jallcom.2019.02.281

Liu, X., Demir, N. K., Wu, Z., & Li, K. (2015). Highly Water-Stable Zirconium Metal-Organic Framework UiO-66 Membranes Supported on Alumina Hollow Fibers for Desalination. *Journal of the American Chemical Society*, *137*(22), 6999–7002. doi:10.1021/jacs.5b02276 PMID:26023819

Liu, Y., Chi, X., Han, Q., Du, Y., Huang, J., Lin, X., & Liu, Y. (2019). Metal-Organic Framework-derived Hierarchical Co3O4@MnCo2O4.5 Nanocubes with enhanced Electrocatalytic Activity for Na-O_2 Batteries. *Nanoscale*, *11*(12), 5285–5294. doi:10.1039/C8NR08818G PMID:30843015

Liu, Y., Xu, X., Xia, Q., Yuan, G., He, Q., & Cui, Y. (2010). Multiple topological isomerism of three-connected networks in silver-based metal–organoboron frameworks. *Chemical Communications (Cambridge)*, *46*(15), 2608–2610. doi:10.1039/b923365b PMID:20449323

Li, W. (2018, September). Metal–Organic Framework Membranes: Production, Modification, and Applications. *Progress in Materials Science*, *100*, 21–63. doi:10.1016/j.pmatsci.2018.09.003

Li, X., Liu, Y., Wang, J., Gascon, J., Li, J., & Van der Bruggen, B. (2017). Metal-Organic Frameworks Based Membranes for Liquid Separation. *Chemical Society Reviews*, *46*(23), 7124–7144. doi:10.1039/C7CS00575J PMID:29110013

Li, Y. P., Zhu, X. H., Li, S. N., Jiang, Y. C., Hu, M. C., & Zhai, Q. G. (2019). Highly selective and sensitive turn off-on fluorescent probes for sensing Al^{3+} ions designed by regulating the excited-state intramolecular proton transfer process in metal-organic frameworks. *ACS Applied Materials & Interfaces*, *11*(12), 11338–11348. doi:10.1021/acsami.8b20410 PMID:30834744

Li, Y., Wee, L. H., Martens, J. A., & Vankelecom, I. F. J. (2017). Interfacial Synthesis of ZIF-8 Membranes with Improved Nanofiltration Performance. *Journal of Membrane Science*, *523*, 561–566. doi:10.1016/j.memsci.2016.09.065

Li, Z. Q., Qiu, L. G., Su, T., Wu, Y., Wang, W., Wu, Z. Y., & Jiang, X. (2009). Ultrasound synthesis of microporous metal-organic framework $Cu_3(BTC)_2$ at ambient temperature and pressure: An efficient and environmentally friendly method. *Materials Letters*, *63*(1), 78–80. doi:10.1016/j.matlet.2008.09.010

Li, Z.-Q., Yang, J.-C., Sui, K.-W., & Yin, N. (2015). Facile synthesis of Metal-Organic Framework MOF-808 for arsenic removal. *Materials Letters*, *160*, 412–414. doi:10.1016/j.matlet.2015.08.004

Lu, C. S., & Lewis, O. (1972). Investigation of film-thickness determination by oscillating quartz resonators with large mass load. *Journal of Applied Physics*, *43*(11), 4385–4390. doi:10.1063/1.1660931

Luo, B., Yuan, L., Chai, Z., Shi, W., & Tang, Q. (2016). U(VI) capture from aqueous solution by highly porous and stable MOFs: UiO-66 and its amine derivative. *Journal of Radioanalytical and Nuclear Chemistry*, *307*(1), 269–276. doi:10.100710967-015-4108-3

Luo, X., Ding, L., & Luo, J. (2015). Adsorptive Removal of Pb(II) Ions from Aqueous Samples with Amino-Functionalization of Metal–Organic Frameworks MIL-101(Cr). *Journal of Chemical & Engineering Data*, *60*(6), 1732–1743. doi:10.1021/je501115m

Lustig, W. P., Mukherjee, S., Rudd, N. D., Desai, A. V., Li, J., & Ghosh, S. K. (2017). Metal-organic frameworks: Functional luminescent and photonic materials for sensing applications. *Chemical Society Reviews*, *46*(11), 3242–3285. doi:10.1039/C6CS00930A PMID:28462954

Lu, T., Zhang, L. C., Sun, M., Deng, D., Su, Y., & Lv, Y. (2016). Amino-functionalized metal-organic frameworks nanoplates-based energy transfer probe for highly selective fluorescence detection of free chlorine. *Analytical Chemistry*, *88*(6), 3413–3420. doi:10.1021/acs.analchem.6b00253 PMID:26905374

Lüthgens, E., Herrig, A., Kastl, K., Steinem, C., Reiss, B., Wegener, J., Pignataro, B., & Janshoff, A. (2003). Adhesion of liposomes: A quartz crystal microbalance study. *Measurement Science & Technology*, *14*(11), 1865–1875. doi:10.1088/0957-0233/14/11/003

Lu, X. Y., Ye, J., Zhang, D., Xie, R., Bogale, R. F., Sun, Y., Zhao, L., Zhao, Q., & Ning, G. (2014). Silver carboxylate metal-organic frameworks with highly antibacterial activity and biocompatibility. *Journal of Inorganic Biochemistry*, *138*, 114–121. doi:10.1016/j.jinorgbio.2014.05.005 PMID:24935093

Ma, A., Ke, F., Jiang, J., Yuan, Q., Luo, Z., Liu, J., & Kumar, A. (2017). Two lanthanide based metal-organic frameworks for highly efficient adsorption and removal of fluoride ion from water. *CrystEngComm*, *19*(16), 2172–2177. doi:10.1039/C7CE00291B

Ma, D., Peh, S. B., Han, G., & Chen, S. B. (2017). Thin-Film Nanocomposite (TFN) Membranes Incorporated with Super-Hydrophilic Metal-Organic Framework (MOF) UiO-66: Toward Enhancement of Water Flux and Salt Rejection. *ACS Applied Materials & Interfaces*, *9*(8), 7523–7534. doi:10.1021/acsami.6b14223 PMID:28186405

Ma, D., Zheng, S. R., Fan, J., Cai, S. L., Dai, Z., Zou, X. Y., Teng, S.-H., & Zhang, W. G. (2020). A new QCM signal enhancement strategy based on streptavidin@ metal-organic framework complex for miRNA detection. *Analytica Chimica Acta*, *1095*, 212–218. doi:10.1016/j.aca.2019.10.034 PMID:31864625

Ma, D., Zheng, S. R., Fan, J., Cai, S., Dai, Z., Zou, X., Teng, S.-H., & Zhang, W. G. (2020). A new amplification strategy for quartz crystal microbalance miRNA sensor based on selective interactions between Metal-organic framework and miRNA. *New Journal of Chemistry*, *44*(5), 1684–1688. doi:10.1039/C9NJ04600C

Madhura, L., Kanchi, S., Sabela, M. I., Singh, S., Bisetty, K., & Inamuddin. (2018). Membrane Technology for Water Purification. *Environmental Chemistry Letters*, *16*(2), 343–365. doi:10.100710311-017-0699-y

Ma, J., Guo, X., Ying, Y., Liu, D., & Zhong, C. (2017). Composite Ultrafiltration Membrane Tailored by MOF@GO with Highly Improved Water Purification Performance. *Chemical Engineering Journal*, *313*, 890–898. doi:10.1016/j.cej.2016.10.127

Makhetha, T. A., & Moutloali, R. M. (2018). Antifouling Properties of Cu(Tpa)@GO/PES Composite Membranes and Selective Dye Rejection. *Journal of Membrane Science*, *554*(January), 195–210. doi:10.1016/j.memsci.2018.03.003

Maleki, A., Hayati, B., Naghizadeh, M., & Joo, S. W. (2015). Adsorption of hexavalent chromium by metal organic frameworks from aqueous solution. *Journal of Industrial and Engineering Chemistry*, *28*, 211–216. doi:10.1016/j.jiec.2015.02.016

Maranescu, B., Lupa, L., & Visa, A. (2017). Heavy metal removal from waste waters by phosphonate Metal Organic Frameworks. *Pure and Applied Chemistry*. Advance online publication. doi:10.1515/pac-2017-0307

Maroofi, S. M., & Mahmoodi, N. M. (2019). Zeolitic Imidazolate Framework-Polyvinylpyrrolidone-Polyethersulfone Composites Membranes: From Synthesis to the Detailed Pollutant Removal from Wastewater Using Cross Flow System. *Colloids and Surfaces. A, Physicochemical and Engineering Aspects*, *572*(March), 211–220. doi:10.1016/j.colsurfa.2019.03.093

Marzo, A. M. L., Pons, J., & Merkoci, A. (2013). All-integrated and highly sensitive paper based device with sample treatment platform for Cd^{2+} immunodetection in drinking/tap waters. *Analytical Chemistry*, *85*(7), 3532–3538. doi:10.1021/ac3034536 PMID:23445438

Massoudinejad, M., Ghaderpoori, M., Shahsavani, A., & Amini, M. M. (2016). Adsorption of fluoride over a metal organic framework Uio-66 functionalized with amine groups and optimization with response surface methodology. *Journal of Molecular Liquids*, *221*, 279–286. doi:10.1016/j.molliq.2016.05.087

Massoudinejad, M., Shahsavani, A., Kamarehie, B., Jafari, A., Ghaderpoori, M., Amini, M. M., & Ghaderpoury, A. (2018). Highly efficient adsorption of fluoride from aqueous solutions by metal organic frameworks: Modeling, isotherms, and kinetics. *Fluoride*, *51*(4), 355–365.

Matilainen, A., Vepsäläinen, M., & Sillanpää, M. (2010). Natural Organic Matter Removal by Coagulation during Drinking Water Treatment: A Review. *Advances in Colloid and Interface Science*, *159*(2), 189–197. doi:10.1016/j.cis.2010.06.007 PMID:20633865

Ma, X. H., Yang, Z., Yao, Z.-K., Xu, Z.-L., & Tang, C. Y. (2016, July). A Facile Preparation of Novel Positively Charged MOF/Chitosan Nanofiltration Membranes. *Journal of Membrane Science*, *525*, 269–276. doi:10.1016/j.memsci.2016.11.015

Ma, Z., Yuan, T., Fan, Y., Wang, L., Duan, Z., Du, W., Zhang, D., & Xu, J. (2020). A benzene vapor sensor based on a metal-organic framework-modified quartz crystal microbalance. *Sensors and Actuators. B, Chemical*, *311*, 127365. doi:10.1016/j.snb.2019.127365

McDonald, T. M., Mason, J. A., Kong, X., Bloch, E. D., Gygi, D., Dani, A., Crocellà, V., Giordanino, F., Odoh, S. O., Drisdell, W. S., Vlaisavljevich, B., Dzubak, A. L., Poloni, R., Schnell, S. K., Planas, N., Lee, K., Pascal, T., Wan, L. F., Prendergast, D., ... Long, J. R. (2015). Cooperative insertion of CO2 in diamine-appended metal-organic frameworks. *Nature*, *519*(7543), 303–308. doi:10.1038/nature14327 PMID:25762144

Meenakshi, S., Sundaram, C. S., & Sukumar, R. (2008). Enhanced fluoride sorption by mechanochemically activated kaolinites. *Journal of Hazardous Materials*, *153*(1-2), 164–172. doi:10.1016/j.jhazmat.2007.08.031 PMID:17897780

Mehtab, T., Yasin, G., Arif, M., Shakeel, M., Korai, R. M., Nadeem, M., Muhammad, N., & Lu, X. (2019). Metal-Organic Frameworks for Energy Storage Devices: Batteries and Supercapacitors. *Journal of Energy Storage*, *21*, 632–646. doi:10.1016/j.est.2018.12.025

Melek, Y., & Aytun, O. U. (2007, August). An energy benchmarking model based on artificial neural network method utilizing US Commercial Buildings Energy Consumption Survey (CBECS) database. *International Journal of Energy Research*, *31*, 135–147. doi:10.1002/er

Meng, Y., Shu, L., Liu, L., Wu, Y., Xie, L.-H., Zhao, M.-J., & Li, J.-R. (2019). A High-Flux Mixed Matrix Nanofiltration Membrane with Highly Water-Dispersible MOF Crystallites as Filler. *Journal of Membrane Science*, *591*(August), 117360. doi:10.1016/j.memsci.2019.117360

Mermer, Ö., Okur, S., Sümer, F., Özbek, C., Sayın, S., & Yılmaz, M. (2012). *Gas sensing properties of carbon nanotubes modified with calixarene molecules measured by QCM techniques*. Academic Press.

Milner, P. J., Siegelman, R. L., Forse, A. C., Gonzalez, M. I., Runčevski, T., Martell, J. D., Reimer, J. A., & Long, J. R. (2017). A diaminopropane-appended metal–organic framework enabling efficient CO2 capture from coal flue gas via a mixed adsorption mechanism. *Journal of the American Chemical Society*, *139*(38), 13541–13553. doi:10.1021/jacs.7b07612 PMID:28906108

Mohammadnezhad, F., Feyzi, M., & Zinadini, S. (2019). A Novel Ce-MOF/PES Mixed Matrix Membrane; Synthesis, Characterization and Antifouling Evaluation. *Journal of Industrial and Engineering Chemistry*, *71*, 99–111. doi:10.1016/j.jiec.2018.09.032

Mon, M., Bruno, R., Ferrando-Soria, J., Armentano, D., & Pardo, E. (2018). Metal-organic framework technologies for water remediation: Towards a sustainable ecosystem. *Journal of Materials Chemistry. A, Materials for Energy and Sustainability*, *6*(12), 4912–4947. doi:10.1039/C8TA00264A

Muller, U., Schubert, M., Teich, F., & Puetter, H. (2006). Metal-organic frameworks- prospective industrial applications. *Journal of Materials Chemistry*, *16*(7), 626–636. doi:10.1039/B511962F

Murrieta-Rico, F. N., Petranovskii, V., Sergiyenko, O., Mercorelli, P., Antúnez-Garcia, J., de Dios Sanchez-Lopez, J., & Yocupicio-Gaxiola, R. I. (2019, October). Experimental analysis of measurement process for a QCM using the pulse coincidence method. In *IECON 2019-45th Annual Conference of the IEEE Industrial Electronics Society* (Vol. 1, pp. 4657-4662). IEEE. 10.1109/IECON.2019.8927337

Murrieta-Rico, F. N., Mercorelli, P., Sergiyenko, O. Y., Petranovskii, V., Hernandez-Balbuena, D., & Tyrsa, V. (2015). Mathematical modelling of molecular adsorption in zeolite coated frequency domain sensors. *IFAC-PapersOnLine*, *48*(1), 41–46. doi:10.1016/j.ifacol.2015.05.060

Nasrollahpour, A., & Moradi, S. E. (2017). Hexavalent chromium removal from water by ionic liquid modified metal-organic frameworks adsorbent. *Microporous and Mesoporous Materials*, *243*, 47–55. doi:10.1016/j.micromeso.2017.02.006

Nath, D., Mohamedali, M., Henni, A., & Ibrahim, H. (2018). Composites of Metal-Organic Frameworks (MOFs): Synthesis and Applications in Separation and Catalysis. *Metal-Organic Frameworks*, (iii), 251–295. doi:10.1002/9783527809097.ch9

Navarro, M., Benito, J., Paseta, L., Gascón, I., Coronas, J., & Téllez, C. (2018). Thin-Film Nanocomposite Membrane with the Minimum Amount of MOF by the Langmuir-Schaefer Technique for Nanofiltration. *ACS Applied Materials & Interfaces*, *10*(1), 1278–1287. doi:10.1021/acsami.7b17477 PMID:29243908

Ngeno, E. C., Orata, F. O., Baraza, L. D., Shikuku, V. O., & Kimosop, S. J. (2016). Adsorption of Caffeine and Ciprofloxacin onto Pyrolitically Derived Water Hyacinth Biochar: Isothermal, Kinetic and Thermodynamic Studies. *Journal of Chemistry and Chemical Engineering*, *10*(4). Advance online publication. doi:10.17265/1934-7375/2016.04.006

Nuijs, A. L. N. V., Tarcomnicu, I., & Covaci, A. (2011). Application of hydrophilic interaction chromatography for the analysis of polar contaminants in food and environmental samples. *Journal of Chromatography. A*, *1218*(35), 5964–5974. doi:10.1016/j.chroma.2011.01.075 PMID:21316059

O'Keeffe, M., Peskov, M. A., Ramsden, S. J., & Yaghi, O. M. (2008). The reticular chemistry structure resource (RCSR) database of, and symbols for, crystal nets. *Accounts of Chemical Research*, *41*(12), 1782–1789. doi:10.1021/ar800124u PMID:18834152

Odum, H. T. (2000). *Heavy Metals in the Environment Using Wetlands for Their Removal*. Lewis Publishers.

Okada, K., Sawai, S., Ikigaki, K., Tokudome, Y., Falcaro, P., & Takahashi, M. (2017). Electrochemical sensing and catalysis using Cu 3 (BTC) 2 coating electrodes from Cu (OH) 2 films. *CrystEngComm*, *19*(29), 4194–4200. doi:10.1039/C7CE00416H

Okur, S., Zhang, Z., Sarheed, M., Nick, P., Lemmer, U., & Heinke, L. (2020). Towards a MOF e-Nose: A SURMOF sensor array for detection and discrimination of plant oil scents and their mixtures. *Sensors and Actuators. B, Chemical*, *306*, 127502. doi:10.1016/j.snb.2019.127502

Osada, M., Sasaki, I., Nishioka, M., Sadakata, M., & Okubo, T. (1998). Synthesis of a faujasite thin layer and its application for SO2 sensing at elevated temperatures. *Microporous and Mesoporous Materials*, *23*(5-6), 287–294. doi:10.1016/S1387-1811(98)00125-5

Osman, S., Senthil, R. A., Pan, J., & Sun, Y. (2019). A Novel Coral structured Porous-like Amorphous Carbon derived from Zinc-based Fluorinated Metal-Organic Framework as Superior Cathode Material for High Performance Supercapacitors. *Journal of Power Sources*, *414*, 401–411. doi:10.1016/j.jpowsour.2019.01.026

Ostermann, R., Cravillon, J., Weidmann, C., & Smarsly, B. M. (2011). *Metal – organic framework nanofibers via electrospinning w z*. doi:10.1039/C0CC02271C

Pachfule, P., Das, R., Poddar, P., & Banerjee, R. (2011). Solvothermal synthesis, structure, and properties of metal organic framework isomers derived from a partially fluorinated link. *Crystal Growth & Design*, *11*(4), 1215–1222. doi:10.1021/cg101414x

Pan, B. J., Pan, B. C., Zhang, W. M., Lv, L., Zhang, Q. X., & Zheng, S. R. (2009). Development of polymeric and polymer-based hybrid adsorbents for pollutants removal from waters. *Chemical Engineering Journal*, *151*(1-3), 19–29. doi:10.1016/j.cej.2009.02.036

Pankajakshan, A., Kuznetsov, D., & Mandal, S. (2019). Ultrasensitive detection of Hg (II) ions in aqueous medium using zinc-based metal-organic framework. *Inorganic Chemistry*, *58*(2), 1377–1381. doi:10.1021/acs.inorgchem.8b02898 PMID:30628772

Paseta, L., Antorán, D., Coronas, J., & Téllez, C. (2019). 110th Anniversary: Polyamide/Metal-Organic Framework Bilayered Thin Film Composite Membranes for the Removal of Pharmaceutical Compounds from Water. *Industrial & Engineering Chemistry Research*, *58*(10), 4222–4230. doi:10.1021/acs.iecr.8b06017

Patrick, G. L. (2009). *Introduction to medicinal chemistry. 4*. Oxford University.

Pelgrift, R. Y., & Friedman, A. J. (2013). Nanotechnology as a therapeutic tool to combat microbial resistance. *Advanced Drug Delivery Reviews*, *65*(13-14), 1803–1815. doi:10.1016/j.addr.2013.07.011 PMID:23892192

Peng, Y., Huang, H., Zhang, Y., Kang, C., Chen, S., Song, L., Liu, D., & Zhong, C. (2018). A versatile MOF-based trap for heavy metal ion capture and dispersion. *Nature Communications*, *9*(187), 1–9. doi:10.103841467-017-02600-2 PMID:29335517

Perez, E. V., Balkus, K. J. Jr, Ferraris, J. P., & Musselman, I. H. (2009). Mixed-matrix membranes containing MOF-5 for gas separations. *Journal of Membrane Science*, *328*(1–2), 165–173. doi:10.1016/j.memsci.2008.12.006

Petit, C., & Bandosz, T. J. (2009). MOF-graphite oxide composites: Combining the uniqueness of graphene layers and metal-organic frameworks. *Advanced Materials*, *21*(46), 4753–4757. doi:10.1002/adma.200901581

Petit, C., & Bandosz, T. J. (2011). Synthesis, characterization, and ammonia adsorption properties of mesoporous metal-organic framework (MIL(Fe))-graphite oxide composites: Exploring the limits of materials fabrication. *Advanced Functional Materials*, *21*(11), 2108–2117. doi:10.1002/adfm.201002517

Petit, C., Burress, J., & Bandosz, T. J. (2011). The synthesis and characterization of copper-based metal-organic framework/graphite oxide composites. *Carbon*, *49*(2), 563–572. doi:10.1016/j.carbon.2010.09.059

Petrovic, M., Farre, M., Alda, M. L. D., Perez, S., Postigo, C., Kock, M., Radjenovic, J., Gros, M., & Barcelo, D. (2010). Recent trends in the liquid chromatography-mass spectrometry analysis of organic contaminants in environmental samples. *Journal of Chromatography. A*, *1217*(25), 4004–4017. doi:10.1016/j.chroma.2010.02.059 PMID:20334864

Pichon, A., & James, S. L. (2008). An array-based study of reactivity under solvent-free mechanochemical conditions-insights and trends. *CrystEngComm*, *10*(12), 1839–1847. doi:10.1039/b810857a

Pradeep, T., & Anshup, T. (2009). Noble metal nanoparticles for water purification: A critical review. *Thin Solid Films*, *517*(24), 6441–6478. doi:10.1016/j.tsf.2009.03.195

Qin, J. S., Du, D. Y., Guan, W., Bo, X. J., Li, Y. F., Guo, L. P., Su, Z.-M., Wang, Y.-Y., Lan, Y.-Q., & Zhou, H. C. (2015). Ultrastable Polymolybdate-Based Metal-Organic Frameworks as Highly Active Electrocatalysts for Hydrogen Generation from Water. *Journal of the American Chemical Society*, *137*(22), 7169–7177. doi:10.1021/jacs.5b02688 PMID:25933041

Qiu, M., & He, C. (2018, December). Efficient Removal of Heavy Metal Ions by Forward Osmosis Membrane with a Polydopamine Modified Zeolitic Imidazolate Framework Incorporated Selective Layer. *Journal of Hazardous Materials*, *367*, 339–347. doi:10.1016/j.jhazmat.2018.12.096 PMID:30599406

Qiu, X., Xu, X. Y., Chen, X., Wu, Y., & Guo, H. (2018). Preparation of a molecularly imprinted sensor based on quartz crystal microbalance for specific recognition of sialic acid in human urine. *Analytical and Bioanalytical Chemistry*, *410*(18), 4387–4395. doi:10.100700216-018-1094-7 PMID:29736700

Quan, B., Jin, A., Yu, S.-H., Kang, S. M., Jeong, J., Abruña, H. D., Jin, L., Piao, Y., & Sung, Y.-E. (2018). Solvothermal-derived S-doped graphene as an anode material for sodium-ion batteries. *Advancement of Science*, *5*(5), 1700880. doi:10.1002/advs.201700880 PMID:29876213

Que, E. L., Domaille, D. W., & Chang, C. J. (2008). Metals in neurobiology: Probing their chemistry and biology with molecular imaging. *Chemical Reviews*, *108*(5), 1517–1549. doi:10.1021/cr078203u PMID:18426241

Ragab, D., Gomaa, H. G., Sabouni, R., Salem, M., Ren, M., & Zhu, J. (2016). Micropollutants Removal from Water Using Microfiltration Membrane Modified with ZIF-8 Metal Organic Frameworks (MOFs). *Chemical Engineering Journal*, *300*, 273–279. doi:10.1016/j.cej.2016.04.033

Rao, Z., Feng, K., Tang, B., & Wu, P. (2017). Surface Decoration of Amino-Functionalized Metal-Organic Framework/Graphene Oxide Composite onto Polydopamine-Coated Membrane Substrate for Highly Efficient Heavy Metal Removal. *ACS Applied Materials & Interfaces*, *9*(3), 2594–2605. doi:10.1021/acsami.6b15873 PMID:28035795

Rascio, N., & Izzo, F. N. (2011). Heavy metal hyperaccumulating plants: How and why do they do it? And what makes them so interesting? *Plant Science*, *180*(2), 169–181. doi:10.1016/j.plantsci.2010.08.016 PMID:21421358

Rehman, M. A., Yusoff, I., & Alias, Y. (2015). Fluoride adsorption by doped and un-doped magnetic ferrites CuCeFeO: Preparation, characterization, optimization and modeling for effectual remediation technologies. *Journal of Hazardous Materials*, *299*, 316–324. doi:10.1016/j.jhazmat.2015.06.030 PMID:26143194

Ren, Y., Li, T., Zhang, W., Wang, S., Shi, M., Shan, C., Zhang, W., Guan, X., Lv, L., Hua, M., & Pan, B. (2019). MIL-PVDF Blend Ultrafiltration Membranes with Ultrahigh MOF Loading for Simultaneous Adsorption and Catalytic Oxidation of Methylene Blue. *Journal of Hazardous Materials*, *365*, 312–321. doi:10.1016/j.jhazmat.2018.11.013 PMID:30447639

Ren, Z., Luo, J., & Wan, Y. (2018). Highly Permeable Biocatalytic Membrane Prepared by 3D Modification: Metal-Organic Frameworks Ameliorate Its Stability for Micropollutants Removal. *Chemical Engineering Journal*, *348*, 389–398. doi:10.1016/j.cej.2018.04.203

Ricco, R., Konstas, K., Styles, M. J., Richardson, J. J., Babarao, R., Suzuki, K., Scopece, P., & Falcaro, P. (2015). Lead(II) uptake by aluminium based magnetic framework composites (MFCs) in water. *Journal of Materials Chemistry. A, Materials for Energy and Sustainability*, *3*(39), 19822–19831. doi:10.1039/C5TA04154F

Rieth, A. J., Wright, A. M., & Dincă, M. (2019). Kinetic stability of metal–organic frameworks for corrosive and coordinating gas capture. *Nature Reviews. Materials*, *4*(11), 708–725. doi:10.103841578-019-0140-1

Rivera, J. M., Rincón, S., Youssef, C. B., & Zepeda, A. (2016). Highly efficient adsorption of aqueous Pb(II) with mesoporous Metal-Organic Framework-5: An equilibrium and kinetic study. *Journal of Nanomaterials*, *2016*, 1–9. Advance online publication. doi:10.1155/2016/8095737

Rojas-Mayorga, C. K., Bonilla-Petriciolet, A., Sánchez-Ruiz, F. J., Moreno-Pérez, J., Reynel-Ávila, H. E., Aguayo-Villarreal, I. A., & Mendoza-Castillo, D. I. (2015). Breakthrough curve modeling of liquid-phase adsorption of fluoride ions on aluminum doped bone char using micro-columns: Effectiveness of data fitting approaches. *Journal of Molecular Liquids*, *208*, 114–121. doi:10.1016/j.molliq.2015.04.045

Rosi, N. L., Eddaoudi, M., Kim, J., O'Keeffe, M., & Yaghi, O. M. (2002). Advances in the chemistry of metal-organic frameworks. *CrystEngComm*, *4*(68), 401–404. doi:10.1039/B203193K

Rosseinsky, M. J. (2014). Perspective: Metal-Organic Frameworks – Opportunities and Challenges. *APL Materials*, *2*(12), 124001. doi:10.1063/1.4904880

Rouhani, F., Morsali, A., & Retailleau, P. (2018). Simple one-pot preparation of a rapid response AIE fluorescent metal-organic framework. *ACS Applied Materials & Interfaces*, *10*(42), 36259–36266. doi:10.1021/acsami.8b12404 PMID:30259725

Roushani, M., Saedi, Z., & Baghelani, Y. M. (2017). Removal of cadmium ions from aqueous solutions using TMU-16-NH_2 metal organic framework. *Environmental Nanotechnology, Monitoring & Management*, *7*, 89–96. doi:10.1016/j.enmm.2017.01.003

Ruan, H., Guo, C., Yu, H., Shen, J., Gao, C., Sotto, A., & Van der Bruggen, B. (2016). Fabrication of a Mil-53(Al) Nanocomposite Membrane and Potential Application in Desalination of Dye Solutions. *Industrial & Engineering Chemistry Research*, *55*(46), 12099–12110. doi:10.1021/acs.iecr.6b03201

Rui, K., Wang, X., Du, M., Zhang, Y., Wang, Q., Ma, Z., Zhang, Q., Li, D., Huang, X., Sun, G., Zhu, J., & Huang, W. (2018). Dual-Function Metal–Organic Framework-Based Wearable Fibers for Gas Probing and Energy Storage. *ACS Applied Materials & Interfaces*, *10*(3), 2837–2842. doi:10.1021/acsami.7b16761 PMID:29286235

Saleem, H., Rafique, U., & Davies, R. P. (2016). Investigations on post-synthetically modified UiO-66-NH2 for the adsorptive removal of heavy metal ions from aqueous solution. *Microporous and Mesoporous Materials*, *221*, 238–244. doi:10.1016/j.micromeso.2015.09.043

Samanta, P., Desai, A. V., Sharma, S., Chandra, P., & Ghosh, S. K. (2018). Selective recognition of Hg^{2+} ion in water by a functionalized metal-organic framework (MOF) based chemodosimeter. *Inorganic Chemistry*, *57*(5), 2360–2364. doi:10.1021/acs.inorgchem.7b02426 PMID:29461048

Sancet, M. P. A. (2013). Surface anchored metal-organic frameworks as stimulus responsive antifouling coatings. *Biointerphases*, *8*. PMID:24706148

Sasaki, I., Tsuchiya, H., Nishioka, M., Sadakata, M., & Okubo, T. (2002). Gas sensing with zeolite-coated quartz crystal microbalances—Principal component analysis approach. *Sensors and Actuators. B, Chemical*, *86*(1), 26–33. doi:10.1016/S0925-4005(02)00132-6

Sauerbrey, G. (1959). Verwendung von Schwingquarzen zur Wägung dünner Schichten und zur Mikrowägung. *Zeitschrift für Physik*, *155*(2), 206–222. doi:10.1007/BF01337937

Sava Gallis, D. F., Butler, K. S., Agola, J. O., Pearce, C., & McBride, A. (2019). Antibacterial Countermeasures via MOF – Supported Sustained Therapeutic Release. *ACS Applied Materials & Interfaces*. Advance online publication. doi:10.1021/acsami.8b21698 PMID:30682243

Seo, P. W., Khan, N. A., Hasan, Z., & Jhung, S. H. (2016). Adsorptive Removal of Artificial Sweeteners from Water Using Metal–Organic Frameworks Functionalized with Urea or Melamine. *ACS Applied Materials & Interfaces*, *8*(43), 29799–29807. doi:10.1021/acsami.6b11115 PMID:27723294

Shekhah, O., Liu, J., Fischer, R. A., & Wöll, C. (2011). MOF Thin Films: Existing and Future Applications. *Chemical Society Reviews*, *40*(2), 1081–1106. doi:10.1039/c0cs00147c PMID:21225034

Shekhah, O., Wang, H., Kowarik, S., Schreiber, F., Paulus, M., Tolan, M., Sternemann, C., Evers, F., Zacher, D., Fischer, R. A., & Wöll, C. (2007). Step-by-step route for the synthesis of metal-organic frameworks. *Journal of the American Chemical Society*, *129*(49), 15118–15119. doi:10.1021/ja076210u PMID:18020338

Shen, C., Zhao, Y., Li, W., Yang, Y., Liu, R., & Morgen, D. (2019). Global Profile of Heavy Metals and Semimetals Adsorption Using Drinking Water Treatment Residual. *Chemical Engineering Journal*, *372*, 1019–1027. doi:10.1016/j.cej.2019.04.219

Sheng, D., Zhu, L., Xu, C., Xiao, C., Wang, Y., Wang, Y., Chen, L., Diwu, J., Chen, J., Chai, Z., Albrecht-Schmitt, T. E., & Wang, S. (2017). Efficient and selective uptake of TcO4- by a cationic metal-organic framework material with open Ag^+ sites. *Environmental Science & Technology*, *51*(6), 3471–3479. doi:10.1021/acs.est.7b00339 PMID:28211267

Shen, J., Mkongo, G., Abbt-Braun, G., Ceppi, S. L., Richards, B. S., & Schäfer, A. I. (2015). Renewable energy powered membrane technology: Fluoride removal in a rural community in northern Tanzania. *Separation and Purification Technology*, *149*, 349–361. doi:10.1016/j.seppur.2015.05.027

Shi, C., Wang, X., Gao, Y., Rong, H., Song, Y., Liu, H.-J., & Liu, Q. (2017). Nickel Metal-Organic Framework Nanoparticles as Electrode Materials for Li-Ion Batteries and Supercapacitors. *Journal of Solid State Electrochemistry*, *21*(8), 2415–2423. doi:10.100710008-017-3591-6

Shi, P. F., Hu, H. C., Zhang, Z. Y., Xiong, G., & Zhao, B. (2015). Heterometal-organic frameworks as highly sensitive and highly selective luminescent probes to detect I^- ions in aqueous solutions. *Chemical Communications (Cambridge)*, *51*(19), 3985–3988. doi:10.1039/C4CC09081K PMID:25620743

Shi, Z., Xu, C., Guan, H., Li, L., Fan, L., Wang, Y., Liu, L., Meng, Q., & Zhang, R. (2018). Magnetic metal organic frameworks (MOFs) composite for removal of lead and malachite green in wastewater. *Colloids and Surfaces. A, Physicochemical and Engineering Aspects*, *539*, 382–390. doi:10.1016/j.colsurfa.2017.12.043

Shooto, N. D., Dikio, E. D., Wankasi, D., & Sikhwivhilu, L. (2015). Synthesis, morphology and lead ion adsorption properties of metal organic frameworks of copper and cobalt. *Chemical Sciences Journal*, *6*(4), 113.

Siegel, F. R. (2002). *Environmental Geochemistry of Potentially Toxic Metals*. Springer., doi:10.1007/978-3-662-04739-2

Silva, I. M. P., Carvalho, M. A., Oliveira, C. S., Profirio, D. M., Ferreira, R. B., Corbi, P. P., & Formiga, A. L. B. (2016). Enhanced performance of a metal-organic framework analogue to MIL-101(Cr) containing amine groups for ibuprofen and nimesulide controlled release. *Inorganic Chemistry Communications*, *70*, 47–50. doi:10.1016/j.inoche.2016.05.020

Si, X., Jiao, C., Li, F., Zhang, J., Wang, S., Liu, S., Li, Z., Sun, L., Xu, F., Gabelica, Z., & Schick, C. (2011). High and selective CO_2 uptake, H_2 storage and methanol sensing on the amine-decorated 12-connected MOF CAU-1. *Energy & Environmental Science*, *4*(11), 4522–4527. doi:10.1039/c1ee01380g

Sobha, K., Poornima, A., Harini, P., & Veeraiah, K. A. (2007). study on biochemical changes in the fresh water fish, catla catla (hamilton) exposed to the heavy metal toxicant cadmium chloride, *Kathmandu Uni. J. Sci. Eng. Techno*, *1*(4), 1–11.

Song, Z., Qiu, F., Zaia, E. W., Wang, Z., Kunz, M., Guo, J., Brady, M., Mi, B., & Urban, J. J. (2017). Dual-channel, molecular-sieving core/shell ZIF@ MOF architectures as engineered fillers in hybrid membranes for highly selective CO2 separation. *Nano Letters*, *17*(11), 6752–6758. doi:10.1021/acs.nanolett.7b02910 PMID:29072837

Son, W. J., Kim, J., Kim, J., & Ahn, W. S. (2008). Sonochemical synthesis of MOF-5. *ChemComm.*, *2008*, 6336–6338. PMID:19048147

Son, W. J., Kim, J., Kim, J., & Ahn, W. S. (2008). Sonochemical synthesis of MOF-5. *Chemical Communications*, (47), 6336–6338. doi:10.1039/b814740j PMID:19048147

Sorribas, S., Gorgojo, P., Téllez, C., Coronas, J., & Livingston, A. G. (2013). High Flux Thin Film Nanocomposite Membranes Based on Metal-Organic Frameworks for Organic Solvent Nanofiltration. *Journal of the American Chemical Society*, *135*(40), 15201–15208. doi:10.1021/ja407665w PMID:24044635

Stern, B. R., Solioz, M., Krewski, D., Aggett, P., Aw, T. C., Baker, S., Crump, K., Dourson, M., Haber, L., Hertzberg, R., Keen, C., Meek, B., Rudenko, L., Schoeny, R., Slob, W., & Starr, T. (2007). Copper and human health: biochemistry, genetics, and strategies for modelling doseresponse relationships. *J. Toxicol. and Env. Heal. B*, *10*, 157-222.

Stock, N. (2011). Metal-organic frameworks: aluminium-based frameworks. Encyclopedia of Inorganic and Bioinorganic Chemistry, 1-16.

Stock, N., & Biswas, S. (2012). Synthesis of metal-organic frameworks (MOFs): Routes to various MOF topologies, morphologies, and composites. *Chemical Reviews*, *112*(2), 933–969. doi:10.1021/cr200304e PMID:22098087

Streit, H. C., Adlung, M., Shekhah, O., Stammer, X., Arslan, H. K., Zybaylo, O., Ladnorg, T., Gliemann, H., Franzreb, M., Wöll, C., & Wickleder, C. (2012). Surface-Anchored MOF-Based Photonic Antennae. *ChemPhysChem*, *13*(11), 2699–2702. doi:10.1002/cphc.201200262 PMID:22700363

Sun, C.-Y., Qin, C., Wang, X.-L., & Su, Z.-M. (2013). Metal-organic frameworks as potential drug delivery systems. *Expert Opinion on Drug Delivery*, *10*(1), 89–101. doi:10.1517/17425247.2013.741583 PMID:23140545

Sun, D. T., Peng, L., Reeder, W. S., Moosavi, S. M., Tiana, D., Britt, D. K., Oveisi, E., & Queen, W. L. (2018). Rapid, Selective heavy metal removal from water by a Metal-Organic Framework/Polydopamine Composite. *ACS Central Science*, *4*(3), 349–356. doi:10.1021/acscentsci.7b00605 PMID:29632880

Sun, H., Tang, B., & Wu, P. (2017). Development of Hybrid Ultrafiltration Membranes with Improved Water Separation Properties Using Modified Superhydrophilic Metal-Organic Framework Nanoparticles. *ACS Applied Materials & Interfaces*, *9*(25), 21473–21484. doi:10.1021/acsami.7b05504 PMID:28594542

Sun, H., Tang, B., & Wu, P. (2017, October). Hydrophilic Hollow Zeolitic Imidazolate Framework-8 Modified Ultrafiltration Membranes with Significantly Enhanced Water Separation Properties. *Journal of Membrane Science*, *551*, 283–293. doi:10.1016/j.memsci.2018.01.053

Taghizadeh, M., Asgharinezhad, A. A., Pooladi, M., Barzin, M., Abbaszadeh, A., & Tadjarodi, A. (2013). A novel magnetic metal organic framework nanocomposite for extraction and pre concentration of heavy metal ions, and its optimization via experimental design methodology. *Mikrochimica Acta*, *180*(11-12), 1073–1084. doi:10.100700604-013-1010-y

Tahmasebi, E., Masoomi, M. Y., Yamini, Y., & Morsali, A. (2015, January 20). Application of mechanosynthesized azine-decorated zinc(II) metal-organic frameworks for highly efficient removal and extraction of some heavy-metal ions from aqueous samples: A comparative study. *Inorganic Chemistry*, *54*(2), 425–433. doi:10.1021/ic5015384 PMID:25548873

Tang, B., Huang, S., Fang, Y., Hu, J., Malonzo, C., Truhlar, D. G., & Stein, A. (2016). Mechanism of Electrochemical Lithiation of a Metal-Organic Framework without Redox-Active Nodes. *The Journal of Chemical Physics*, *144*(19), 194702. doi:10.1063/1.4948706 PMID:27208960

Tang, D., & Zhang, G. (2016). Efficient removal of fluoride by hierarchical Ce–Fe bimetal oxides adsorbent: Thermodynamics, kinetics and mechanism. *Chemical Engineering Journal*, *283*, 721–729. doi:10.1016/j.cej.2015.08.019

Tang, Y. Y., Ding, C. X., Ng, W. S., & Xie, Y. S. (2013). Synthesis, structures and photoluminescence of Zn (II), Ag (I), Cu (I) and Co (II) coordination polymers of a tetrapyridyl ligand. *RSC Advances*, *3*(39), 18134–18141. doi:10.1039/c3ra43405b

Tan, Q., Chen, X., Wan, H., Zhang, B., Liu, X., Li, L., Wang, C., Gan, Y., Liang, P., Wang, Y., Zhang, J., Wang, H., Miao, L., Jiang, J., van Aken, P. A., & Wang, H. (2020). Metal-Organic Framework-derived High Conductivity Fe_3C with Porous Carbon on Graphene as Advanced Anode Materials for Aqueous Battery-Supercapacitor Hybrid Devices. *Journal of Power Sources*, *448*, 227403. doi:10.1016/j.jpowsour.2019.227403

Tan, Y., Sun, Z., Meng, H., Han, Y., Wu, J., Xu, J., Xu, Y., & Zhang, X. (2019). A New MOFs/Polymer Hybrid Membrane: MIL-68(Al)/PVDF, Fabrication and Application in High-Efficient Removal of p-Nitrophenol and Methylene Blue. *Separation and Purification Technology*, *215*(January), 217–226. doi:10.1016/j.seppur.2019.01.008

Tehrani, A. A., Esrafili, L., Abedi, S., Morsali, A., Carlucci, L., Proserpio, D. M., Wang, J., Junk, P. C., & Liu, T. F. (2017). Urea metal-organic frameworks for nitro-substituted compounds sensing. *Inorganic Chemistry*, *56*(3), 1446–1454. doi:10.1021/acs.inorgchem.6b02518 PMID:28085264

Tella, A. C., & Aaron, I. Y. (2013). ChemInform Abstract: Syntheses and Applications of Metal-Organic Frameworks Materials. *ChemInform*, *44*(11). doi:10.1002/chin.201311277

Tian, D., Li, Y., Chen, R. Y., Chang, Z., Wang, G. Y., & Bu, X. H. (2014). A luminescent metal-organic framework demonstrating ideal detection ability for nitroaromatic explosives. *Journal of Materials Chemistry. A, Materials for Energy and Sustainability*, *2*(5), 1465–1470. doi:10.1039/C3TA13983B

Tomic, E. A. (1965). Thermal Stability of Coordination Polymers. *Journal of Applied Polymer Science*, *9*(11), 3745–3752. doi:10.1002/app.1965.070091121

Tranchemontagn, D. J., Mendoza-Corte, J. L., Keeffe, M., & Yaghi, O. M. (2009). Secondary building units, nets and bonding in the chemistry of metal-organic frameworks. *Chemical Society Reviews*, *38*(5), 1257–1283. doi:10.1039/b817735j PMID:19384437

Trinh, D. X., Thuy, P. N. T., & Taniike, T. (2017). Fabrication of New Composite Membrane Filled with UiO-66 Nanoparticles and Its Application to Nanofiltration. *Separation and Purification Technology*, *177*, 249–256. doi:10.1016/j.seppur.2017.01.004

Tripathy, S. S., Bersillon, J.-L., & Gopal, K. (2006). Removal of fluoride from drinking water by adsorption onto alum-impregnated activated alumina. *Separation and Purification Technology*, *50*(3), 310–317. doi:10.1016/j.seppur.2005.11.036

Tsuge, K., Chishina, Y., Hashiguchi, H., Sasaki, Y., Kato, M., Ishizaka, S., & Kitamura, N. (2016). Luminescent copper (I) complexes with halogenidobridged dimeric core. *Coordination Chemistry Reviews*, *306*, 636–651. doi:10.1016/j.ccr.2015.03.022

Van Assche, T. R., Desmet, G., Ameloot, R., De Vos, D. E., Terryn, H., & Denayer, J. F. (2012). Electrochemical synthesis of thin HKUST-1 layers on copper mesh. *Microporous and Mesoporous Materials*, *158*, 209–213. doi:10.1016/j.micromeso.2012.03.029

Van Goethem, C., Verbeke, R., Hermans, S., Bernstein, R., & Vankelecom, I. F. J. (2016). Controlled Positioning of MOFs in Interfacially Polymerized Thin-Film Nanocomposites. *Journal of Materials Chemistry. A, Materials for Energy and Sustainability*, *4*(42), 16368–16376. doi:10.1039/C6TA05175H

Van Goethem, C., Verbeke, R., Pfanmöller, M., Koschine, T., Dickmann, M., Timpel-Lindner, T., Egger, W., Bals, S., & Vankelecom, I. F. J. (2018, March). The Role of MOFs in Thin-Film Nanocomposite (TFN) Membranes. *Journal of Membrane Science*, *563*, 938–948. doi:10.1016/j.memsci.2018.06.040

Vasudevan, S., Lakshmi, J., & Sozhan, G. (2009). Studies on a Mg-Al-Zn alloy as an anode for the removal of fluoride from drinking water in an electrocoagulation process. *Clean-Soil Air Water*, *37*(4-5), 372–378. doi:10.1002/clen.200900031

Viswanathan, N., & Meenakshi, S. (2009). Role of metal ion incorporation in ion exchange resin on the selectivity of fluoride. *Journal of Hazardous Materials*, *162*(2-3), 920–930. doi:10.1016/j.jhazmat.2008.05.118 PMID:18619734

Vörösmarty, C. J., McIntyre, P. B., Gessner, M. O., Dudgeon, D., Prusevich, A., Green, P., Glidden, S., Bunn, S. E., Sullivan, C. A., Liermann, C. R., & Davies, P. M. (2010). Global threats to human water security and river biodiversity. *Nature*, *467*, 555–561.

Vu, T. A., Le, G. H., Dao, C. D., Dang, L. Q., Nguyen, K. T., Nguyen, Q. K., Dang, P. T., Tran, H. T. K., Duong, Q. T., Nguyen, T. V., & Lee, G. D. (2015). Arsenic removal from aqueous solutions by adsorption using novel MIL-53(Fe) as a highly efficient adsorbent. *RSC Advances*, *5*(7), 5261–5268. doi:10.1039/C4RA12326C

Wan Ngah, W. S., Hanafiah, M. A. K. M. (2008). Removal of heavy metal ions from wastewater by chemically modified plant wastes as adsorbents: A review. *Bioresource Technology*, *99*, 3935–3948. PMID:17681755

Wang, C., & Liu, X. (2015). Superior Removal of Arsenic from Water with Zirconium Metal-Organic Framework UiO-66. *Scientific Reports, 5*, 16613. https://pubmed.ncbi.nlm.nih.gov/26559001

Wang, Hou, Chen, & Meng. (2020). In-Situ Growth of Metal-Organic Framework Film on a Polydopamine-Modified Flexible Substrate for Antibacterial and Forward Osmosis Membranes. *Separation and Purification Technology, 236*, 116239. . doi:10.1016/j.seppur.2019.116239

Wang, C., Liu, X., Chen, J. P., & Li, K. (2015). Superior removal of arsenic from water with zirconium metalorganic framework UiO-66. *Scientific Reports*, *5*(1), 16613. doi:10.1038rep16613 PMID:26559001

Wang, H. H., Zhou, L. J., Wang, Y. L., & Liu, Q. Y. (2016). Terbium-biphenyl-3, 3′-disulfonyl-4, 4′ carboxylate framework with sulfonate sites for luminescent sensing of Cr^{3+}. *Inorganic Chemistry Communications*, *73*, 94–97. doi:10.1016/j.inoche.2016.10.006

Wang, J., Ma, D., Liao, W., Li, S., Huang, M., Liu, H., Wang, Y., Xie, R., & Xu, J. (2017). A hydrostable anionic zinc-organic framework carrier with a **bcu** topology for drug delivery. *CrystEngComm*, *19*(35), 5244–5250. doi:10.1039/C7CE01238A

Wang, K., Gu, J., & Yin, N. (2017). Efficient removal of Pb(II) and Cd(II) using NH2-functionalized Zr-MOFs via rapid microwave-promoted synthesis. *Industrial & Engineering Chemistry Research*, *56*(7), 1880–1887. doi:10.1021/acs.iecr.6b04997

Wang, K., Tao, X. R., Xu, J. Z., & Yin, N. (2016). Novel Chitosan MOF Composite Adsorbent for the Removal of Heavy Metal Ions. *Chemistry Letters*, *45*(12), 1365–1368. doi:10.1246/cl.160718

Wang, L., Duan, S., Fang, M., Liu, J., He, J., Li, J., & Lei, J. (2016). Surface Modification Route to Prepare Novel Polyamide@NH2-MIL-88B Nanocomposite Membranes for Water Treatment. *RSC Advances*, *6*(75), 71250–71261. doi:10.1039/C6RA09080J

Wang, L., Fang, M., Liu, J., He, J., Li, J., & Lei, J. (2015). Layer-by-Layer Fabrication of High-Performance Polyamide/ZIF-8 Nanocomposite Membrane for Nanofiltration Applications. *ACS Applied Materials & Interfaces*, *7*(43), 24082–24093. doi:10.1021/acsami.5b07128 PMID:26485228

Wang, L., Han, Y., Feng, X., Zhou, J., Qi, P., & Wang, B. (2018). Metal-Organic Frameworks for Energy Storage: Batteries and Supercapacitors. *Coordination Chemistry Reviews*, *307*, 361–381. doi:10.1016/j.ccr.2015.09.002

Wang, N., Liu, T., Shen, H., Ji, S., Li, J.-R., & Zhang, R. (2016). Ceramic Tubular MOF Hybrid Membrane Fabricated through in Situ Layer-by-Layer Self-Assembly for Nanofiltration. *AIChE Journal. American Institute of Chemical Engineers*, *62*(2), 538–546. doi:10.1002/aic.15115

Wang, Q. S., Li, J. J., & Li, X. (2018). A luminescent Eu (III)-based metal-organic framework as a highly effective sensor for cation and anion detections. *Sensors and Actuators. B, Chemical*, *258*, 358–364. doi:10.1016/j.snb.2017.11.075

Wang, R., Liu, X., Huang, A., Wang, W., Xiao, Z., Zhang, L., Dai, F., & Sun, D. (2016). Unprecedented solvent-dependent sensitivities in highly efficient detection of metal ions and nitroaromatic compounds by a fluorescent barium metal-organic framework. *Inorganic Chemistry*, *55*(4), 1782–1787. doi:10.1021/acs.inorgchem.5b02693 PMID:26845287

Wang, X., Zhu, H., Sun, T., Liu, Y., Han, T., Lu, J., Dai, H., & Zhai, L. (2019). Synthesis and Study of an Efficient Metal-Organic Framework Adsorbent (MIL-96 (Al)) for Fluoride Removal from Water. *Journal of Nanomaterials*, *2019*, 2019. doi:10.1155/2019/3128179

Wang, Y., Cheng, L., Liu, Z. Y., Wang, X. G., Ding, B., Yin, L., Zhou, B. B., Li, M. S., Wang, J. X., & Zhao, X. J. (2015). An Ideal Detector Composed of Two-Dimensional Cd (II)-Triazole Frameworks for Nitro-Compound Explosives and Potassium Dichromate. *Chemistry (Weinheim an der Bergstrasse, Germany)*, *21*(40), 14171–14178. doi:10.1002/chem.201502167 PMID:26291651

Wang, Y., Yan, J., Wen, N., Xiong, H., Cai, S., He, Q., Hu, Y., Peng, D., Liu, Z., & Liu, Y. (2020, February). Metal-organic frameworks for stimuli-responsive drug delivery. *Biomaterials*, *230*, 119619. Advance online publication. doi:10.1016/j.biomaterials.2019.119619 PMID:31757529

Wang, Z., Tao, H., & Yue, Y. (2019). Metal-Organic-Framework-based Cathodes for Enhancing the Electrochemical Performances of Batteries: A Review. *ChemElectroChem*, *6*(21), 5358–5374. doi:10.1002/celc.201900843

Wei, L. Q., Lu, J. Y., Li, Q. Q., Zhou, Y., Tang, L.-L., & Li, F.-Y. (2017). *, Tang LL, Li FY. A porous Ca-MOF with nano-sized {Ca11} as building unit: Structure, drug loading and release properties. *Inorganic Chemistry Communications*, *78*, 43–47. doi:10.1016/j.inoche.2017.02.010

Wei, T., Zhang, M., Wu, P., Tang, Y. J., Li, S.-L., Shen, F.-C., Wang, X.-L., Zhou, X.-P., & Lan, Y.-Q. (2017). POM-based Metal-Organic Framework/Reduced Graphene Oxide Nanocomposites with Hybrid Behavior of Battery-Supercapacitor for Superior Lithium Storage. *Nano Energy*, *34*, 205–214. doi:10.1016/j.nanoen.2017.02.028

WHO & UNICEF. (2017). *Progress on Drinking Water, Sanitation and Hygiene: 2017 Update and SDG Baseline*. World Health Organization. https://www.unicef.org/publications/files/Progress_on_Drinking_Water_Sanitation_and_Hygiene_2017.pdf

World Health Organization. (2010). *WHO Housing and Health Guidelines*. https://www.who.int/phe%0Ahttp://apps.who.int/bookorders

Wu, D., Liu, J., Jin, J., Cheng, J., Wang, M., Yang, G., & Wang, Y. Y. (2019). New Doubly Interpenetrated MOF with [Zn4O] Clusters and Its Doped Isomorphic MOF: Sensing, Dye, and Gas Adsorption Capacity. *Crystal Growth & Design*, *19*(11), 6774–6783. doi:10.1021/acs.cgd.9b01193

Wu, J. X., & Yan, B. (2017). Eu (III)-functionalized In-MOF (In (OH) bpydc) as fluorescent probe for highly selectively sensing organic small molecules and anions especially for $CHCl_3$ and MnO_4^-. *Journal of Colloid and Interface Science*, *504*, 197–205. doi:10.1016/j.jcis.2017.05.054 PMID:28551513

Wu, M., Ye, H., Zhao, F., & Zeng, B. (2017). High-Quality Metal-Organic Framework ZIF-8 Membrane Supported on Electrodeposited ZnO/2-Methylimidazole Nanocomposite: Efficient Adsorbent for the Enrichment of Acidic Drugs. *Scientific Reports*, *7*(January), 1–9. https://dx.doi.org/10.1038/srep39778

Wu, S., Ge, Y., Wang, Y., Chen, X., Li, F., Xuan, H., & Li, X. (2017). Adsorption of Cr(VI) on nano UiO-66-NH_2 MOFs in water. *Environmental Technology*, 1–12. PMID:28625105

Wu, X., Meng, G., Liu, W., Li, T., Yang, Q., Sun, X., & Liu, J. (2018). Metal-Organic Framework-derived, Zn-doped Porous Carbon Polyhedra with enhanced Activity as Bifunctional Catalysts for Rechargeable Zinc-Air Batteries. *Nano Research*, *11*(1), 163–173. doi:10.100712274-017-1615-2

Wu, Y., Qiu, X., Liang, F., Zhang, Q., Koo, A., Dai, Y., Lei, Y., & Sun, X. (2019). A Metal-Organic-Framework-derived Bifunctional Catalysts for Hybrid Sodium-Air Batteries. *Applied Catalysis B: Environmental*, *241*, 407–414. doi:10.1016/j.apcatb.2018.09.063

Wu, Y., Xu, G., Wei, F., Song, Q., Tang, T., Wang, X., & Hu, Q. (2016). Determination of Hg(II) in tea and mushroom samples based on metal-organic frameworks as solid phase extraction sorbents. *Microporous and Mesoporous Materials*, *235*, 204–210. doi:10.1016/j.micromeso.2016.08.010

Xia, H., Zhang, J., Yang, Z., Guo, S., Guo, S., & Xu, Q. (2017). 2D MOF nanoflake-assembled spherical microstructures for enhanced supercapacitor and electrocatalysis performances. *Nano-Micro Letters*, *9*(4), 43. doi:10.100740820-017-0144-6 PMID:30393738

Xie, A. (2019, July). Photo-Fenton Self-Cleaning PVDF/NH2-MIL-88B(Fe) Membranes towards Highly-Efficient Oil/Water Emulsion Separation. *Journal of Membrane Science*, *595*, 117499. https://doi.org/10.1016/j.memsci.2019.117499

Xie, A. (2020). Graphene Oxide/Fe(III)-Based Metal-Organic Framework Membrane for Enhanced Water Purification Based on Synergistic Separation and Photo-Fenton Processes. *Applied Catalysis B: Environmental*, *264*(July), 118548. https://doi.org/10.1016/j.apcatb.2019.118548

Xiong, P., Zhao, X., & Xu, Y. (2018). Nitrogen-doped Carbon Nanotubes derived from Metal-Organic Frameworks for Potassium-Ion battery Anodes. *ChemSusChem*, *11*(1), 202–208. doi:10.1002/cssc.201701759 PMID:28994244

Xu, F. (2018). *Phosphorus Removal and Recovery From Anaerobic Digestion Residues*. Elsevier. https://linkinghub.elsevier.com/retrieve/pii/S2468012518300038

Xu, F., Sun, L., Huang, P., Sun, Y., Zheng, Q., Zou, Y., Chu, H., Yan, E., Zhang, H., Wang, J., & Du, Y. (2018). A pyridine vapor sensor based on metal-organic framework-modified quartz crystal microbalance. *Sensors and Actuators. B, Chemical*, *254*, 872–877. doi:10.1016/j.snb.2017.07.026

Xu, G. W., Wu, Y. P., Dong, W. W., Zhao, J., Wu, X. Q., Li, D. S., & Zhang, Q. (2017). A multifunctional Tb-MOF for highly discriminative sensing of Eu3+/Dy3+ and as a catalyst support of Ag nanoparticles. *Small*, *13*(22), 1602996. doi:10.1002mll.201602996 PMID:28418186

Xu, G., Nie, P., Dou, H., Ding, B., Li, L., & Zhang, X. (2017). Exploring Metal Organic Frameworks for Energy Storage in Batteries and Supercapacitors. *Materials Today*, *20*(4), 191–209. doi:10.1016/j.mattod.2016.10.003

Xu, H., Cao, C. S., & Zhao, B. (2015). A water-stable lanthanide-organic framework as a recyclable luminescent probe for detecting pollutant phosphorus anions. *Chemical Communications (Cambridge)*, *51*(51), 10280–10283. doi:10.1039/C5CC02596F PMID:25940110

Xu, H., Gao, J. K., Qian, X. F., Wang, J. P., He, H. J., Cui, Y. J., Yang, Y., Wang, Z. Y., & Qian, G. D. J. (2016). Metal-organic framework nanosheets for fast-response and highly sensitive luminescent sensing of Fe^{3+}. *Journal of Materials Chemistry. A, Materials for Energy and Sustainability*, *4*(28), 10900–10905. doi:10.1039/C6TA03065C

Xu, H., Xiao, Y. Q., Rao, X. T., Dou, Z. S., Li, W. F., Cui, Y. J., Wang, Z. Y., & Qian, G. D. (2011). A metal-organic framework for selectively sensing of PO_4^{3-} anion in aqueous solution. *Journal of Alloys and Compounds*, *509*(5), 2552–2554. doi:10.1016/j.jallcom.2010.11.087

Xu, X. Y., & Yan, B. (2016). Eu (III) functionalized Zr-based metalorganic framework as excellent fluorescent probe for Cd^{2+} detection in aqueous environment. *Sensors and Actuators. B, Chemical*, *222*, 347–353. doi:10.1016/j.snb.2015.08.082

Xu, Y., Gao, X., Wang, Q., Wang, X., Ji, Z., & Gao, C. (2016). Highly Stable MIL-101(Cr) Doped Water Permeable Thin Film Nanocomposite Membranes for Water Treatment. *RSC Advances*, *6*(86), 82669–82675.

Xu, Y., Gao, X., Wang, X., Wang, Q., Ji, Z., & Wang, X. (2016). Highly and Stably Water Permeable Thin Film Nanocomposite Membranes Doped with MIL-101 (Cr) Nanoparticles for Reverse Osmosis Application. *Materials (Basel)*, *9*(11).

Xu, Y., Li, Q., Xue, H., & Pang, H. (2018). Metal-Organic Frameworks for Direct Electrochemical Applications. *Coordination Chemistry Reviews*, *376*, 292–318. doi:10.1016/j.ccr.2018.08.010

Yaghi, O. M. (2003). Reticular Synthesis and the Design of New Materials. *Nature*, *423*(6941), 705–14. doi:10.1038/nature01650

Yaghi, O. M., Li, G., & Li, H. (1995). Selective Binding and Removal of Guests in a Microporous Metal–Organic Framework. *Nature*, *378*(6558), 703–706. https://dx.doi.org/10.1038/378703a0

Yaghi, O. M., & Li, H. (1995). Hydrothermal synthesis of a metal-organic framework containing large rectangular channels. *Journal of the American Chemical Society*, *117*(41), 10401–10402. doi:10.1021/ja00146a033

Yaghi, O. M., Li, H., Davis, C., Richardson, D., & Groy, T. L. (1998). Synthetic Strategies, Structure Patterns, and Emerging Properties in the Chemistry of Modular Porous Solids. *Accounts of Chemical Research*, *31*(8), 474–484. doi:10.1021/ar970151f

Yahaya, N. Z. S. (2020). Synthesis and Characterizations of MIL-140B-Al2O3/YSZ Ceramic Membrane Using Solvothermal Method for Seawater Desalination. *Journal of the Australian Ceramic Society*, *56*(1), 291–300.

Yamagiwa, H., Sato, S., Fukawa, T., Ikehara, T., Maeda, R., Mihara, T., & Kimura, M. (2014). Detection of volatile organic compounds by weight-detectable sensors coated with metal-organic frameworks. *Scientific Reports*, *4*(1), 1–6. PMID:25175808

Yang, Y. (2005). *High-efficiency solution processable self-organization of polymer blends*. doi:10.1038/nmat1500

Yang, C. X., Ren, H. B., & Yan, X. P. (2013). Fluorescent metal-organic framework MIL-53(Al) for highly selective and sensitive detection of Fe^{3+} in aqueous solution. *Analytical Chemistry*, *85*(15), 7441–7446. doi:10.1021/ac401387z PMID:23826852

Yang, F. (2018). Metal-Organic Frameworks Supported on Nanofiber for Desalination by Direct Contact Membrane Distillation. *ACS Applied Materials & Interfaces*, *10*(13), 11251–11260.

Yang, F., Li, W., & Tang, B. (2018). Facile synthesis of amorphous UiO-66 (Zr-MOF) for supercapacitor application. *Journal of Alloys and Compounds*, *733*, 8–14. doi:10.1016/j.jallcom.2017.10.129

Yang, H., Cui, W., Han, Y., & Wang, B. (2018). Porous Nanocomposite derived from Zn, Ni-Bimetallic Metal-Organic Framework as an Anode Material for Lithium-Ion Batteries. *Chinese Chemical Letters*, *29*(6), 842–844. doi:10.1016/j.cclet.2017.09.024

Yang, J.-C., & Yin, X.-B. (2017). $CoFe_2O_4$@MIL-100(Fe) hybrid magnetic nanoparticles exhibit fast and selective adsorption of arsenic with high adsorption capacity. *Scientific Reports*, *7*(1), 40955. doi:10.1038rep40955 PMID:28102334

Yang, J., Wang, Z., Li, Y. S., Zhuang, Q. X., Zhao, W. R., & Gu, J. (2016). Porphyrin MOFs for reversible fluorescent and colorimetric sensing of mercury (II) ions in aqueous phase. *RSC Advances*, *6*(74), 69807–69814. doi:10.1039/C6RA13766K

Yang, L., Lian, C., Li, X., Han, Y., Yang, L., Cai, T., & Shao, C. (2017). Highly selective bifunctional luminescent sensor toward nitrobenzene and Cu^{2+} ion based on microporous metal-organic frameworks: Synthesis, structures, and properties. *ACS Applied Materials & Interfaces*, *9*(20), 17208–17217. doi:10.1021/acsami.7b01758 PMID:28467044

Yang, L., Wang, Z., & Zhang, J. (2017). Zeolite Imidazolate Framework Hybrid Nanofiltration (NF) Membranes with Enhanced Permselectivity for Dye Removal. *Journal of Membrane Science*, *532*(March), 76–86. https://dx.doi.org/10.1016/j.memsci.2017.03.014

Yang, Q. X., Zhao, Q. Q., Ren, S. S., Lu, Q. Q., Guo, X. M., & Chen, Z. J. (2016). Fabrication of core-shell Fe_3O_4@ MIL-100(Fe) magnetic microspheres for the removal of Cr(VI) in aqueous solution. *Journal of Solid State Chemistry*, *244*, 25–30. doi:10.1016/j.jssc.2016.09.010

Yang, S., Zou, Q., Wang, T., & Zhang, L. (2019). Effects of GO and MOF@GO on the Permeation and Antifouling Properties of Cellulose Acetate Ultrafiltration Membrane. *Journal of Membrane Science*, *569*, 48–59. https://doi.org/10.1016/j.memsci.2018.09.068

Yang, W., Bai, Z., Shi, W., Yuan, L., Tian, T., Chai, Z., Wang, H., & Sun, Z. (2013). MOF-76: From a luminescent probe to highly efficient UVI sorption material. *Chemical Communications (Cambridge)*, *49*(88), 10415–10417. doi:10.1039/c3cc44983a PMID:24079004

Yang, Z. R., Wang, M. M., & Yin, X. B. (2017). Boric-acid functional lanthanide metal-organic frameworks for selective ratiometric fluorescence detection of fluoride ions. *Analytical Chemistry*, *89*(3), 1930–1936. doi:10.1021/acs.analchem.6b04421 PMID:28208304

Yao, Z. Q., Li, G. Y., Xu, J., Hu, L., & Bu, X. H. (2018). A water-stable luminescent Zn (II) metalorganic framework as chemosensor for high-efficiency detection of Cr^{VI}- anions ($Cr_2O_7^{2-}$ and CrO_4^{2-}) in aqueous solution. *Chemistry (Weinheim an der Bergstrasse, Germany)*, *24*(13), 3192–3198. doi:10.1002/chem.201705328 PMID:29210125

Yin, N., Wang, K., Wang, L., & Li, Z. (2016). Amino-Functionalized MOFs Combining Ceramic Membrane Ultrafiltration for Pb (II) Removal. *Chemical Engineering Journal*, *306*, 619–628. https://dx.doi.org/10.1016/j.cej.2016.07.064

Yin, N., Wang, K., Xia, Y., & Li, Z. (2018). Novel melamine modified metal-organic frameworks for remarkably high removal of heavy metal Pb(II). *Desalination*, *430*, 120–127. doi:10.1016/j.desal.2017.12.057

Yi, P., & Chen, K. L. (2014). Release kinetics of multiwalled carbon nanotubes deposited on silica surfaces: Quartz crystal microbalance with dissipation (QCM-D) measurements and modeling. *Environmental Science & Technology*, *48*(8), 4406–4413. doi:10.1021/es405471u PMID:24654566

Yuan, Y. J., Van der Werff, M. J., & Xu, W. L. (2011). A field-programmable-gate-array-based high-speed transceiver for a quartz crystal microbalance induced bond-rupture sensor. *Measurement Science & Technology*, *22*(4), 045201. doi:10.1088/0957-0233/22/4/045201

Yu, C. X., Hu, F. L., Song, J. G., Zhang, J. L., Liu, S. S., Wang, B. X., Meng, H., Liu, L.-L., & Ma, L. F. (2020). Ultrathin two-dimensional metal-organic framework nanosheets decorated with tetra-pyridyl calix [4] arene: Design, synthesis and application in pesticide detection. *Sensors and Actuators. B, Chemical*, *310*, 127819. doi:10.1016/j.snb.2020.127819

Yu, D., Shao, Q., Song, Q., Cui, J., Zhang, Y., Wu, B., Ge, L., Wang, Y., Zhang, Y., Qin, Y., Vajtai, R., Ajayan, P. M., Wang, H., Xu, T., & Wu, Y. (2020). A solvent-assisted ligand exchange approach enables metal-organic frameworks with diverse and complex architectures. *Nature Communications*, *11*(1), 927. doi:10.103841467-020-14671-9 PMID:32066754

Yusuf, A. S., Popoola, L. T., & Babatunde, E. O. (2019). Adsorption of cadmium ion from aqueous solutions by copper-based Metal Organic Framework: Equilibrium modeling and kinetic studies. *Applied Water Science*, *9*(4), 106. doi:10.100713201-019-0991-z

Zeng, X., Yang, Z., Meng, J., Chen, L., Chen, H., & Qin, H. (2019). The Cube-like ZnO/C Composites derived from Metal Organic Framework-5 as Anodic Material with High Electrochemical Performance for Ni-Zn Rechargeable Battery. *Journal of Power Sources*, *438*, 226986. doi:10.1016/j.jpowsour.2019.226986

Zhang, H., Nai, J., Yu, L., & Lou, X. W. D. (2017). Metal-Organic-Framework based materials as platforms for renewable energy and environment applications. *Joule*, *1*(1), 77–107. doi:10.1016/j.joule.2017.08.008

Zhang, K., Lee, T. H., Cha, J. H., Jang, H. W., Choi, J.-W., Mahmoudi, M., & Shokouhimehr, M. (2019). Metal-Organic Framework-derived Metal Oxide Nanoparticles@Reduced Graphene Oxide Composites as Cathode Materials for Rechargeable Aluminium-Ion Batteries. *Scientific Reports*, *9*(1), 13739. doi:10.103841598-019-50156-6 PMID:31551435

Zhang, L., Zhang, Y., Huang, S., Yuan, Y., Li, H., Jin, Z., Wu, J., Liao, Q., Hu, L., Lu, J., Ruan, S., & Zeng, Y.-J. (2018). Co_3O_4/Ni-based MOFs on Carbon Cloth for Flexible Alkaline Battery-Supercapacitor Hybrid Devices and near-infrared Photocatalytic Hydrogen Evolution. *Electrochimica Acta*, *281*, 189–197. doi:10.1016/j.electacta.2018.05.162

Zhang, M. (2018). Insights into the Use of Metal–Organic Framework As High-Performance Anticorrosion Coatings. *ACS Applied Materials & Interfaces*, *10*(3), 2259–2263. https://dx.doi.org/10.1021/acsami.7b18713

Zhang, M., Feng, G., Song, Z., Zhou, Y. P., Chao, H. Y., Yuan, D., Tan, T. T., Guo, Z., Hu, Z., Tang, B. Z., Liu, B., & Zhao, D. (2014). Two-dimensional metal-organic framework with wide channels and responsive turn-on fluorescence for the chemical sensing of volatile organic compounds. *Journal of the American Chemical Society*, *136*(20), 7241–7244. doi:10.1021/ja502643p PMID:24824627

Zhang, N., Yang, X., Yu, X., Jia, Y., Wang, J., Kong, L., Jim, Z., Sun, B., Luo, T., & Liu, J. (2014). Al-1,2,5-benzenetricarboxylic metal-organic frameworks: A promising adsorbent for defluoridation of water with pH insensitivity and low aluminium residual. *Chemical Engineering Journal*, *252*, 220–229. doi:10.1016/j.cej.2014.04.090

Zhang, N., Yuan, L., Guo, W., Luo, S., Chai, Z., & Shi, W. (2017). Extending the use of highly porous and functionalized MOFs to Th(IV) capture. *ACS Applied Materials & Interfaces*, *9*(30), 25216–25224. doi:10.1021/acsami.7b04192 PMID:28699737

Zhang, P. (2018). Ultrathin Reduced Graphene Oxide/MOF Nanofiltration Membrane with Improved Purification Performance at Low Pressure. *Chemosphere*, *204*, 378–389. https://doi.org/10.1016/j.chemosphere.2018.04.064

Zhang, Q., Wang, J., Kirillov, A. M., Dou, W., Xu, C., Yang, L., Fang, R., & Liu, W. (2018). Multifunctional Ln-MOF luminescent probe for efficient sensing of Fe^{3+}, Ce^{3+} and Acetone. *ACS Applied Materials & Interfaces*, *10*(28), 23976–23986. doi:10.1021/acsami.8b06103 PMID:29920195

Zhang, Q., Yu, J., Cai, J., Zhang, L., Cui, Y., Yang, Y., Chen, B., & Qian, G. (2015). A porous Zrcluster-based cationic metal-organic framework for highly efficient $Cr_2O_7^{2-}$ removal from water. *Chemical Communications (Cambridge)*, *51*(79), 14732–14734. doi:10.1039/C5CC05927E PMID:26291500

Zhang, R. (2014). Coordination-Driven in Situ Self-Assembly Strategy for the Preparation of Metal-Organic Framework Hybrid Membranes. *Angewandte Chemie International Edition*, *53*(37), 9775–9779.

Zhang, S. (2020). Water-Soluble MOF Nanoparticles Modified Polyethersulfone Membrane for Improving Flux and Molecular Retention. *Applied Surface Science*, *505*, 144553. https://doi.org/10.1016/j.apsusc.2019.144553

Zhang, X., Wang, M., Zhu, G., Li, D., Yan, D., Lu, T., & Pan, L. (2017). Porous Cake-like TiO_2 derived from Metal-Organic Frameworks as Superior Anode Material for Sodium Ion Batteries. *Ceramics International, 43*(2), 2898–2402. doi:10.1016/j.ceramint.2016.11.028

Zhang, Y., & Wang, J. (2018). A water-stable and biofriendly 2D anionic zinc(II) metal–organic framework for drug delivery. *Inorganica Chimica Acta, 477*, 8–14. doi:10.1016/j.ica.2018.01.015

Zhang, Z., & Awaga, K. (2016). Redox-Active Metal-Organic Frameworks as Electrode Materials for Batteries. *MRS Bulletin, 41*(11), 883–889. doi:10.1557/mrs.2016.245

Zhao, B., Chen, X. Y., Chen, Z., Shi, W., Cheng, P., Yan, S. P., & Liao, D. Z. (2009). A porous 3D heterometal-organic framework containing both lanthanide and high-spin Fe (II) ions. *Chemical Communications, 21*(21), 3113–3115. doi:10.1039/b821496d PMID:19462104

Zhao, D., Liu, X. H., Zhao, Y., Wang, P., Liu, Y., Azam, M., Ai-Resayes, S. I., Lu, Y., & Sun, W. Y. (2017). Luminescent Cd (II)-organic frameworks with chelating NH_2 sites for selective detection of Fe (III) and antibiotics. *Journal of Materials Chemistry. A, Materials for Energy and Sustainability, 5*(30), 15797–15807. doi:10.1039/C7TA03849F

Zhao, D., Wan, X. Y., Song, H. J., Hao, L. Y., Su, Y. Y., & Lv, Y. (2014). Metal-organic frameworks (MOFs) combined with ZnO quantum dots as a fluorescent sensing platform for phosphate. *Sensors and Actuators. B, Chemical, 197*, 50–57. doi:10.1016/j.snb.2014.02.070

Zhao, H., Ni, J., Zhang, J. J., Liu, S. Q., Sun, Y. J., Zhou, H., Li, Y. Q., & Duan, C. Y. (2018). A trichromatic MOF composite for multidimensional ratiometric luminescent sensing. *Chemical Science (Cambridge), 9*(11), 2918–2926. doi:10.1039/C8SC00021B PMID:29732075

Zhao, M., Lu, Q., Ma, Q., & Zhang, H. (2017). Two-Dimensional Metal–Organic Framework Nanosheets. *Small Methods, 1*(1–2), 1600030. Advance online publication. doi:10.1002mtd.201600030

Zhao, W., Peng, J., Wang, W., Jin, B., Chen, T., Liu, S., Zhao, Q., & Huang, W. (2019). Interlayer Hydrogen-bonded Metal Porphyrin Frameworks/MXene Hybrid Film with High Capacitance for Flexible All-Solid-State Supercapacitors. *Small, 15*(18), 1901351. doi:10.1002mll.201901351 PMID:30957989

Zhao, X. (2015). Recent Advances in the Fabrication and Structure-Specific Applications of Graphene-Based Inorganic Hybrid Membranes. *Nanoscale, 7*(12), 5080–5093.

Zhao, X. L., Tian, D., Gao, Q., Sun, H. W., Xu, J., & Bu, X. H. (2016). A chiral lanthanide metal-organic framework for selective sensing of Fe (II) ions. *Dalton Transactions (Cambridge, England), 45*(3), 1040–1046. doi:10.1039/C5DT03283K PMID:26649625

Zhao, X. Q., Zhao, B., Shi, B. W., & Cheng, P. (2009). Structures and luminescent properties of a series of Ln-Ag heterometallic coordination polymers. *CrystEngComm, 11*(7), 1261–1269. doi:10.1039/b900430k

Zhao, X., Liu, D., Huang, H., Zhang, W., Yang, Q., & Zhong, C. (2014). The stability and defluoridation performance of MOFs in fluoride solutions. *Microporous and Mesoporous Materials, 185*, 72–78. doi:10.1016/j.micromeso.2013.11.002

Zheng, Y. H., Tan, C. L., & Wang, Q. M. (2012). A luminescent lanthanide complex-based anion sensor with electron-donating methoxy groups for monitoring multiple anions in environmental and biological processes. *Spectrochimica Acta. Part A: Molecular and Biomolecular Spectroscopy, 96*, 387–394. doi:10.1016/j.saa.2012.05.064 PMID:22722072

Zheng, Y., Li, X., & Dutta, P. K. (2012). Exploitation of unique properties of zeolites in the development of gas sensors. *Sensors (Basel), 12*(4), 5170–5194. doi:10.3390120405170 PMID:22666081

Zhong, M., Kong, L., Li, N., Liu, Y.-Y., Zhu, J., & Bu, X.-H. (2019). Synthesis of MOF-derived Nanostructures and Their Applications as Anodes in Lithium and Sodium Ion Batteries. *Coordination Chemistry Reviews*, *388*, 172–201. doi:10.1016/j.ccr.2019.02.029

Zhou, M., Zhao, H., Ko, F., Servati, P., Bahi, A., Soltanian, S., Ge, F., Zhao, Y., & Cai, Z. (2019). Metal Organic Frameworks-derived Porous $NiCo_2S_4$ Nanorods and N-doped Carbon for High-Performance Battery-Supercapacitor Hybrid Device. *Journal of Power Sources*, *440*, 227146. doi:10.1016/j.jpowsour.2019.227146

Zhou, S., Wei, Y., & Hou, J. (2017). Self-Sacrificial Template Strategy Coupled with Smart in Situ Seeding for Highly Oriented Metal-Organic Framework Layers: From Films to Membranes. *Chemistry of Materials*, *29*(17), 7103–7107.

Zhou, S., Wei, Y., & Zhuang, L. (2017). Introduction of Metal Precursors by Electrodeposition for the in Situ Growth of Metal-Organic Framework Membranes on Porous Metal Substrates. *Journal of Materials Chemistry. A, Materials for Energy and Sustainability*, *5*(5), 1948–1951.

Zhou, W., Begum, S., Wang, Z., Krolla, P., Wagner, D., Bräse, S., & Tsotsalas, M. (2018). High Antimicrobial Activity of Metal–Organic Framework-Templated Porphyrin Polymer Thin Films. *ACS Applied Materials & Interfaces*, *10*(2), 1528–1533. doi:10.1021/acsami.7b14866 PMID:29243917

Zhou, Y., & Yan, B. (2016). A responsive MOF nanocomposite for decoding volatile organic compounds. *Chemical Communications (Cambridge)*, *52*(11), 2265–2268. doi:10.1039/C5CC09029F PMID:26727287

Zhu, B.-J., Yu, X.-Y., Jia, Y., Peng, F.-M., Sun, B., Zhang, M.-Y., Luo, T., Liu, J.-H., & Huang, X.-J. (2012). Iron and 1,3,5-Benzenetricarboxylic Metal Organic Coordination polymers prepared by Solvothermal method and their application in efficient As(V) removal from aqueous solutions. *The Journal of Physical Chemistry C*, *116*(15), 8601–8607. doi:10.1021/jp212514a

Zhu, J. (2017). Elevated Performance of Thin Film Nanocomposite Membranes Enabled by Modified Hydrophilic MOFs for Nanofiltration. *ACS Applied Materials & Interfaces*, *9*(2), 1975–1986.

Zhu, K., Chen, C., Xu, H., Gao, Y., Tan, X., Alsaedi, A., & Hayat, T. (2017). Cr(VI) reduction and immobilization by core double shell structured magnetic polydopamine@zeolitic idazolate frameworks-8 microspheres. *ACS Sustainable Chemistry & Engineering*, *5*(8), 6795–6802. doi:10.1021/acssuschemeng.7b01036

Zhu, L., Xiao, C., Dai, X., Li, J., Gui, D., Sheng, D., Chen, L., Zhou, R., Chai, Z., Albrecht-Schmitt, T. E., & Wang, S. (2017). Exceptional perrhenate/pertechnetate uptake and subsequent immobilization by a low-dimensional cationic coordination polymer: Overcoming the hofmeister bias selectivity. *Environmental Science & Technology Letters*, *4*(7), 316–322. doi:10.1021/acs.estlett.7b00165

Zhu, X.-H., Yang, C.-X., & Yan, X.-P. (2018). Metal-organic framework-801 for efficient removal of fluoride from water. *Microporous and Mesoporous Materials*, *259*, 163–170. doi:10.1016/j.micromeso.2017.10.001

Zou, F., Yu, R., Li, R., & Li, W. (2013). Microwave-Assisted Synthesis of HKUST-1 and Functionalized HKUST-1-@ $H_3PW_{12}O_{40}$: Selective Adsorption of Heavy Metal Ions in Water Analyzed with Synchrotron Radiation. *ChemPhysChem*, *14*(12), 2835–2832. doi:10.1002/cphc.201300215 PMID:23818161

Zuo, J., & Chung, T. S. (2016). Metal-Organic Framework-Functionalized Alumina Membranes for Vacuum Membrane Distillation. *Water (Switzerland)*, *8*(12), 1–15.

Zuo, W., Li, R., Zhou, C., Li, Y., Xia, J., & Liu, J. (2017). Battery-Supercapacitor Hybrid Devices: Recent Progress and Future Prospects. *Advancement of Science*, *4*(7), 1600539. doi:10.1002/advs.201600539 PMID:28725528

Related References

To continue our tradition of advancing academic research, we have compiled a list of recommended IGI Global readings. These references will provide additional information and guidance to further enrich your knowledge and assist you with your own research and future publications.

Abed, S., Khir, T., & Ben Brahim, A. (2016). Thermodynamic and Energy Study of a Regenerator in Gas Turbine Cycle and Optimization of Performances. *International Journal of Energy Optimization and Engineering*, *5*(2), 25–44. doi:10.4018/IJEOE.2016040102

Abu Bakar, W. A., Abdullah, W. N., Ali, R., & Mokhtar, W. N. (2016). Polymolybdate Supported Nano Catalyst for Desulfurization of Diesel. In T. Saleh (Ed.), *Applying Nanotechnology to the Desulfurization Process in Petroleum Engineering* (pp. 263–280). Hershey, PA: IGI Global. doi:10.4018/978-1-4666-9545-0.ch009

Addo-Tenkorang, R., Helo, P., & Kantola, J. (2016). Engineer-To-Order Product Development: A Communication Network Analysis for Supply-Chain's Sustainable Competitive Advantage. In R. Addo-Tenkorang, J. Kantola, P. Helo, & A. Shamsuzzoha (Eds.), *Supply Chain Strategies and the Engineer-to-Order Approach* (pp. 43–59). Hershey, PA: IGI Global. doi:10.4018/978-1-5225-0021-6.ch003

Adebiyi, I. D., Popoola, P. A., & Pityana, S. (2016). Mitigation of Wear Damage by Laser Surface Alloying Technique. In E. Akinlabi, R. Mahamood, & S. Akinlabi (Eds.), *Advanced Manufacturing Techniques Using Laser Material Processing* (pp. 172–196). Hershey, PA: IGI Global. doi:10.4018/978-1-5225-0329-3.ch007

Ahmad, W. (2016). Sulfur in Petroleum: Petroleum Desulfurization Techniques. In T. Saleh (Ed.), *Applying Nanotechnology to the Desulfurization Process in Petroleum Engineering* (pp. 1–52). Hershey, PA: IGI Global. doi:10.4018/978-1-4666-9545-0.ch001

Ahmed, I., Ahmad, N., Mehmood, I., Haq, I. U., Hassan, M., & Khan, M. U. (2016). Applications of Nanotechnology in Transportation Engineering. In A. Khitab & W. Anwar (Eds.), *Advanced Research on Nanotechnology for Civil Engineering Applications* (pp. 180–207). Hershey, PA: IGI Global. doi:10.4018/978-1-5225-0344-6.ch006

Aikhuele, D. (2018). A Study of Product Development Engineering and Design Reliability Concerns. *International Journal of Applied Industrial Engineering*, *5*(1), 79–89. doi:10.4018/IJAIE.2018010105

Al-Najar, B. T., & Bououdina, M. (2016). Bioinspired Nanoparticles for Efficient Drug Delivery System. In M. Bououdina (Ed.), *Emerging Research on Bioinspired Materials Engineering* (pp. 69–103). Hershey, PA: IGI Global. doi:10.4018/978-1-4666-9811-6.ch003

Al-Shebeeb, O. A., Rangaswamy, S., Gopalakrishan, B., & Devaru, D. G. (2017). Evaluation and Indexing of Process Plans Based on Electrical Demand and Energy Consumption. *International Journal of Manufacturing, Materials, and Mechanical Engineering*, *7*(3), 1–19. doi:10.4018/IJMMME.2017070101

Alexakis, H., & Makris, N. (2016). Validation of the Discrete Element Method for the Limit Stability Analysis of Masonry Arches. In V. Sarhosis, K. Bagi, J. Lemos, & G. Milani (Eds.), *Computational Modeling of Masonry Structures Using the Discrete Element Method* (pp. 292–325). Hershey, PA: IGI Global. doi:10.4018/978-1-5225-0231-9.ch012

AlMegren, H. A., Gonzalez-Cortes, S., Huang, Y., Chen, H., Qian, Y., Alkinany, M., ... Xiao, T. (2016). Preparation of Deep Hydrodesulfurzation Catalysts for Diesel Fuel using Organic Matrix Decomposition Method. In H. Al-Megren & T. Xiao (Eds.), *Petrochemical Catalyst Materials, Processes, and Emerging Technologies* (pp. 216–253). Hershey, PA: IGI Global. doi:10.4018/978-1-4666-9975-5.ch009

Alshammari, A., Kalevaru, V. N., Bagabas, A., & Martin, A. (2016). Production of Ethylene and its Commercial Importance in the Global Market. In H. Al-Megren & T. Xiao (Eds.), *Petrochemical Catalyst Materials, Processes, and Emerging Technologies* (pp. 82–115). Hershey, PA: IGI Global. doi:10.4018/978-1-4666-9975-5.ch004

Amel, M. (2016). Synthesis, Characterizations, and Biological Effects Study of Some Quinoline Family. In M. Bououdina (Ed.), *Emerging Research on Bioinspired Materials Engineering* (pp. 160–196). Hershey, PA: IGI Global. doi:10.4018/978-1-4666-9811-6.ch006

Amna, T., Haasan, M. S., Khil, M., & Hwang, I. (2016). Impact of Electrospun Biomimetic Extracellular Environment on Proliferation and Intercellular Communication of Muscle Precursor Cells: An Overview – Intercellular Communication of Muscle Precursor Cells with Extracellular Environment. In M. Bououdina (Ed.), *Emerging Research on Bioinspired Materials Engineering* (pp. 247–265). Hershey, PA: IGI Global. doi:10.4018/978-1-4666-9811-6.ch009

Amuda, M. O., Lawal, T. F., & Akinlabi, E. T. (2017). Research Progress on Rheological Behavior of AA7075 Aluminum Alloy During Hot Deformation. *International Journal of Materials Forming and Machining Processes*, *4*(1), 53–96. doi:10.4018/IJMFMP.2017010104

An, M., & Qin, Y. (2016). Challenges of Railway Safety Risk Assessment and Maintenance Decision Making. In B. Rai (Ed.), *Handbook of Research on Emerging Innovations in Rail Transportation Engineering* (pp. 173–211). Hershey, PA: IGI Global. doi:10.4018/978-1-5225-0084-1.ch009

Anil, M., Ayyildiz-Tamis, D., Tasdemir, S., Sendemir-Urkmez, A., & Gulce-Iz, S. (2016). Bioinspired Materials and Biocompatibility. In M. Bououdina (Ed.), *Emerging Research on Bioinspired Materials Engineering* (pp. 294–322). Hershey, PA: IGI Global. doi:10.4018/978-1-4666-9811-6.ch011

Armutlu, H. (2018). Intelligent Biomedical Engineering Operations by Cloud Computing Technologies. In U. Kose, G. Guraksin, & O. Deperlioglu (Eds.), *Nature-Inspired Intelligent Techniques for Solving Biomedical Engineering Problems* (pp. 297–317). Hershey, PA: IGI Global. doi:10.4018/978-1-5225-4769-3.ch015

Arokiyaraj, S., Saravanan, M., Bharanidharan, R., Islam, V. I., Bououdina, M., & Vincent, S. (2016). Green Synthesis of Metallic Nanoparticles Using Plant Compounds and Their Applications: Metallic Nanoparticles Synthesis Using Plants. In M. Bououdina (Ed.), *Emerging Research on Bioinspired Materials Engineering* (pp. 1–34). Hershey, PA: IGI Global. doi:10.4018/978-1-4666-9811-6.ch001

Atik, M., Sadek, M., & Shahrour, I. (2017). Single-Run Adaptive Pushover Procedure for Shear Wall Structures. In V. Plevris, G. Kremmyda, & Y. Fahjan (Eds.), *Performance-Based Seismic Design of Concrete Structures and Infrastructures* (pp. 59–83). Hershey, PA: IGI Global. doi:10.4018/978-1-5225-2089-4.ch003

Aydin, A., Akyol, E., Gungor, M., Kaya, A., & Tasdelen, S. (2018). Geophysical Surveys in Engineering Geology Investigations With Field Examples. In N. Ceryan (Ed.), *Handbook of Research on Trends and Digital Advances in Engineering Geology* (pp. 257–280). Hershey, PA: IGI Global. doi:10.4018/978-1-5225-2709-1.ch007

Azevedo, N. M., Lemos, J. V., & Rocha de Almeida, J. (2016). Discrete Element Particle Modelling of Stone Masonry. In V. Sarhosis, K. Bagi, J. Lemos, & G. Milani (Eds.), *Computational Modeling of Masonry Structures Using the Discrete Element Method* (pp. 146–170). Hershey, PA: IGI Global. doi:10.4018/978-1-5225-0231-9.ch007

Bamufleh, H. S., Noureldin, M. M., & El-Halwagi, M. M. (2016). Sustainable Process Integration in the Petrochemical Industries. In H. Al-Megren & T. Xiao (Eds.), *Petrochemical Catalyst Materials, Processes, and Emerging Technologies* (pp. 150–163). Hershey, PA: IGI Global. doi:10.4018/978-1-4666-9975-5.ch006

Banerjee, S., Gautam, R. K., Gautam, P. K., Jaiswal, A., & Chattopadhyaya, M. C. (2016). Recent Trends and Advancement in Nanotechnology for Water and Wastewater Treatment: Nanotechnological Approach for Water Purification. In A. Khitab & W. Anwar (Eds.), *Advanced Research on Nanotechnology for Civil Engineering Applications* (pp. 208–252). Hershey, PA: IGI Global. doi:10.4018/978-1-5225-0344-6.ch007

Bas, T. G. (2017). Nutraceutical Industry with the Collaboration of Biotechnology and Nutrigenomics Engineering: The Significance of Intellectual Property in the Entrepreneurship and Scientific Research Ecosystems. In T. Bas & J. Zhao (Eds.), *Comparative Approaches to Biotechnology Development and Use in Developed and Emerging Nations* (pp. 1–17). Hershey, PA: IGI Global. doi:10.4018/978-1-5225-1040-6.ch001

Beale, R., & André, J. (2017). *Design Solutions and Innovations in Temporary Structures*. Hershey, PA: IGI Global. doi:10.4018/978-1-5225-2199-0

Behnam, B. (2017). Simulating Post-Earthquake Fire Loading in Conventional RC Structures. In P. Samui, S. Chakraborty, & D. Kim (Eds.), *Modeling and Simulation Techniques in Structural Engineering* (pp. 425–444). Hershey, PA: IGI Global. doi:10.4018/978-1-5225-0588-4.ch015

Ben Hamida, I., Salah, S. B., Msahli, F., & Mimouni, M. F. (2018). Distribution Network Reconfiguration Using SPEA2 for Power Loss Minimization and Reliability Improvement. *International Journal of Energy Optimization and Engineering*, *7*(1), 50–65. doi:10.4018/IJEOE.2018010103

Benjamin, S. R., de Lima, F., & Rathoure, A. K. (2016). Genetically Engineered Microorganisms for Bioremediation Processes: GEMs for Bioremediaton. In A. Rathoure & V. Dhatwalia (Eds.), *Toxicity and Waste Management Using Bioremediation* (pp. 113–140). Hershey, PA: IGI Global. doi:10.4018/978-1-4666-9734-8.ch006

Bhaskar, S. V., & Kudal, H. N. (2017). Effect of TiCN and AlCrN Coating on Tribological Behaviour of Plasma-nitrided AISI 4140 Steel. *International Journal of Surface Engineering and Interdisciplinary Materials Science*, *5*(2), 1–17. doi:10.4018/IJSEIMS.2017070101

Bhowmik, S., Sahoo, P., Acharyya, S. K., Dhar, S., & Chattopadhyay, J. (2016). Effect of Microstructure Degradation on Fracture Toughness of 20MnMoNi55 Steel in DBT Region. *International Journal of Manufacturing, Materials, and Mechanical Engineering*, *6*(3), 11–27. doi:10.4018/IJMMME.2016070102

Bhutto, A. W., Abro, R., Abbas, T., Yu, G., & Chen, X. (2016). Desulphurization of Fuel Oils Using Ionic Liquids. In H. Al-Megren & T. Xiao (Eds.), *Petrochemical Catalyst Materials, Processes, and Emerging Technologies* (pp. 254–284). Hershey, PA: IGI Global. doi:10.4018/978-1-4666-9975-5.ch010

Bhuyan, D. (2018). Designing of a Twin Tube Shock Absorber: A Study in Reverse Engineering. In K. Kumar & J. Davim (Eds.), *Design and Optimization of Mechanical Engineering Products* (pp. 83–104). Hershey, PA: IGI Global. doi:10.4018/978-1-5225-3401-3.ch005

Bouloudenine, M., & Bououdina, M. (2016). Toxic Effects of Engineered Nanoparticles on Living Cells. In M. Bououdina (Ed.), *Emerging Research on Bioinspired Materials Engineering* (pp. 35–68). Hershey, PA: IGI Global. doi:10.4018/978-1-4666-9811-6.ch002

Brunetti, A., Sellaro, M., Drioli, E., & Barbieri, G. (2016). Membrane Engineering and its Role in Oil Refining and Petrochemical Industry. In H. Al-Megren & T. Xiao (Eds.), *Petrochemical Catalyst Materials, Processes, and Emerging Technologies* (pp. 116–149). Hershey, PA: IGI Global. doi:10.4018/978-1-4666-9975-5.ch005

Bügler, M., & Borrmann, A. (2016). Simulation Based Construction Project Schedule Optimization: An Overview on the State-of-the-Art. In F. Miranda & C. Abreu (Eds.), *Handbook of Research on Computational Simulation and Modeling in Engineering* (pp. 482–507). Hershey, PA: IGI Global. doi:10.4018/978-1-4666-8823-0.ch016

Calderon, F. A., Giolo, E. G., Frau, C. D., Rengel, M. G., Rodriguez, H., Tornello, M., ... Gallucci, R. (2018). Seismic Microzonation and Site Effects Detection Through Microtremors Measures: A Review. In N. Ceryan (Ed.), *Handbook of Research on Trends and Digital Advances in Engineering Geology* (pp. 326–349). Hershey, PA: IGI Global. doi:10.4018/978-1-5225-2709-1.ch009

Carmona-Murillo, J., & Valenzuela-Valdés, J. F. (2016). Motivation on Problem Based Learning. In D. Fonseca & E. Redondo (Eds.), *Handbook of Research on Applied E-Learning in Engineering and Architecture Education* (pp. 179–203). Hershey, PA: IGI Global. doi:10.4018/978-1-4666-8803-2.ch009

Ceryan, N. (2016). A Review of Soft Computing Methods Application in Rock Mechanic Engineering. In P. Samui (Ed.), *Handbook of Research on Advanced Computational Techniques for Simulation-Based Engineering* (pp. 1–70). Hershey, PA: IGI Global. doi:10.4018/978-1-4666-9479-8.ch001

Ceryan, N., & Can, N. K. (2018). Prediction of The Uniaxial Compressive Strength of Rocks Materials. In N. Ceryan (Ed.), *Handbook of Research on Trends and Digital Advances in Engineering Geology* (pp. 31–96). Hershey, PA: IGI Global. doi:10.4018/978-1-5225-2709-1.ch002

Ceryan, S. (2018). Weathering Indices Used in Evaluation of the Weathering State of Rock Material. In N. Ceryan (Ed.), *Handbook of Research on Trends and Digital Advances in Engineering Geology* (pp. 132–186). Hershey, PA: IGI Global. doi:10.4018/978-1-5225-2709-1.ch004

Chandrasekaran, S., Silva, B., Patil, A., Oo, A. M., & Campbell, M. (2016). Evaluating Engineering Students' Perceptions: The Impact of Team-Based Learning Practices in Engineering Education. *International Journal of Quality Assurance in Engineering and Technology Education*, *5*(4), 42–59. doi:10.4018/IJQAETE.2016100103

Chen, H., Padilla, R. V., & Besarati, S. (2017). Supercritical Fluids and Their Applications in Power Generation. In L. Chen & Y. Iwamoto (Eds.), *Advanced Applications of Supercritical Fluids in Energy Systems* (pp. 369–402). Hershey, PA: IGI Global. doi:10.4018/978-1-5225-2047-4.ch012

Chen, L. (2017). Principles, Experiments, and Numerical Studies of Supercritical Fluid Natural Circulation System. In L. Chen & Y. Iwamoto (Eds.), *Advanced Applications of Supercritical Fluids in Energy Systems* (pp. 136–187). Hershey, PA: IGI Global. doi:10.4018/978-1-5225-2047-4.ch005

Clementi, F., Di Sciascio, G., Di Sciascio, S., & Lenci, S. (2017). Influence of the Shear-Bending Interaction on the Global Capacity of Reinforced Concrete Frames: A Brief Overview of the New Perspectives. In V. Plevris, G. Kremmyda, & Y. Fahjan (Eds.), *Performance-Based Seismic Design of Concrete Structures and Infrastructures* (pp. 84–111). Hershey, PA: IGI Global. doi:10.4018/978-1-5225-2089-4.ch004

Cortés-Polo, D., Calle-Cancho, J., Carmona-Murillo, J., & González-Sánchez, J. (2017). Future Trends in Mobile-Fixed Integration for Next Generation Networks: Classification and Analysis. *International Journal of Vehicular Telematics and Infotainment Systems*, *1*(1), 33–53. doi:10.4018/IJVTIS.2017010103

Cui, X., Zeng, S., Li, Z., Zheng, Q., Yu, X., & Han, B. (2018). Advanced Composites for Civil Engineering Infrastructures. In K. Kumar & J. Davim (Eds.), *Composites and Advanced Materials for Industrial Applications* (pp. 212–248). Hershey, PA: IGI Global. doi:10.4018/978-1-5225-5216-1.ch010

Dalgıç, S., & Kuşku, İ. (2018). Geological and Geotechnical Investigations in Tunneling. In N. Ceryan (Ed.), *Handbook of Research on Trends and Digital Advances in Engineering Geology* (pp. 482–529). Hershey, PA: IGI Global. doi:10.4018/978-1-5225-2709-1.ch014

de la Varga, D., Soto, M., Arias, C. A., van Oirschot, D., Kilian, R., Pascual, A., & Álvarez, J. A. (2017). Constructed Wetlands for Industrial Wastewater Treatment and Removal of Nutrients. In Á. Val del Río, J. Campos Gómez, & A. Mosquera Corral (Eds.), *Technologies for the Treatment and Recovery of Nutrients from Industrial Wastewater* (pp. 202–230). Hershey, PA: IGI Global. doi:10.4018/978-1-5225-1037-6.ch008 del Valle-Zermeño, R., Chimenos, J. M., & Formosa, J. (2016). Flue Gas Desulfurization: Processes and Technologies. In T. Saleh (Ed.), *Applying Nanotechnology to the Desulfurization Process in Petroleum Engineering* (pp. 337–377). Hershey, PA: IGI Global. doi:10.4018/978-1-4666-9545-0.ch011

Delgado, J. M., Henriques, A. A., & Delgado, R. M. (2016). Structural Non-Linear Models and Simulation Techniques: An Efficient Combination for Safety Evaluation of RC Structures. In F. Miranda & C. Abreu (Eds.), *Handbook of Research on Computational Simulation and Modeling in Engineering* (pp. 540–584). Hershey, PA: IGI Global. doi:10.4018/978-1-4666-8823-0.ch018

Delgado, P. S., Arêde, A., Pouca, N. V., & Costa, A. (2016). Numerical Modeling of RC Bridges for Seismic Risk Analysis. In F. Miranda & C. Abreu (Eds.), *Handbook of Research on Computational Simulation and Modeling in Engineering* (pp. 457–481). Hershey, PA: IGI Global. doi:10.4018/978-1-4666-8823-0.ch015

Deng, Y., & Liu, S. (2016). Catalysis with Room Temperature Ionic Liquids Mediated Metal Nanoparticles. In H. Al-Megren & T. Xiao (Eds.), *Petrochemical Catalyst Materials, Processes, and Emerging Technologies* (pp. 285–329). Hershey, PA: IGI Global. doi:10.4018/978-1-4666-9975-5.ch011

Deperlioglu, O. (2018). Intelligent Techniques Inspired by Nature and Used in Biomedical Engineering. In U. Kose, G. Guraksin, & O. Deperlioglu (Eds.), *Nature-Inspired Intelligent Techniques for Solving Biomedical Engineering Problems* (pp. 51–77). Hershey, PA: IGI Global. doi:10.4018/978-1-5225-4769-3.ch003

Dias, G. L., Magalhães, R. R., Ferreira, D. D., & Vitoriano, F. A. (2016). The Use of a Robotic Arm for Displacement Measurements in a Cantilever beam. *International Journal of Manufacturing, Materials, and Mechanical Engineering, 6*(3), 45–57. doi:10.4018/IJMMME.2016070104

Dimitratos, N., Villa, A., Chan-Thaw, C. E., Hammond, C., & Prati, L. (2016). Valorisation of Glycerol to Fine Chemicals and Fuels. In H. Al-Megren & T. Xiao (Eds.), *Petrochemical Catalyst Materials, Processes, and Emerging Technologies* (pp. 352–384). Hershey, PA: IGI Global. doi:10.4018/978-1-4666-9975-5.ch013

Dixit, A. (2018). Application of Silica-Gel-Reinforced Aluminium Composite on the Piston of Internal Combustion Engine: Comparative Study of Silica-Gel-Reinforced Aluminium Composite Piston With Aluminium Alloy Piston. In K. Kumar & J. Davim (Eds.), *Composites and Advanced Materials for Industrial Applications* (pp. 63–98). Hershey, PA: IGI Global. doi:10.4018/978-1-5225-5216-1.ch004

Drei, A., Milani, G., & Sincraian, G. (2016). Application of DEM to Historic Masonries, Two Case-Studies in Portugal and Italy: Aguas Livres Aqueduct and Arch-Tympana of a Church. In V. Sarhosis, K. Bagi, J. Lemos, & G. Milani (Eds.), *Computational Modeling of Masonry Structures Using the Discrete Element Method* (pp. 326–366). Hershey, PA: IGI Global. doi:10.4018/978-1-5225-0231-9.ch013

Dutta, S., Roy, P. K., & Nandi, D. (2016). Optimal Allocation of Static Synchronous Series Compensator Controllers using Chemical Reaction Optimization for Reactive Power Dispatch. *International Journal of Energy Optimization and Engineering, 5*(3), 43–62. doi:10.4018/IJEOE.2016070103

Dutta, S., Roy, P. K., & Nandi, D. (2016). Quasi Oppositional Teaching-Learning based Optimization for Optimal Power Flow Incorporating FACTS. *International Journal of Energy Optimization and Engineering*, *5*(2), 64–84. doi:10.4018/IJEOE.2016040104

Eloy, S., Dias, M. S., Lopes, P. F., & Vilar, E. (2016). Digital Technologies in Architecture and Engineering: Exploring an Engaged Interaction within Curricula. In D. Fonseca & E. Redondo (Eds.), *Handbook of Research on Applied E-Learning in Engineering and Architecture Education* (pp. 368–402). Hershey, PA: IGI Global. doi:10.4018/978-1-4666-8803-2.ch017

Elsayed, A. M., Dakkama, H. J., Mahmoud, S., Al-Dadah, R., & Kaialy, W. (2017). Sustainable Cooling Research Using Activated Carbon Adsorbents and Their Environmental Impact. In T. Kobayashi (Ed.), *Applied Environmental Materials Science for Sustainability* (pp. 186–221). Hershey, PA: IGI Global. doi:10.4018/978-1-5225-1971-3.ch009

Ercanoglu, M., & Sonmez, H. (2018). General Trends and New Perspectives on Landslide Mapping and Assessment Methods. In N. Ceryan (Ed.), *Handbook of Research on Trends and Digital Advances in Engineering Geology* (pp. 350–379). Hershey, PA: IGI Global. doi:10.4018/978-1-5225-2709-1.ch010

Erinosho, M. F., Akinlabi, E. T., & Pityana, S. (2016). Enhancement of Surface Integrity of Titanium Alloy with Copper by Means of Laser Metal Deposition Process. In E. Akinlabi, R. Mahamood, & S. Akinlabi (Eds.), *Advanced Manufacturing Techniques Using Laser Material Processing* (pp. 60–91). Hershey, PA: IGI Global. doi:10.4018/978-1-5225-0329-3.ch004

Farag, H., & Kishida, M. (2016). Kinetic Models for Complex Parallel–Consecutive Reactions Assessment of Reaction Network and Product Selectivity. In H. Al-Megren & T. Xiao (Eds.), *Petrochemical Catalyst Materials, Processes, and Emerging Technologies* (pp. 330–351). Hershey, PA: IGI Global. doi:10.4018/978-1-4666-9975-5.ch012

Faroz, S. A., Pujari, N. N., Rastogi, R., & Ghosh, S. (2017). Risk Analysis of Structural Engineering Systems Using Bayesian Inference. In P. Samui, S. Chakraborty, & D. Kim (Eds.), *Modeling and Simulation Techniques in Structural Engineering* (pp. 390–424). Hershey, PA: IGI Global. doi:10.4018/978-1-5225-0588-4.ch014

Fernando, P. R., Hamigah, T., Disne, S., Wickramasingha, G. G., & Sutharshan, A. (2018). The Evaluation of Engineering Properties of Low Cost Concrete Blocks by Partial Doping of Sand with Sawdust: Low Cost Sawdust Concrete Block. *International Journal of Strategic Engineering*, *1*(2), 26–42. doi:10.4018/IJoSE.2018070103

Fragiadakis, M., Stefanou, I., & Psycharis, I. N. (2016). Vulnerability Assessment of Damaged Classical Multidrum Columns. In V. Sarhosis, K. Bagi, J. Lemos, & G. Milani (Eds.), *Computational Modeling of Masonry Structures Using the Discrete Element Method* (pp. 235–253). Hershey, PA: IGI Global. doi:10.4018/978-1-5225-0231-9.ch010

Gaines, T. W., Williams, K. R., & Wagener, K. B. (2016). ADMET: Functionalized Polyolefins. In H. Al-Megren & T. Xiao (Eds.), *Petrochemical Catalyst Materials, Processes, and Emerging Technologies* (pp. 1–21). Hershey, PA: IGI Global. doi:10.4018/978-1-4666-9975-5.ch001

Garg, H. (2016). Bi-Criteria Optimization for Finding the Optimal Replacement Interval for Maintaining the Performance of the Process Industries. In P. Vasant, G. Weber, & V. Dieu (Eds.), *Handbook of Research on Modern Optimization Algorithms and Applications in Engineering and Economics* (pp. 643–675). Hershey, PA: IGI Global. doi:10.4018/978-1-4666-9644-0.ch025

Gaspar, P. D., Dinho da Silva, P., Gonçalves, J. P., & Carneiro, R. (2016). Computational Modelling and Simulation to Assist the Improvement of Thermal Performance and Energy Efficiency in Industrial Engineering Systems: Application to Cold Stores. In F. Miranda & C. Abreu (Eds.), *Handbook of Research on Computational Simulation and Modeling in Engineering* (pp. 1–68). Hershey, PA: IGI Global. doi:10.4018/978-1-4666-8823-0.ch001

Ge, H., Tang, M., & Wen, X. (2016). Ni/ZnO Nano Sorbent for Reactive Adsorption Desulfurization of Refinery Oil Streams. In T. Saleh (Ed.), *Applying Nanotechnology to the Desulfurization Process in Petroleum Engineering* (pp. 216–239). Hershey, PA: IGI Global. doi:10.4018/978-1-4666-9545-0.ch007

Ghosh, S., Mitra, S., Ghosh, S., & Chakraborty, S. (2017). Seismic Reliability Analysis in the Framework of Metamodelling Based Monte Carlo Simulation. In P. Samui, S. Chakraborty, & D. Kim (Eds.), *Modeling and Simulation Techniques in Structural Engineering* (pp. 192–208). Hershey, PA: IGI Global. doi:10.4018/978-1-5225-0588-4.ch006

Gil, M., & Otero, B. (2017). Learning Engineering Skills through Creativity and Collaboration: A Game-Based Proposal. In R. Alexandre Peixoto de Queirós & M. Pinto (Eds.), *Gamification-Based E-Learning Strategies for Computer Programming Education* (pp. 14–29). Hershey, PA: IGI Global. doi:10.4018/978-1-5225-1034-5.ch002

Gill, J., Ayre, M., & Mills, J. (2017). Revisioning the Engineering Profession: How to Make It Happen! In M. Gray & K. Thomas (Eds.), *Strategies for Increasing Diversity in Engineering Majors and Careers* (pp. 156–175). Hershey, PA: IGI Global. doi:10.4018/978-1-5225-2212-6.ch008

Gopal, S., & Al-Hazmi, M. H. (2016). Advances in Catalytic Technologies for Selective Oxidation of Lower Alkanes. In H. Al-Megren & T. Xiao (Eds.), *Petrochemical Catalyst Materials, Processes, and Emerging Technologies* (pp. 22–52). Hershey, PA: IGI Global. doi:10.4018/978-1-4666-9975-5.ch002

Goyal, N., Ram, M., Bhardwaj, A., & Kumar, A. (2016). Thermal Power Plant Modelling with Fault Coverage Stochastically. *International Journal of Manufacturing, Materials, and Mechanical Engineering*, *6*(3), 28–44. doi:10.4018/IJMMME.2016070103

Goyal, N., Ram, M., & Kumar, P. (2017). Welding Process under Fault Coverage Approach for Reliability and MTTF. In M. Ram & J. Davim (Eds.), *Mathematical Concepts and Applications in Mechanical Engineering and Mechatronics* (pp. 222–245). Hershey, PA: IGI Global. doi:10.4018/978-1-5225-1639-2.ch011

Gray, M., & Lundy, C. (2017). Engineering Study Abroad: High Impact Strategy for Increasing Access. In M. Gray & K. Thomas (Eds.), *Strategies for Increasing Diversity in Engineering Majors and Careers* (pp. 42–59). Hershey, PA: IGI Global. doi:10.4018/978-1-5225-2212-6.ch003

Guha, D., Roy, P. K., & Banerjee, S. (2016). Application of Modified Biogeography Based Optimization in AGC of an Interconnected Multi-Unit Multi-Source AC-DC Linked Power System. *International Journal of Energy Optimization and Engineering*, *5*(3), 1–18. doi:10.4018/IJEOE.2016070101

Guha, D., Roy, P. K., & Banerjee, S. (2016). Grey Wolf Optimization to Solve Load Frequency Control of an Interconnected Power System: GWO Used to Solve LFC Problem. *International Journal of Energy Optimization and Engineering*, *5*(4), 62–83. doi:10.4018/IJEOE.2016100104

Gupta, A. K., Dey, A., & Mukhopadhyay, A. K. (2016). Micromechanical and Finite Element Modeling for Composites. In S. Datta & J. Davim (Eds.), *Computational Approaches to Materials Design: Theoretical and Practical Aspects* (pp. 101–162). Hershey, PA: IGI Global. doi:10.4018/978-1-5225-0290-6.ch005

Guraksin, G. E. (2018). Internet of Things and Nature-Inspired Intelligent Techniques for the Future of Biomedical Engineering. In U. Kose, G. Guraksin, & O. Deperlioglu (Eds.), *Nature-Inspired Intelligent Techniques for Solving Biomedical Engineering Problems* (pp. 263–282). Hershey, PA: IGI Global. doi:10.4018/978-1-5225-4769-3.ch013

Hansman, C. A. (2016). Developing Mentoring Programs in Engineering and Technology Education. *International Journal of Quality Assurance in Engineering and Technology Education*, *5*(2), 1–15. doi:10.4018/IJQAETE.2016040101

Hasan, U., Chegenizadeh, A., & Nikraz, H. (2016). Nanotechnology Future and Present in Construction Industry: Applications in Geotechnical Engineering. In A. Khitab & W. Anwar (Eds.), *Advanced Research on Nanotechnology for Civil Engineering Applications* (pp. 141–179). Hershey, PA: IGI Global. doi:10.4018/978-1-5225-0344-6.ch005

Hejazi, T., & Akbari, L. (2017). A Multiresponse Optimization Model for Statistical Design of Processes with Discrete Variables. In M. Ram & J. Davim (Eds.), *Mathematical Concepts and Applications in Mechanical Engineering and Mechatronics* (pp. 17–37). Hershey, PA: IGI Global. doi:10.4018/978-1-5225-1639-2.ch002

Hejazi, T., & Hejazi, A. (2017). Monte Carlo Simulation for Reliability-Based Design of Automotive Complex Subsystems. In M. Ram & J. Davim (Eds.), *Mathematical Concepts and Applications in Mechanical Engineering and Mechatronics* (pp. 177–200). Hershey, PA: IGI Global. doi:10.4018/978-1-5225-1639-2.ch009

Hejazi, T., & Poursabbagh, H. (2017). Reliability Analysis of Engineering Systems: An Accelerated Life Testing for Boiler Tubes. In M. Ram & J. Davim (Eds.), *Mathematical Concepts and Applications in Mechanical Engineering and Mechatronics* (pp. 154–176). Hershey, PA: IGI Global. doi:10.4018/978-1-5225-1639-2.ch008

Henao, J., & Sotelo, O. (2018). Surface Engineering at High Temperature: Thermal Cycling and Corrosion Resistance. In A. Pakseresht (Ed.), *Production, Properties, and Applications of High Temperature Coatings* (pp. 131–159). Hershey, PA: IGI Global. doi:10.4018/978-1-5225-4194-3.ch006

Huirache-Acuña, R., Alonso-Nuñez, G., Rivera-Muñoz, E. M., Gutierrez, O., & Pawelec, B. (2016). Trimetallic Sulfide Catalysts for Hydrodesulfurization. In T. Saleh (Ed.), *Applying Nanotechnology to the Desulfurization Process in Petroleum Engineering* (pp. 240–262). Hershey, PA: IGI Global. doi:10.4018/978-1-4666-9545-0.ch008

Ilori, O. O., Adetan, D. A., & Umoru, L. E. (2017). Effect of Cutting Parameters on the Surface Residual Stress of Face-Milled Pearlitic Ductile Iron. *International Journal of Materials Forming and Machining Processes*, *4*(1), 38–52. doi:10.4018/IJMFMP.2017010103

Imam, M. H., Tasadduq, I. A., Ahmad, A., Aldosari, F., & Khan, H. (2017). Automated Generation of Course Improvement Plans Using Expert System. *International Journal of Quality Assurance in Engineering and Technology Education*, *6*(1), 1–12. doi:10.4018/IJQAETE.2017010101

Injeti, S. K., & Kumar, T. V. (2018). A WDO Framework for Optimal Deployment of DGs and DSCs in a Radial Distribution System Under Daily Load Pattern to Improve Techno-Economic Benefits. *International Journal of Energy Optimization and Engineering*, *7*(2), 1–38. doi:10.4018/IJEOE.2018040101

Ishii, N., Anami, K., & Knisely, C. W. (2018). *Dynamic Stability of Hydraulic Gates and Engineering for Flood Prevention*. Hershey, PA: IGI Global. doi:10.4018/978-1-5225-3079-4

J., J., Chowdhury, S., Goyal, P., Samui, P., & Dalkiliç, Y. (2016). Determination of Bearing Capacity of Shallow Foundation Using Soft Computing. In P. Saxena, D. Singh, & M. Pant (Eds.), *Problem Solving and Uncertainty Modeling through Optimization and Soft Computing Applications* (pp. 292-328). Hershey, PA: IGI Global. doi:10.4018/978-1-4666-9885-7.ch014

Jagan, J., Gundlapalli, P., & Samui, P. (2016). Utilization of Classification Techniques for the Determination of Liquefaction Susceptibility of Soils. In S. Bhattacharyya, P. Banerjee, D. Majumdar, & P. Dutta (Eds.), *Handbook of Research on Advanced Hybrid Intelligent Techniques and Applications* (pp. 124–160). Hershey, PA: IGI Global. doi:10.4018/978-1-4666-9474-3.ch005

Jayapalan, S. (2018). A Review of Chemical Treatments on Natural Fibers-Based Hybrid Composites for Engineering Applications. In K. Kumar & J. Davim (Eds.), *Composites and Advanced Materials for Industrial Applications* (pp. 16–37). Hershey, PA: IGI Global. doi:10.4018/978-1-5225-5216-1.ch002

Jeet, K., & Dhir, R. (2016). Software Module Clustering Using Bio-Inspired Algorithms. In P. Vasant, G. Weber, & V. Dieu (Eds.), *Handbook of Research on Modern Optimization Algorithms and Applications in Engineering and Economics* (pp. 445–470). Hershey, PA: IGI Global. doi:10.4018/978-1-4666-9644-0.ch017

Joshi, S. D., & Talange, D. B. (2016). Fault Tolerant Control for a Fractional Order AUV System. *International Journal of Energy Optimization and Engineering*, *5*(2), 1–24. doi:10.4018/IJEOE.2016040101

Julião, D., Ribeiro, S., de Castro, B., Cunha-Silva, L., & Balula, S. S. (2016). Polyoxometalates-Based Nanocatalysts for Production of Sulfur-Free Diesel. In T. Saleh (Ed.), *Applying Nanotechnology to the Desulfurization Process in Petroleum Engineering* (pp. 426–458). Hershey, PA: IGI Global. doi:10.4018/978-1-4666-9545-0.ch014

Kamthan, P. (2016). On the Nature of Collaborations in Agile Software Engineering Course Projects. *International Journal of Quality Assurance in Engineering and Technology Education*, *5*(2), 42–59. doi:10.4018/IJQAETE.2016040104

Karaman, O., Celik, C., & Urkmez, A. S. (2016). Self-Assembled Biomimetic Scaffolds for Bone Tissue Engineering. In M. Bououdina (Ed.), *Emerging Research on Bioinspired Materials Engineering* (pp. 104–132). Hershey, PA: IGI Global. doi:10.4018/978-1-4666-9811-6.ch004

Karkalos, N. E., Markopoulos, A. P., & Dossis, M. F. (2017). Optimal Model Parameters of Inverse Kinematics Solution of a 3R Robotic Manipulator Using ANN Models. *International Journal of Manufacturing, Materials, and Mechanical Engineering*, 7(3), 20–40. doi:10.4018/IJMMME.2017070102

Kesimal, A., Karaman, K., Cihangir, F., & Ercikdi, B. (2018). Excavatability Assessment of Rock Masses for Geotechnical Studies. In N. Ceryan (Ed.), *Handbook of Research on Trends and Digital Advances in Engineering Geology* (pp. 231–256). Hershey, PA: IGI Global. doi:10.4018/978-1-5225-2709-1.ch006

Khanh, D. V., Vasant, P. M., Elamvazuthi, I., & Dieu, V. N. (2016). Multi-Objective Optimization of Two-Stage Thermo-Electric Cooler Using Differential Evolution: MO Optimization of TEC Using DE. In F. Miranda & C. Abreu (Eds.), *Handbook of Research on Computational Simulation and Modeling in Engineering* (pp. 139–170). Hershey, PA: IGI Global. doi:10.4018/978-1-4666-8823-0.ch004

Kim, D., Hassan, M. K., Chang, S., & Bigdeli, Y. (2016). Nonlinear Vibration Control of 3D Irregular Structures Subjected to Seismic Loads. In P. Samui (Ed.), *Handbook of Research on Advanced Computational Techniques for Simulation-Based Engineering* (pp. 103–119). Hershey, PA: IGI Global. doi:10.4018/978-1-4666-9479-8.ch003

Knoflacher, H. (2017). The Role of Engineers and Their Tools in the Transport Sector after Paradigm Change: From Assumptions and Extrapolations to Science. In H. Knoflacher & E. Ocalir-Akunal (Eds.), *Engineering Tools and Solutions for Sustainable Transportation Planning* (pp. 1–29). Hershey, PA: IGI Global. doi:10.4018/978-1-5225-2116-7.ch001

Kose, U. (2018). Towards an Intelligent Biomedical Engineering With Nature-Inspired Artificial Intelligence Techniques. In U. Kose, G. Guraksin, & O. Deperlioglu (Eds.), *Nature-Inspired Intelligent Techniques for Solving Biomedical Engineering Problems* (pp. 1–26). Hershey, PA: IGI Global. doi:10.4018/978-1-5225-4769-3.ch001

Kostić, S. (2018). A Review on Enhanced Stability Analyses of Soil Slopes Using Statistical Design. In N. Ceryan (Ed.), *Handbook of Research on Trends and Digital Advances in Engineering Geology* (pp. 446–481). Hershey, PA: IGI Global. doi:10.4018/978-1-5225-2709-1.ch013

Kumar, A., Patil, P. P., & Prajapati, Y. K. (2018). *Advanced Numerical Simulations in Mechanical Engineering*. Hershey, PA: IGI Global. doi:10.4018/978-1-5225-3722-9

Kumar, G. R., Rajyalakshmi, G., & Manupati, V. K. (2017). Surface Micro Patterning of Aluminium Reinforced Composite through Laser Peening. *International Journal of Manufacturing, Materials, and Mechanical Engineering*, 7(4), 15–27. doi:10.4018/IJMMME.2017100102

Kumari, N., & Kumar, K. (2018). Fabrication of Orthotic Calipers With Epoxy-Based Green Composite. In K. Kumar & J. Davim (Eds.), *Composites and Advanced Materials for Industrial Applications* (pp. 157–176). Hershey, PA: IGI Global. doi:10.4018/978-1-5225-5216-1.ch008

Kuppusamy, R. R. (2018). Development of Aerospace Composite Structures Through Vacuum-Enhanced Resin Transfer Moulding Technology (VERTMTy): Vacuum-Enhanced Resin Transfer Moulding. In K. Kumar & J. Davim (Eds.), *Composites and Advanced Materials for Industrial Applications* (pp. 99–111). Hershey, PA: IGI Global. doi:10.4018/978-1-5225-5216-1.ch005

Lemos, J. V. (2016). The Basis for Masonry Analysis with UDEC and 3DEC. In V. Sarhosis, K. Bagi, J. Lemos, & G. Milani (Eds.), *Computational Modeling of Masonry Structures Using the Discrete Element Method* (pp. 61–89). Hershey, PA: IGI Global. doi:10.4018/978-1-5225-0231-9.ch003

Loy, J., Howell, S., & Cooper, R. (2017). Engineering Teams: Supporting Diversity in Engineering Education. In M. Gray & K. Thomas (Eds.), *Strategies for Increasing Diversity in Engineering Majors and Careers* (pp. 106–129). Hershey, PA: IGI Global. doi:10.4018/978-1-5225-2212-6.ch006

Macher, G., Armengaud, E., Kreiner, C., Brenner, E., Schmittner, C., Ma, Z., ... Krammer, M. (2018). Integration of Security in the Development Lifecycle of Dependable Automotive CPS. In N. Druml, A. Genser, A. Krieg, M. Menghin, & A. Hoeller (Eds.), *Solutions for Cyber-Physical Systems Ubiquity* (pp. 383–423). Hershey, PA: IGI Global. doi:10.4018/978-1-5225-2845-6.ch015

Maghsoodlou, S., & Poreskandar, S. (2016). Controlling Electrospinning Jet Using Microscopic Model for Ideal Tissue Engineering Scaffolds. *International Journal of Chemoinformatics and Chemical Engineering*, *5*(2), 1–16. doi:10.4018/IJCCE.2016070101

Mahendramani, G., & Lakshmana Swamy, N. (2018). Effect of Weld Groove Area on Distortion of Butt Welded Joints in Submerged Arc Welding. *International Journal of Manufacturing, Materials, and Mechanical Engineering*, *8*(2), 33–44. doi:10.4018/IJMMME.2018040103

Maiti, S. (2016). Engineered Gellan Polysaccharides in the Design of Controlled Drug Delivery Systems. In M. Bououdina (Ed.), *Emerging Research on Bioinspired Materials Engineering* (pp. 266–293). Hershey, PA: IGI Global. doi:10.4018/978-1-4666-9811-6.ch010

Majumdar, J. D., Weisheit, A., & Manna, I. (2016). Laser Surface Processing for Tailoring of Properties by Optimization of Microstructure. In E. Akinlabi, R. Mahamood, & S. Akinlabi (Eds.), *Advanced Manufacturing Techniques Using Laser Material Processing* (pp. 121–171). Hershey, PA: IGI Global. doi:10.4018/978-1-5225-0329-3.ch006

Maldonado-Macías, A. A., García-Alcaraz, J. L., Hernández-Arellano, J. L., & Cortes-Robles, G. (2016). An Ergonomic Compatibility Perspective on the Selection of Advanced Manufacturing Technology: A Case Study for CNC Vertical Machining Centers. In G. Alor-Hernández, C. Sánchez-Ramírez, & J. García-Alcaraz (Eds.), *Handbook of Research on Managerial Strategies for Achieving Optimal Performance in Industrial Processes* (pp. 137–165). Hershey, PA: IGI Global. doi:10.4018/978-1-5225-0130-5.ch008

Mamaghani, I. H. (2016). Application of Discrete Finite Element Method for Analysis of Unreinforced Masonry Structures. In V. Sarhosis, K. Bagi, J. Lemos, & G. Milani (Eds.), *Computational Modeling of Masonry Structures Using the Discrete Element Method* (pp. 440–458). Hershey, PA: IGI Global. doi:10.4018/978-1-5225-0231-9.ch017

Mansor, M. R., Sapuan, S. M., Salim, M. A., Akop, M. Z., Musthafah, M. T., & Shaharuzaman, M. A. (2016). Concurrent Design of Green Composites. In D. Verma, S. Jain, X. Zhang, & P. Gope (Eds.), *Green Approaches to Biocomposite Materials Science and Engineering* (pp. 48–75). Hershey, PA: IGI Global. doi:10.4018/978-1-5225-0424-5.ch003

Mansouri, I., & Esmaeili, E. (2016). Nanotechnology Applications in the Construction Industry. In A. Khitab & W. Anwar (Eds.), *Advanced Research on Nanotechnology for Civil Engineering Applications* (pp. 111–140). Hershey, PA: IGI Global. doi:10.4018/978-1-5225-0344-6.ch004

Manzoor, A. (2016). MOOCs for Enhancing Engineering Education. In D. Fonseca & E. Redondo (Eds.), *Handbook of Research on Applied E-Learning in Engineering and Architecture Education* (pp. 204–223). Hershey, PA: IGI Global. doi:10.4018/978-1-4666-8803-2.ch010

Martin, A., Kalevaru, V. N., & Radnik, J. (2016). Palladium in Heterogeneous Oxidation Catalysis. In H. Al-Megren & T. Xiao (Eds.), *Petrochemical Catalyst Materials, Processes, and Emerging Technologies* (pp. 53–81). Hershey, PA: IGI Global. doi:10.4018/978-1-4666-9975-5.ch003

Melnyczuk, J. M., & Palchoudhury, S. (2016). Introduction to Bio-Inspired Hydrogel and Their Application: Hydrogels. In M. Bououdina (Ed.), *Emerging Research on Bioinspired Materials Engineering* (pp. 133–159). Hershey, PA: IGI Global. doi:10.4018/978-1-4666-9811-6.ch005

Mitra-Kirtley, S., Mullins, O. C., & Pomerantz, A. E. (2016). Sulfur and Nitrogen Chemical Speciation in Crude Oils and Related Carbonaceous Materials. In T. Saleh (Ed.), *Applying Nanotechnology to the Desulfurization Process in Petroleum Engineering* (pp. 53–83). Hershey, PA: IGI Global. doi:10.4018/978-1-4666-9545-0.ch002

Moalosi, R., Uziak, J., & Oladiran, M. T. (2016). Using Blended Learning Approach to Deliver Courses in An Engineering Programme. *International Journal of Quality Assurance in Engineering and Technology Education*, *5*(1), 23–39. doi:10.4018/IJQAETE.2016010103

Mohammadzadeh, S., & Kim, Y. (2017). Nonlinear System Identification of Smart Buildings. In P. Samui, S. Chakraborty, & D. Kim (Eds.), *Modeling and Simulation Techniques in Structural Engineering* (pp. 328–347). Hershey, PA: IGI Global. doi:10.4018/978-1-5225-0588-4.ch011

Mohanty, I., & Bhattacherjee, D. (2016). Artificial Neural Network and Its Application in Steel Industry. In S. Datta & J. Davim (Eds.), *Computational Approaches to Materials Design: Theoretical and Practical Aspects* (pp. 267–300). Hershey, PA: IGI Global. doi:10.4018/978-1-5225-0290-6.ch010

Mohebkhah, A., & Sarhosis, V. (2016). Discrete Element Modeling of Masonry-Infilled Frames. In V. Sarhosis, K. Bagi, J. Lemos, & G. Milani (Eds.), *Computational Modeling of Masonry Structures Using the Discrete Element Method* (pp. 200–234). Hershey, PA: IGI Global. doi:10.4018/978-1-5225-0231-9.ch009

Molina, G. J., Aktaruzzaman, F., Soloiu, V., & Rahman, M. (2017). Design and Testing of a Jet-Impingement Instrument to Study Surface-Modification Effects by Nanofluids. *International Journal of Surface Engineering and Interdisciplinary Materials Science*, *5*(2), 43–61. doi:10.4018/IJSEIMS.2017070104

Montalvan-Sorrosa, D., de los Cobos-Vasconcelos, D., & Gonzalez-Sanchez, A. (2016). Nanotechnology Applied to the Biodesulfurization of Fossil Fuels and Spent Caustic Streams. In T. Saleh (Ed.), *Applying Nanotechnology to the Desulfurization Process in Petroleum Engineering* (pp. 378–389). Hershey, PA: IGI Global. doi:10.4018/978-1-4666-9545-0.ch012

Montillet, J., Yu, K., Bonenberg, L. K., & Roberts, G. W. (2016). Optimization Algorithms in Local and Global Positioning. In P. Vasant, G. Weber, & V. Dieu (Eds.), *Handbook of Research on Modern Optimization Algorithms and Applications in Engineering and Economics* (pp. 1–53). Hershey, PA: IGI Global. doi:10.4018/978-1-4666-9644-0.ch001

Moreira, F., & Ferreira, M. J. (2016). Teaching and Learning Requirements Engineering Based on Mobile Devices and Cloud: A Case Study. In D. Fonseca & E. Redondo (Eds.), *Handbook of Research on Applied E-Learning in Engineering and Architecture Education* (pp. 237–262). Hershey, PA: IGI Global. doi:10.4018/978-1-4666-8803-2.ch012

Mukherjee, A., Saeed, R. A., Dutta, S., & Naskar, M. K. (2017). Fault Tracking Framework for Software-Defined Networking (SDN). In C. Singhal & S. De (Eds.), *Resource Allocation in Next-Generation Broadband Wireless Access Networks* (pp. 247–272). Hershey, PA: IGI Global. doi:10.4018/978-1-5225-2023-8.ch011

Mukhopadhyay, A., Barman, T. K., & Sahoo, P. (2018). Electroless Nickel Coatings for High Temperature Applications. In K. Kumar & J. Davim (Eds.), *Composites and Advanced Materials for Industrial Applications* (pp. 297–331). Hershey, PA: IGI Global. doi:10.4018/978-1-5225-5216-1.ch013

Náprstek, J., & Fischer, C. (2017). Dynamic Stability and Post-Critical Processes of Slender Auto-Parametric Systems. In V. Plevris, G. Kremmyda, & Y. Fahjan (Eds.), *Performance-Based Seismic Design of Concrete Structures and Infrastructures* (pp. 128–171). Hershey, PA: IGI Global. doi:10.4018/978-1-5225-2089-4.ch006

Nautiyal, L., Shivach, P., & Ram, M. (2018). Optimal Designs by Means of Genetic Algorithms. In M. Ram & J. Davim (Eds.), *Soft Computing Techniques and Applications in Mechanical Engineering* (pp. 151–161). Hershey, PA: IGI Global. doi:10.4018/978-1-5225-3035-0.ch007

Nazir, R. (2017). Advanced Nanomaterials for Water Engineering and Treatment: Nano-Metal Oxides and Their Nanocomposites. In T. Saleh (Ed.), *Advanced Nanomaterials for Water Engineering, Treatment, and Hydraulics* (pp. 84–126). Hershey, PA: IGI Global. doi:10.4018/978-1-5225-2136-5.ch005

Nogueira, A. F., Ribeiro, J. C., Fernández de Vega, F., & Zenha-Rela, M. A. (2018). Evolutionary Approaches to Test Data Generation for Object-Oriented Software: Overview of Techniques and Tools. In M. Khosrow-Pour, D.B.A. (Ed.), Incorporating Nature-Inspired Paradigms in Computational Applications (pp. 162-194). Hershey, PA: IGI Global. doi:10.4018/978-1-5225-5020-4.ch006

Nunes, J. F., Moreira, P. M., & Tavares, J. M. (2016). Human Motion Analysis and Simulation Tools: A Survey. In F. Miranda & C. Abreu (Eds.), *Handbook of Research on Computational Simulation and Modeling in Engineering* (pp. 359–388). Hershey, PA: IGI Global. doi:10.4018/978-1-4666-8823-0.ch012

Ogunlaja, A. S., & Tshentu, Z. R. (2016). Molecularly Imprinted Polymer Nanofibers for Adsorptive Desulfurization. In T. Saleh (Ed.), *Applying Nanotechnology to the Desulfurization Process in Petroleum Engineering* (pp. 281–336). Hershey, PA: IGI Global. doi:10.4018/978-1-4666-9545-0.ch010

Ong, P., & Kohshelan, S. (2016). Performances of Adaptive Cuckoo Search Algorithm in Engineering Optimization. In P. Vasant, G. Weber, & V. Dieu (Eds.), *Handbook of Research on Modern Optimization Algorithms and Applications in Engineering and Economics* (pp. 676–699). Hershey, PA: IGI Global. doi:10.4018/978-1-4666-9644-0.ch026

Osho, M. B. (2018). Industrial Enzyme Technology: Potential Applications. In S. Bharati & P. Chaurasia (Eds.), *Research Advancements in Pharmaceutical, Nutritional, and Industrial Enzymology* (pp. 375–394). Hershey, PA: IGI Global. doi:10.4018/978-1-5225-5237-6.ch017

Padmaja, P., & Marutheswar, G. (2017). Certain Investigation on Secured Data Transmission in Wireless Sensor Networks. *International Journal of Mobile Computing and Multimedia Communications, 8*(1), 48–61. doi:10.4018/IJMCMC.2017010104

Paixão, S. M., Silva, T. P., Arez, B. F., & Alves, L. (2016). Advances in the Reduction of the Costs Inherent to Fossil Fuels' Biodesulfurization towards Its Potential Industrial Application. In T. Saleh (Ed.), *Applying Nanotechnology to the Desulfurization Process in Petroleum Engineering* (pp. 390–425). Hershey, PA: IGI Global. doi:10.4018/978-1-4666-9545-0.ch013

Palmer, S., & Hall, W. (2017). An Evaluation of Group Work in First-Year Engineering Design Education. In R. Tucker (Ed.), *Collaboration and Student Engagement in Design Education* (pp. 145–168). Hershey, PA: IGI Global. doi:10.4018/978-1-5225-0726-0.ch007

Panneer, R. (2017). Effect of Composition of Fibers on Properties of Hybrid Composites. *International Journal of Manufacturing, Materials, and Mechanical Engineering, 7*(4), 28–43. doi:10.4018/IJMMME.2017100103

Parker, J. (2016). Hubble's Expanding Universe: A Model for Quality in Technology Infused engineering and Technology Education. *International Journal of Quality Assurance in Engineering and Technology Education, 5*(2), 16–29. doi:10.4018/IJQAETE.2016040102

Paul, S., & Roy, P. (2018). Optimal Design of Power System Stabilizer Using a Novel Evolutionary Algorithm. *International Journal of Energy Optimization and Engineering, 7*(3), 24–46. doi:10.4018/IJEOE.2018070102

Pavaloiu, A. (2018). Artificial Intelligence Ethics in Biomedical-Engineering-Oriented Problems. In U. Kose, G. Guraksin, & O. Deperlioglu (Eds.), *Nature-Inspired Intelligent Techniques for Solving Biomedical Engineering Problems* (pp. 219–231). Hershey, PA: IGI Global. doi:10.4018/978-1-5225-4769-3.ch010

Peña, F. (2016). A Semi-Discrete Approach for the Numerical Simulation of Freestanding Blocks. In V. Sarhosis, K. Bagi, J. Lemos, & G. Milani (Eds.), *Computational Modeling of Masonry Structures Using the Discrete Element Method* (pp. 416–439). Hershey, PA: IGI Global. doi:10.4018/978-1-5225-0231-9.ch016

Penchovsky, R., & Traykovska, M. (2016). Synthetic Approaches to Biology: Engineering Gene Control Circuits, Synthesizing, and Editing Genomes. In M. Bououdina (Ed.), *Emerging Research on Bioinspired Materials Engineering* (pp. 323–351). Hershey, PA: IGI Global. doi:10.4018/978-1-4666-9811-6.ch012

Pieroni, A., & Iazeolla, G. (2016). Engineering QoS and Energy Saving in the Delivery of ICT Services. In P. Vasant & N. Voropai (Eds.), *Sustaining Power Resources through Energy Optimization and Engineering* (pp. 208–226). Hershey, PA: IGI Global. doi:10.4018/978-1-4666-9755-3.ch009

Pioro, I., Mahdi, M., & Popov, R. (2017). Application of Supercritical Pressures in Power Engineering. In L. Chen & Y. Iwamoto (Eds.), *Advanced Applications of Supercritical Fluids in Energy Systems* (pp. 404–457). Hershey, PA: IGI Global. doi:10.4018/978-1-5225-2047-4.ch013

Plaksina, T., & Gildin, E. (2017). Rigorous Integrated Evolutionary Workflow for Optimal Exploitation of Unconventional Gas Assets. *International Journal of Energy Optimization and Engineering*, *6*(1), 101–122. doi:10.4018/IJEOE.2017010106

Puppala, A. J., Bheemasetti, T. V., Zou, H., Yu, X., Pedarla, A., & Cai, G. (2016). Spatial Variability Analysis of Soil Properties using Geostatistics. In P. Samui (Ed.), *Handbook of Research on Advanced Computational Techniques for Simulation-Based Engineering* (pp. 195–226). Hershey, PA: IGI Global. doi:10.4018/978-1-4666-9479-8.ch008

Ramdani, N., & Azibi, M. (2018). Polymer Composite Materials for Microelectronics Packaging Applications: Composites for Microelectronics Packaging. In K. Kumar & J. Davim (Eds.), *Composites and Advanced Materials for Industrial Applications* (pp. 177–211). Hershey, PA: IGI Global. doi:10.4018/978-1-5225-5216-1.ch009

Ramesh, M., Garg, R., & Subrahmanyam, G. V. (2017). Investigation of Influence of Quenching and Annealing on the Plane Fracture Toughness and Brittle to Ductile Transition Temperature of the Zinc Coated Structural Steel Materials. *International Journal of Surface Engineering and Interdisciplinary Materials Science*, *5*(2), 33–42. doi:10.4018/IJSEIMS.2017070103

Razavi, A. M., & Ahmad, R. (2016). Agile Software Development Challenges in Implementation and Adoption: Focusing on Large and Distributed Settings – Past Experiences, Emergent Topics. In I. Ghani, D. Jawawi, S. Dorairaj, & A. Sidky (Eds.), *Emerging Innovations in Agile Software Development* (pp. 175–207). Hershey, PA: IGI Global. doi:10.4018/978-1-4666-9858-1.ch010

Reccia, E., Cecchi, A., & Milani, G. (2016). FEM/DEM Approach for the Analysis of Masonry Arch Bridges. In V. Sarhosis, K. Bagi, J. Lemos, & G. Milani (Eds.), *Computational Modeling of Masonry Structures Using the Discrete Element Method* (pp. 367–392). Hershey, PA: IGI Global. doi:10.4018/978-1-5225-0231-9.ch014

Ro, H. K., & McIntosh, K. (2016). Constructing Conducive Environment for Women of Color in Engineering Undergraduate Education. In U. Thomas & J. Drake (Eds.), *Critical Research on Sexism and Racism in STEM Fields* (pp. 23–48). Hershey, PA: IGI Global. doi:10.4018/978-1-5225-0174-9.ch002

Rodulfo-Baechler, S. M. (2016). Dual Role of Perovskite Hollow Fiber Membrane in the Methane Oxidation Reactions. In H. Al-Megren & T. Xiao (Eds.), *Petrochemical Catalyst Materials, Processes, and Emerging Technologies* (pp. 385–430). Hershey, PA: IGI Global. doi:10.4018/978-1-4666-9975-5.ch014

Rudolf, S., Biryuk, V. V., & Volov, V. (2018). Vortex Effect, Vortex Power: Technology of Vortex Power Engineering. In V. Kharchenko & P. Vasant (Eds.), *Handbook of Research on Renewable Energy and Electric Resources for Sustainable Rural Development* (pp. 500–533). Hershey, PA: IGI Global. doi:10.4018/978-1-5225-3867-7.ch021

Sah, A., Bhadula, S. J., Dumka, A., & Rawat, S. (2018). A Software Engineering Perspective for Development of Enterprise Applications. In A. Elçi (Ed.), *Handbook of Research on Contemporary Perspectives on Web-Based Systems* (pp. 1–23). Hershey, PA: IGI Global. doi:10.4018/978-1-5225-5384-7.ch001

Sahoo, P., & Roy, S. (2017). Tribological Behavior of Electroless Ni-P, Ni-P-W and Ni-P-Cu Coatings: A Comparison. *International Journal of Surface Engineering and Interdisciplinary Materials Science, 5*(1), 1–15. doi:10.4018/IJSEIMS.2017010101

Sahoo, S. (2018). Laminated Composite Hypar Shells as Roofing Units: Static and Dynamic Behavior. In K. Kumar & J. Davim (Eds.), *Composites and Advanced Materials for Industrial Applications* (pp. 249–269). Hershey, PA: IGI Global. doi:10.4018/978-1-5225-5216-1.ch011

Sahu, H., & Hungyo, M. (2018). Introduction to SDN and NFV. In A. Dumka (Ed.), *Innovations in Software-Defined Networking and Network Functions Virtualization* (pp. 1–25). Hershey, PA: IGI Global. doi:10.4018/978-1-5225-3640-6.ch001

Saikia, P., Bharadwaj, S. K., & Miah, A. T. (2016). Peroxovanadates and Its Bio-Mimicking Relation with Vanadium Haloperoxidases. In M. Bououdina (Ed.), *Emerging Research on Bioinspired Materials Engineering* (pp. 197–219). Hershey, PA: IGI Global. doi:10.4018/978-1-4666-9811-6.ch007

Saladino, R., Botta, G., & Crucianelli, M. (2016). Advances in Nanotechnology Transition Metal Catalysts in Oxidative Desulfurization (ODS) Processes: Nanotechnology Applied to ODS Processing. In T. Saleh (Ed.), *Applying Nanotechnology to the Desulfurization Process in Petroleum Engineering* (pp. 180–215). Hershey, PA: IGI Global. doi:10.4018/978-1-4666-9545-0.ch006

Saleh, T. A., Danmaliki, G. I., & Shuaib, T. D. (2016). Nanocomposites and Hybrid Materials for Adsorptive Desulfurization. In T. Saleh (Ed.), *Applying Nanotechnology to the Desulfurization Process in Petroleum Engineering* (pp. 129–153). Hershey, PA: IGI Global. doi:10.4018/978-1-4666-9545-0.ch004

Saleh, T. A., Shuaib, T. D., Danmaliki, G. I., & Al-Daous, M. A. (2016). Carbon-Based Nanomaterials for Desulfurization: Classification, Preparation, and Evaluation. In T. Saleh (Ed.), *Applying Nanotechnology to the Desulfurization Process in Petroleum Engineering* (pp. 154–179). Hershey, PA: IGI Global. doi:10.4018/978-1-4666-9545-0.ch005

Salem, A. M., & Shmelova, T. (2018). Intelligent Expert Decision Support Systems: Methodologies, Applications, and Challenges. In T. Shmelova, Y. Sikirda, N. Rizun, A. Salem, & Y. Kovalyov (Eds.), *Socio-Technical Decision Support in Air Navigation Systems: Emerging Research and Opportunities* (pp. 215–242). Hershey, PA: IGI Global. doi:10.4018/978-1-5225-3108-1.ch007

Samal, M. (2017). FE Analysis and Experimental Investigation of Cracked and Un-Cracked Thin-Walled Tubular Components to Evaluate Mechanical and Fracture Properties. In P. Samui, S. Chakraborty, & D. Kim (Eds.), *Modeling and Simulation Techniques in Structural Engineering* (pp. 266–293). Hershey, PA: IGI Global. doi:10.4018/978-1-5225-0588-4.ch009

Samal, M., & Balakrishnan, K. (2017). Experiments on a Ring Tension Setup and FE Analysis to Evaluate Transverse Mechanical Properties of Tubular Components. In P. Samui, S. Chakraborty, & D. Kim (Eds.), *Modeling and Simulation Techniques in Structural Engineering* (pp. 91–115). Hershey, PA: IGI Global. doi:10.4018/978-1-5225-0588-4.ch004

Santhanakumar, M., Adalarasan, R., & Rajmohan, M. (2016). An Investigation in Abrasive Waterjet Cutting of Al6061/SiC/Al2O3 Composite Using Principal Component Based Response Surface Methodology. *International Journal of Manufacturing, Materials, and Mechanical Engineering*, *6*(4), 30–47. doi:10.4018/IJMMME.2016100103

Sareen, N., & Bhattacharya, S. (2016). Cleaner Energy Fuels: Hydrodesulfurization and Beyond. In T. Saleh (Ed.), *Applying Nanotechnology to the Desulfurization Process in Petroleum Engineering* (pp. 84–128). Hershey, PA: IGI Global. doi:10.4018/978-1-4666-9545-0.ch003

Sarhosis, V. (2016). Micro-Modeling Options for Masonry. In V. Sarhosis, K. Bagi, J. Lemos, & G. Milani (Eds.), *Computational Modeling of Masonry Structures Using the Discrete Element Method* (pp. 28–60). Hershey, PA: IGI Global. doi:10.4018/978-1-5225-0231-9.ch002

Sarhosis, V., Oliveira, D. V., & Lourenco, P. B. (2016). On the Mechanical Behavior of Masonry. In V. Sarhosis, K. Bagi, J. Lemos, & G. Milani (Eds.), *Computational Modeling of Masonry Structures Using the Discrete Element Method* (pp. 1–27). Hershey, PA: IGI Global. doi:10.4018/978-1-5225-0231-9.ch001

Satyam, N. (2016). Liquefaction Modelling of Granular Soils using Discrete Element Method. In P. Samui (Ed.), *Handbook of Research on Advanced Computational Techniques for Simulation-Based Engineering* (pp. 381–441). Hershey, PA: IGI Global. doi:10.4018/978-1-4666-9479-8.ch015

Sawant, S. (2018). Deep Learning and Biomedical Engineering. In U. Kose, G. Guraksin, & O. Deperlioglu (Eds.), *Nature-Inspired Intelligent Techniques for Solving Biomedical Engineering Problems* (pp. 283–296). Hershey, PA: IGI Global. doi:10.4018/978-1-5225-4769-3.ch014

Sezgin, H., & Berkalp, O. B. (2018). Textile-Reinforced Composites for the Automotive Industry. In K. Kumar & J. Davim (Eds.), *Composites and Advanced Materials for Industrial Applications* (pp. 129–156). Hershey, PA: IGI Global. doi:10.4018/978-1-5225-5216-1.ch007

Shah, M. Z., Gazder, U., Bhatti, M. S., & Hussain, M. (2018). Comparative Performance Evaluation of Effects of Modifier in Asphaltic Concrete Mix. *International Journal of Strategic Engineering*, *1*(2), 13–25. doi:10.4018/IJoSE.2018070102

Shah, V. S., Shah, H. R., & Samui, P. (2016). Application of Meta-Models (MPMR and ELM) for Determining OMC, MDD and Soaked CBR Value of Soil. In S. Bhattacharyya, P. Banerjee, D. Majumdar, & P. Dutta (Eds.), *Handbook of Research on Advanced Hybrid Intelligent Techniques and Applications* (pp. 454–482). Hershey, PA: IGI Global. doi:10.4018/978-1-4666-9474-3.ch015

Sharma, N., & Kumar, K. (2018). Fabrication of Porous NiTi Alloy Using Organic Binders. In K. Kumar & J. Davim (Eds.), *Composites and Advanced Materials for Industrial Applications* (pp. 38–62). Hershey, PA: IGI Global. doi:10.4018/978-1-5225-5216-1.ch003

Sharma, T. K. (2016). Application of Shuffled Frog Leaping Algorithm in Software Project Scheduling. In P. Saxena, D. Singh, & M. Pant (Eds.), *Problem Solving and Uncertainty Modeling through Optimization and Soft Computing Applications* (pp. 225–238). Hershey, PA: IGI Global. doi:10.4018/978-1-4666-9885-7.ch011

Shivach, P., Nautiyal, L., & Ram, M. (2018). Applying Multi-Objective Optimization Algorithms to Mechanical Engineering. In M. Ram & J. Davim (Eds.), *Soft Computing Techniques and Applications in Mechanical Engineering* (pp. 287–301). Hershey, PA: IGI Global. doi:10.4018/978-1-5225-3035-0.ch014

Shmelova, T. (2018). Stochastic Methods for Estimation and Problem Solving in Engineering: Stochastic Methods of Decision Making in Aviation. In S. Kadry (Ed.), *Stochastic Methods for Estimation and Problem Solving in Engineering* (pp. 139–160). Hershey, PA: IGI Global. doi:10.4018/978-1-5225-5045-7.ch006

Shukla, R., Anapagaddi, R., Singh, A. K., Allen, J. K., Panchal, J. H., & Mistree, F. (2016). Integrated Computational Materials Engineering for Determining the Set Points of Unit Operations for Production of a Steel Product Mix. In S. Datta & J. Davim (Eds.), *Computational Approaches to Materials Design: Theoretical and Practical Aspects* (pp. 163–191). Hershey, PA: IGI Global. doi:10.4018/978-1-5225-0290-6.ch006

Siero González, L. R., & Romo Vázquez, A. (2017). Didactic Sequences Teaching Mathematics for Engineers With Focus on Differential Equations. In M. Ramírez-Montoya (Ed.), *Handbook of Research on Driving STEM Learning With Educational Technologies* (pp. 129–151). Hershey, PA: IGI Global. doi:10.4018/978-1-5225-2026-9.ch007

Singh, R., & Dutta, S. (2018). Visible Light Active Nanocomposites for Photocatalytic Applications. In K. Kumar & J. Davim (Eds.), *Composites and Advanced Materials for Industrial Applications* (pp. 270–296). Hershey, PA: IGI Global. doi:10.4018/978-1-5225-5216-1.ch012

Singh, R., & Lou, H. H. (2016). Safety and Efficiency Enhancement in LNG Terminals. In H. Al-Megren & T. Xiao (Eds.), *Petrochemical Catalyst Materials, Processes, and Emerging Technologies* (pp. 164–176). Hershey, PA: IGI Global. doi:10.4018/978-1-4666-9975-5.ch007

Sözbilir, H., Özkaymak, Ç., Uzel, B., & Sümer, Ö. (2018). Criteria for Surface Rupture Microzonation of Active Faults for Earthquake Hazards in Urban Areas. In N. Ceryan (Ed.), *Handbook of Research on Trends and Digital Advances in Engineering Geology* (pp. 187–230). Hershey, PA: IGI Global. doi:10.4018/978-1-5225-2709-1.ch005

Stanciu, I. (2018). Stochastic Methods in Microsystems Engineering. In S. Kadry (Ed.), *Stochastic Methods for Estimation and Problem Solving in Engineering* (pp. 161–176). Hershey, PA: IGI Global. doi:10.4018/978-1-5225-5045-7.ch007

Strebkov, D., Nekrasov, A., Trubnikov, V., & Nekrasov, A. (2018). Single-Wire Resonant Electric Power Systems for Renewable-Based Electric Grid. In V. Kharchenko & P. Vasant (Eds.), *Handbook of Research on Renewable Energy and Electric Resources for Sustainable Rural Development* (pp. 449–474). Hershey, PA: IGI Global. doi:10.4018/978-1-5225-3867-7.ch019

Subburaman, D., Jagan, J., Dalkiliç, Y., & Samui, P. (2016). Reliability Analysis of Slope Using MPMR, GRNN and GPR. In F. Miranda & C. Abreu (Eds.), *Handbook of Research on Computational Simulation and Modeling in Engineering* (pp. 208–224). Hershey, PA: IGI Global. doi:10.4018/978-1-4666-8823-0.ch007

Sun, J., Wan, S., Lin, J., & Wang, Y. (2016). Advances in Catalytic Conversion of Syngas to Ethanol and Higher Alcohols. In H. Al-Megren & T. Xiao (Eds.), *Petrochemical Catalyst Materials, Processes, and Emerging Technologies* (pp. 177–215). Hershey, PA: IGI Global. doi:10.4018/978-1-4666-9975-5.ch008

Tüdeş, Ş., Kumlu, K. B., & Ceryan, S. (2018). Integration Between Urban Planning and Natural Hazards For Resilient City. In N. Ceryan (Ed.), *Handbook of Research on Trends and Digital Advances in Engineering Geology* (pp. 591–630). Hershey, PA: IGI Global. doi:10.4018/978-1-5225-2709-1.ch017

Tyukhov, I., Rezk, H., & Vasant, P. (2016). Modern Optimization Algorithms and Applications in Solar Photovoltaic Engineering. In P. Vasant & N. Voropai (Eds.), *Sustaining Power Resources through Energy Optimization and Engineering* (pp. 390–445). Hershey, PA: IGI Global. doi:10.4018/978-1-4666-9755-3.ch016

Ulamis, K. (2018). Soil Liquefaction Assessment by Anisotropic Cyclic Triaxial Test. In N. Ceryan (Ed.), *Handbook of Research on Trends and Digital Advances in Engineering Geology* (pp. 631–664). Hershey, PA: IGI Global. doi:10.4018/978-1-5225-2709-1.ch018

Umar, M. A., Tenuche, S. S., Yusuf, S. A., Abdulsalami, A. O., & Kufena, A. M. (2016). Usability Engineering in Agile Software Development Processes. In I. Ghani, D. Jawawi, S. Dorairaj, & A. Sidky (Eds.), *Emerging Innovations in Agile Software Development* (pp. 208–221). Hershey, PA: IGI Global. doi:10.4018/978-1-4666-9858-1.ch011

Üzüm, O., & Çakır, Ö. A. (2016). A Bio-Inspired Phenomena in Cementitious Materials: Self-Healing. In M. Bououdina (Ed.), *Emerging Research on Bioinspired Materials Engineering* (pp. 220–246). Hershey, PA: IGI Global. doi:10.4018/978-1-4666-9811-6.ch008

Valente, M., & Milani, G. (2017). Seismic Assessment and Retrofitting of an Under-Designed RC Frame Through a Displacement-Based Approach. In V. Plevris, G. Kremmyda, & Y. Fahjan (Eds.), *Performance-Based Seismic Design of Concrete Structures and Infrastructures* (pp. 36–58). Hershey, PA: IGI Global. doi:10.4018/978-1-5225-2089-4.ch002

Vasant, P. (2018). A General Medical Diagnosis System Formed by Artificial Neural Networks and Swarm Intelligence Techniques. In U. Kose, G. Guraksin, & O. Deperlioglu (Eds.), *Nature-Inspired Intelligent Techniques for Solving Biomedical Engineering Problems* (pp. 130–145). Hershey, PA: IGI Global. doi:10.4018/978-1-5225-4769-3.ch006

Vergara, D., Lorenzo, M., & Rubio, M. (2016). On the Use of Virtual Environments in Engineering Education. *International Journal of Quality Assurance in Engineering and Technology Education*, *5*(2), 30–41. doi:10.4018/IJQAETE.2016040103

Verrollot, J., Tolonen, A., Harkonen, J., & Haapasalo, H. J. (2018). Challenges and Enablers for Rapid Product Development. *International Journal of Applied Industrial Engineering*, *5*(1), 25–49. doi:10.4018/IJAIE.2018010102

Wagner, C., & Ryan, C. (2016). Physical and Digital Integration Strategies of Electronic Device Supply Chains and Their Applicability to ETO Supply Chains. In R. Addo-Tenkorang, J. Kantola, P. Helo, & A. Shamsuzzoha (Eds.), *Supply Chain Strategies and the Engineer-to-Order Approach* (pp. 224–245). Hershey, PA: IGI Global. doi:10.4018/978-1-5225-0021-6.ch011

Wang, Z., Wu, P., Lan, L., & Ji, S. (2016). Preparation, Characterization and Desulfurization of the Supported Nickel Phosphide Catalysts. In H. Al-Megren & T. Xiao (Eds.), *Petrochemical Catalyst Materials, Processes, and Emerging Technologies* (pp. 431–458). Hershey, PA: IGI Global. doi:10.4018/978-1-4666-9975-5.ch015

Yardimci, A. G., & Karpuz, C. (2018). Fuzzy Rock Mass Rating: Soft-Computing-Aided Preliminary Stability Analysis of Weak Rock Slopes. In N. Ceryan (Ed.), *Handbook of Research on Trends and Digital Advances in Engineering Geology* (pp. 97–131). Hershey, PA: IGI Global. doi:10.4018/978-1-5225-2709-1.ch003

Zhang, L., Ding, S., Sun, S., Han, B., Yu, X., & Ou, J. (2016). Nano-Scale Behavior and Nano-Modification of Cement and Concrete Materials. In A. Khitab & W. Anwar (Eds.), *Advanced Research on Nanotechnology for Civil Engineering Applications* (pp. 28–79). Hershey, PA: IGI Global. doi:10.4018/978-1-5225-0344-6.ch002

Zindani, D., & Kumar, K. (2018). Industrial Applications of Polymer Composite Materials. In K. Kumar & J. Davim (Eds.), *Composites and Advanced Materials for Industrial Applications* (pp. 1–15). Hershey, PA: IGI Global. doi:10.4018/978-1-5225-5216-1.ch001

Zindani, D., Maity, S. R., & Bhowmik, S. (2018). A Decision-Making Approach for Material Selection of Polymeric Composite Bumper Beam. In K. Kumar & J. Davim (Eds.), *Composites and Advanced Materials for Industrial Applications* (pp. 112–128). Hershey, PA: IGI Global. doi:10.4018/978-1-5225-5216-1.ch006

About the Contributors

Joel Antúnez-García was born in Ensenada B. C., México, in 1975. He received the B. Sc. degree in Physics from Universidad Autónoma de Baja California (UABC), México, in 1999. The M. Sc. from Centro de Investigación Científica y de Educación Superior de Ensenada (CICESE), México, in 2004. The Ph. D. in Physical-Industrial Engineering from Universidad Autónoma de Nuevo Léon (UANL), Méxio, in 2010. From 2012 to 2013, he did a postdoctoral stay at Centro de Nanociencias y Nanotecnología at UNAM, working on DFT calculations to obtain different zeolites' electronic properties. From 2013-2015 he worked as a professor at Centro de Enseñanza Técnica y Superior (CETYS university). From 2016 to date, he has been involved in the theoretical study of bi-and tri-metallic catalysts based on MoS2 compounds and zeolites.

Paran Jyoti Borpatra was born in Lakhimpur district of Assam, India. He obtained his Ph.D degree (2020) under the supervision of Dr. Pranjal Kumar Baruah at Gauhati University, Assam, working in the area of organic synthesis especially C-H activation reactions, multicomponent reactions. Dr. Borpatra completed a two years research project as JRF under 'DBT's Twinning Program for the NE' titled "De Novo Rationally Designed Curcumin Derived Analogues for Increased Bioavailability and Its Metal Analogues for Anticancer and Protective Studies" under the guidance of Dr. Pranjal Kumar Baruah. His current research area is C–H functionalization and development of novel methodologies for heterocyclic molecules. He has also co-authored a number of publications in journals with good impact factors.

Cleiton Gonçalves Dias obtained recently his title of Bachelor of Pharmacy from Universidade Vale do Rio Doce (MG, Brazil), having worked with Dr Marcus Vinícius Dias-Souza in researches centered in drug-nutrient interactions. He is also one of the pharmacists partners of the Reference Center for Drug Information (CRIMED, Pitagoras College, Ipatinga, MG, Brazil).

Marcus Vinícius Dias-Souza is a Pharmacist with Specialist degree in Clinical Pharmacology, Master of Science degree in Immunopathology and PhD in Microbiology, the last one obtained at the Federal University of Minas Gerais, Brazil. He is Editor-in-Chief of JAPHAC Journal, and currently, he is professor of pharmaceutical chemistry, organic chemistry, pharmacognosy and pathophysiology-pharmacotherapeutics related disciplines at Pitágoras College (Ipatinga, MG, Brazil). He is head of the Integrated Pharmacology and Drug Interactions Research Group (GPqFAR, Brazil) and of the MONARCHY study group of pain treatment (***M***anagement of ***O***poiods, ***N***saids, ***A***nalgesics and other d***R***ugs for ***CH***ronic pain in ***Y***oung adults), and is the coordinator of the Reference Center for Drug Information (CRIMED) at Pitágoras College

Mintu Maan Dutta was born in Nazira, Sibsagar, Assam, India. He obtained his B.Sc. degree from Cotton College in 2012 and M.Sc degree with specialization in Physical Chemistry in 2014, from Gauhati University, Assam. He obtained his Ph.D. degree in 2020 from Department of Chemistry, Gauhati University under the supervision of Professor Prodeep Phukan. His PhD. thesis entitled "Surface functionalized magnetic nanoparticles as catalyst for organic transformations". He has co-authored number of publications in reputed journals that includes Dalton Transactions, Catalysis Communications, Chemistry Select, etc. His current research interest includes heterogeneous catalysis, magnetic nanoparticles, organic transformations. Presently he is working as Assistant Professor in Arya Vidyapeeth College, Assam.

Pankaj Indurkar is Scientist in Membrane Science and Separation Technology division of CSIR-Central Salt and Marine Chemicals Research Institute.

Ashish Kumar is presently heading chemistry department in School of Civil and School of Mechanical Engineering. He has been a member of review boards of various journals of Springer, Bentham, Taylor and Francis etc and conferences in India and abroad. He has also authored and reviewed diverse chapters in refereed books with Springer nature and Elsevier, Wiley etc. He is editor/Editorial Board member of more than 15 international Journals. He also published one book on Engineering chemistry. He also evaluated Ph.D thesis's from India and Abroad. He himself guided more than 20 (Ph.D / Masters and UG) students. He was recipient of Research Excellence Award from LPU for last FOUR consecutive years.

Fabian N. Murrieta-Rico received B.Eng. and M.Eng. degrees from Mexicali Institute of Technology (ITM) in 2004 and 2013 respectively. In 2017, he received his PhD in Materials Physics at Ensenada Center for Scientific Research and Higher Education (CICESE). He has worked as an automation engineer, systems designer, as a university professor, and as postdoctoral researcher at Facultad de Ingeniería, Arquitectura y Diseño from Universidad Autónoma de Baja California (UABC). Currently he works as professor of bioengineering topics at UABC, a postdoctoral researcher at the Centro de Nanociencias y Nanotecnología from Universidad Nacional Autónoma de México. His research has been published in different journals and presented at international conferences since 2009. He has served as reviewer for different journals, some of them include IEEE Transactions on Industrial Electronics, IEEE Transactions on Instrumentation, Measurement and Sensor Review. His research interests are focused on the field of time and frequency metrology, the design of wireless sensor networks, automated systems and highly sensitive chemical detectors.

Khemnath Patir obtained his M.Sc Degree in Chemistry (2013) and Ph.D Degree (2020) from Gauhati University-Guwahati, Assam, India. Currently, he is an Assistant Professor in the Department of Chemistry, Behali Degree College-Borgang, Biswanath, Assam, India. Formerly, he was a guest faculty in the Department of Chemistry, Bodoland University-Kokrajhar, Assam, India for a period of one year (2019-2020). His research area includes luminescent materials, photocatalysis, adsorption and nanotechnology. He has four scientific papers in reputed journals, ACS sustainable chemistry and engineering, Chemistry-An Asian Journal, Nanoscale Advances, Acta Crystal B.

Vitalii Petranovskii received his PhD in Physical chemistry from Moscow Institute of Crystallography in 1988. In 1993-1994 he worked as invited scientist at National Institute of Materials and Chemical Research, Japan. Since 1995, he is working at "Centro de Nanociencias y Nanotecnología, Universidad

Nacional Autónoma de México" (2006-2014 – as the Nanocatalysis department chair). His research interests include synthesis and properties of nanoparticles supported over zeolite matrices. He is a member of Mexican Academy of Sciences, International Zeolite Association, and Mendeleev Russian Chemical Society. He has published over 150 papers in peer-reviewed journals and 7 invited book chapters. Also he is co-author of monograph "Clusters and matrix isolated cluster superstructures", SPb, 1995.

Shveta Sharma is currently perusing her Ph.D under supervision of Prof. Ashish Kumar and Also working as assistant professor in Government college Himachal Pradesh.

Victor Shikuku is a lecturer of Physical Chemistry at the Department of Physical Sciences, Kaimosi Friends University College, Kenya. He holds a Ph.D. in Chemistry from Maseno University (Kenya), with expertise in adsorption science. Dr. Shikuku also has post-doctoral research experience at Trier University, Germany funded by the Alexander Humbolt foundation, and other research visits at CSIR-NML (India) and Federal University of Santa Maria (Brazil) funded by the CV Raman Fellowship and EXCEED-SWINDON, respectively. He has an interest in the development of low-cost materials for application in environmental remediation.

Patrick Ssebugere is a Lecturer of Physical Chemistry at the Department of Chemistry, Makerere University. He has been studying the prevalence, behavior and sources of persistent organic pollutants (POPs) in sub-Saharan Africa since 2007. Currently his research focuses on development of wastewater treatment techniques, sustainable energy production and, studying the prenatal exposure to undesirable chemical compounds and child health. He has trained in courses such as occupational toxicology, food toxicology and safety, health risk assessment and management of toxic chemicals.

Abhinay Thakur is currently perusing his Ph.D under supervision of Prof. Ashish Kumar.

Rafael Vargas Bernal received a Bachelor's degree in Communications and Electronics Engineering from the University of Guanajuato in 1995, and the degrees of Master of Science and Doctorate in Science with Specialty in Electronics from the National Institute of Astrophysics, Optics and Electronics (INAOE) in 1997 and 2000, respectively. Since January 2002, he has been a professor-researcher at the Higher Technological Institute of Irapuato (ITESI), Mexico, and particularly since 2006, he has worked in the Department of Materials Engineering where he has established himself as a researcher. He has authored 15 articles in journals, 28 chapters in books, and about 100 conference articles. He is a member of the National System of Researchers (SNI-Mexico). He regularly serves as a reviewer of scientific articles in RSC Advances and Royal Society Open Science as well as Standards in Semiconductor Equipment and Materials International (SEMI). His research interests include two-dimensional materials, nanomaterials, space materials, composite materials, gas sensors, and biosensors.

Anshul Yadav is Scientist in Membrane Science and Separation division of CSIR-Central Salt and Marine Research Institute.

Rosario I. Yocupicio-Gaxiola received the B.S. degree in chemical engineering from Instituto Tecnologico de Los Mochis, in 2007, M. S. degree and Ph.D. in materials science and engineering by Universidad Nacional Autonoma de México (UNAM), in 2011 and 2017, respectively. From 2017 to 2019, he held a postdoc position at UNAM. Actually, he holds a postdoctoral position in Centro de Investigacion Cientıfica y de Educacion de Ensenada (CICESE), Baja California. His research interest includes the synthesis, analysis and porous materials applications, specifically, the study of MOF and bulk and hierarchical zeolites. His research is focused to the use of this kind of materials as catalysts, catalytic supports, sorbents and optical devices.

Index

N

O

P

Q

R

S

T

U

W

Recommended Reference Books

IGI Global's reference books are available in three unique pricing formats: Print Only, E-Book Only, or Print + E-Book.

Shipping fees may apply.

www.igi-global.com

ISBN: 978-1-5225-9276-1
© 2020; 612 pp.
List Price: US$ **345**

ISBN: 978-1-5225-8325-7
© 2019; 169 pp.
List Price: US$ **165**

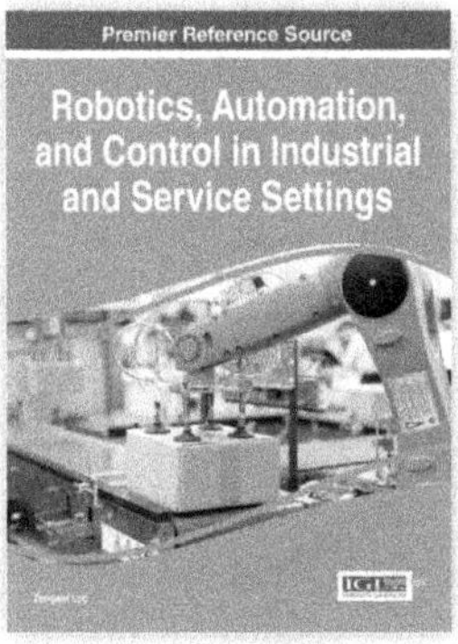

ISBN: 978-1-4666-8693-9
© 2015; 337 pp.
List Price: US$ **215**

ISBN: 978-1-4666-8823-0
© 2016; 824 pp.
List Price: US$ **420**

ISBN: 978-1-4666-8150-7
© 2015; 380 pp.
List Price: US$ **205**

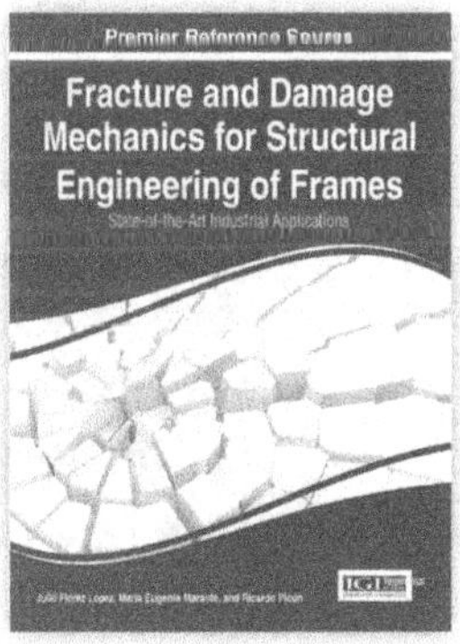

ISBN: 978-1-4666-6379-4
© 2015; 602 pp.
List Price: US$ **225**

Do you want to stay current on the latest research trends, product announcements, news, and special offers?

Join IGI Global's mailing list to receive customized recommendations, exclusive discounts, and more.

Sign up at: **www.igi-global.com/newsletters.**

Publisher of Peer-Reviewed, Timely, and Innovative Academic Research

IGI Global DISSEMINATOR OF KNOWLEDGE

www.igi-global.com Sign up at www.igi-global.com/newsletters facebook.com/igiglobal twitter.com/igiglobal linkedin.com/igiglobal

www.ingramcontent.com/pod-product-compliance
Lightning Source LLC
LaVergne TN
LVHW080529300926
849162LV00007B/1566